Computational Optimizations for
Machine Learning

Computational Optimizations for Machine Learning

Editor

Freddy Gabbay

MDPI • Basel • Beijing • Wuhan • Barcelona • Belgrade • Manchester • Tokyo • Cluj • Tianjin

Editor
Freddy Gabbay
Faculty of Engineering
Ruppin Academic Center
Israel

Editorial Office
MDPI
St. Alban-Anlage 66
4052 Basel, Switzerland

This is a reprint of articles from the Special Issue published online in the open access journal *Mathematics* (ISSN 2227-7390) (available at: https://www.mdpi.com/journal/mathematics/special_issues/Comput_Optim_Mach_Learn).

For citation purposes, cite each article independently as indicated on the article page online and as indicated below:

LastName, A.A.; LastName, B.B.; LastName, C.C. Article Title. *Journal Name* **Year**, *Volume Number*, Page Range.

ISBN 978-3-0365-3186-1 (Hbk)
ISBN 978-3-0365-3187-8 (PDF)

© 2022 by the authors. Articles in this book are Open Access and distributed under the Creative Commons Attribution (CC BY) license, which allows users to download, copy and build upon published articles, as long as the author and publisher are properly credited, which ensures maximum dissemination and a wider impact of our publications.

The book as a whole is distributed by MDPI under the terms and conditions of the Creative Commons license CC BY-NC-ND.

Contents

About the Editor vii

Preface to "Computational Optimizations for Machine Learning" ix

Weijia Shao, Lukas Radke, Fikret Sivrikaya and Sahin Albayrak
Adaptive Online Learning for the Autoregressive Integrated Moving Average Models
Reprinted from: *Mathematics* **2021**, *9*, 1523, doi:10.3390/math9131523 1

Gabriela Czibula, Andrei Mihai, Alexandra-Ioana Albu, Istvan Gergely Czibula, Sorin Burcea and Abdelkader Mezghani
AutoNowP: An Approach Using Deep Autoencoders for Precipitation Nowcasting Based on Weather Radar Reflectivity Prediction
Reprinted from: *Mathematics* **2021**, *9*, 1653, doi:10.3390/math9141653 31

Zhe Wu, David Rincon, Quanquan Gu and Panagiotis D. Christofides
Statistical Machine Learning in Model Predictive Control of Nonlinear Processes
Reprinted from: *Mathematics* **2021**, *9*, 1912, doi:10.3390/math9161912 53

Chaim Baskin, Evgenii Zheltonozhkii, Tal Rozen, Natan Liss, Yoav Chai, Eli Schwartz, Raja Giryes, Alexander M. Bronstein and Avi Mendelson
NICE: Noise Injection and Clamping Estimation for Neural Network Quantization
Reprinted from: *Mathematics* **2021**, *9*, 2144, doi:10.3390/math9172144 91

Alberto Garces-Jimenez, Jose-Manuel Gomez-Pulido, Nuria Gallego-Salvador and Alvaro-Jose Garcia-Tejedor
Genetic and Swarm Algorithms for Optimizing the Control of Building HVAC Systems Using Real Data: A Comparative Study
Reprinted from: *Mathematics* **2021**, *9*, 2181, doi:10.3390/math9182181 103

Ricardo J. Jesus, Mário L. Antunes, Rui A. da Costa, Sergey N. Dorogovtsev, José F. F. Mendes and Rui L. Aguiar
Effect of Initial Configuration of Weights on Training and Function of Artificial Neural Networks
Reprinted from: *Mathematics* **2021**, *9*, 2246, doi:10.3390/math9182246 127

Freddy Gabbay and Gil Shomron
Compression of Neural Networks for Specialized Tasks via Value Locality
Reprinted from: *Mathematics* **2021**, *9*, 2612, doi:10.3390/math9202612 143

Bharathwaj Suresh, Kamlesh Pillai, Gurpreet Singh Kalsi, Avishaii Abuhatzera and Sreenivas Subramoney
Early Prediction of DNN Activation Using Hierarchical Computations
Reprinted from: *Mathematics* **2021**, *9*, 3130, doi:10.3390/math9233130 177

Mehdi Dasineh, Amir Ghaderi, Mohammad Bagherzadeh, Mohammad Ahmadi and Alban Kuriqi
Prediction of Hydraulic Jumps on a Triangular Bed Roughness Using Numerical Modeling and Soft Computing Methods
Reprinted from: *Mathematics* **2021**, *9*, 3135, doi:10.3390/math9233135 195

Ruba Abu Khurma, Ibrahim Aljarah*, Ahmad Sharieh, Mohamed Abd Elaziz, Robertas Damaševičius*, Tomas Krilavičius
A Review of the Modification Strategies of the Nature Inspired Algorithms for Feature Selection Problem
Reprinted from: *Mathematics* **2022**, *10*, 464, doi:10.3390/math10030464 . . . . **219**

About the Editor

Freddy Gabbay received his B.Sc., M.Sc. and Ph.D. in Electrical Engineering from Technion – Israel Institute of Technology, Haifa, Israel. In 1998, he worked as a researcher at Intel's Microprocessor Research Lab. In 1999, he joined Mellanox Technologies and held various positions in leading switch product line architecture and ASIC design. In 2003, he joined Freescale Semiconductor as a senior design manager and led the design of baseband ASIC products. In 2012, he rejoined Mellanox Technologies, where he served as Vice President of Chip Design. Today, he is the Dean of the Engineering Faculty and an associate professor at the Ruppin Academic Center, Emek Hefer, Israel. His research interests include VLSI design, computer architecture, machine learning and domain-specific accelerators. Prof. Gabbay holds 19 patents and is a senior member of IEEE.

Preface to "Computational Optimizations for Machine Learning"

In the recent decade, machine learning has emerged as a powerful tool for an incredible number of applications, such as computer vision, medicine, fintech, autonomous systems, speech recognition, traffic management and social media among many others. Machine learning models provide state-of-the-art and robust accuracy in various applications. Beyond the major impact of machine learning applications on our life and environment, machine learning introduces a revolutionary approach in the method of developing algorithms. While past approaches rely on humans to develop new algorithms, machine learning uses powerful computers to train algorithms based on large datasets. Machine learning can thereby identify complex connections and relations between features that cannot be handled using conventional methods.

The increasing deployment of machine learning algorithms introduces major computational challenges due to the explosive growth in their model size and complexity. These challenges have been further emphasized due to the diversity of hosting computational platforms, from edge devices and cloud systems to high-performance computing. Given that each platform introduces different computational and cost constraints, the need for computational optimizations that are fine-tuned to the application and platform is crucial.

The present book contains the 10 articles accepted for publication among the 15 submissions to the Special Issue "Computational Optimizations for Machine Learning" of the MDPI journal Mathematics.

The 10 articles, which appear in the present book in the order in which they were published in Volume 9 (2021) of the journal, cover a wide range of topics connected to the machine learning computational optimizations theory and applications. These topics include, among others, elements from convolutional neural networks, nature inspired algorithms, neural networks training, quantization, predictive control of nonlinear processes, weather prediction and adaptive online learning.

It is hoped that the book will be interesting and useful to those developing mathematical algorithms and applications in the domain of artificial intelligence and machine learning as well as for those having the appropriate mathematical background and willing to become familiar with recent advances of machine learning computational optimization mathematics, which has nowadays permeated into almost all sectors of human life and activity.

As the Guest Editor of the Special Issue, I am grateful to the authors of the papers for their quality contributions, to the reviewers for their valuable comments toward the improvement of the submitted works and to the administrative staff of the MDPI publications for the support to complete this project. Special thanks are due to the Managing Editor of the Special Issue, Dr. Syna Mu, for his excellent collaboration and valuable assistance.

Freddy Gabbay
Editor

Article

Adaptive Online Learning for the Autoregressive Integrated Moving Average Models

Weijia Shao [1,*], Lukas Friedemann Radke [1], Fikret Sivrikaya [2] and Sahin Albayrak [1,2]

[1] Faculty of Electrical Engineering and Computer Science, Technische Universität Berlin, Ernst-Reuter-Platz 7, 10587 Berlin, Germany; lukas.radke@dai-labor.de (L.F.R.); sahin.albayrak@dai-labor.de (S.A.)
[2] GT-ARC Gemeinnützige GmbH, Ernst-Reuter-Platz 7, 10587 Berlin, Germany; Fikret.Sivrikaya@gt-arc.com
* Correspondence: weijia.shao@campus.tu-berlin.de

Abstract: This paper addresses the problem of predicting time series data using the autoregressive integrated moving average (ARIMA) model in an online manner. Existing algorithms require model selection, which is time consuming and unsuitable for the setting of online learning. Using adaptive online learning techniques, we develop algorithms for fitting ARIMA models without hyperparameters. The regret analysis and experiments on both synthetic and real-world datasets show that the performance of the proposed algorithms can be guaranteed in both theory and practice.

Keywords: ARIMA model; time series analysis; online optimization; online model selection

Citation: Shao, W.; Radke, L.F.; Sivrikaya, F.; Albayrak S. Adaptive Online Learning for the Autoregressive Integrated Moving Average Models. *Mathematics* **2021**, *9*, 1523. https://doi.org/10.3390/math9131523

Academic Editors: Freddy Gabbay, Ioannis K. Argyros and Mihai Postolache

Received: 19 April 2021
Accepted: 24 June 2021
Published: 29 June 2021

Publisher's Note: MDPI stays neutral with regard to jurisdictional claims in published maps and institutional affiliations.

Copyright: © 2021 by the authors. Licensee MDPI, Basel, Switzerland. This article is an open access article distributed under the terms and conditions of the Creative Commons Attribution (CC BY) license (https://creativecommons.org/licenses/by/4.0/).

1. Introduction

The autoregressive integrated moving average (ARIMA) model is an important tool for time series analysis [1], and has been successfully applied to a wide range of domains including the forecasting of household electric consumption [2], scheduling in smart grids [3], finance [4], and environment protection [5]. It specifies that the values of a time series depend linearly on their previous values and error terms. In recent years, online learning (OL) methods have been applied to estimate the univariate [6,7] and multivariate [8,9] ARIMA models for their efficiency and scalability. These methods are based on the fact that any ARIMA model can be approximated by a finite dimensional autoregressive (AR) model, which can be fitted incrementally using online convex optimization algorithms. However, to guarantee accurate predictions, these methods require a proper configuration of hyperparameters, such as the diameter of the decision set, the learning rate, the order of differencing, and the lag of the AR model. Theoretically, these hyperparameters need to be set according to prior knowledge about the data generation, which is impossible to obtain. In practice, the hyperparameters are usually tuned to optimize the goodness of fit on the unseen data, which requires numerical simulation (e.g., cross-validation) on a previously collected dataset. The numerical simulation is notoriously expensive, since it requires multiple training runs for each candidate hyperparameter configuration. Furthermore, a previously collected dataset containing ground truth is needed for validation of the fitted model, which is unsuited for the online setting. Unfortunately, the expensive tuning process needs to be regularly repeated if the statistical properties of the time series change over time in an unforeseen way.

Given a new problem of predicting time series values, it appears that tuning the hyperparameters of the online algorithms can negate the benefits of the online setting. This paper addresses this problem in the online learning framework by proposing new parameter-free algorithms for learning ARIMA models, while their performance can still be guaranteed in both theory and practice. A naive attempt for this would be to directly apply parameter-free online convex optimization (PF-OCO) algorithms to the AR approximation. However, the theoretical performance of the AR approximation and the parameter-free algorithms rely on the bounded gradient vectors of the loss function, which is unreasonable for the widely used squared error with an unbounded domain.

The key contribution of this paper is the design of online learning algorithms for ARIMA models, avoiding regular and expensive hyperparameter tuning without damaging the power of the models. Our algorithms update the model incrementally with a computational complexity that is linearly related to the size of the model parameters and the number of candidate models in each iteration. To obtain a solid theoretical foundation, we first show that, for any locally Lipschitz-continuous function, ARIMA models with fixed order of differencing can be approximated using an AR model of the same order for a large enough lag. Based on this, new algorithms are proposed for learning the AR model adaptively without requiring any prior knowledge about the model parameters. For Lipschitz-continuous loss functions, we apply a new algorithm based on the adaptive follow the regularized leader (FTRL) framework [10] and show that our algorithm achieves a sublinear regret bound depending on the data sequence and the Lipschitz constant. A special treatment on the commonly used squared error is required due to its non-Lipschitz continuity. To obtain a data-dependent regret bound, we combine a polynomial regularizer [11] with the adaptive FTRL framework. Finally, to find the proper order and lag of the AR model in an online manner, multiple AR models are simultaneously maintained, and an adaptive hedge algorithm is applied to aggregate their predictions. In the previous attempts [12,13] to solve this online model selection (OMS) problem, the exponentiated gradient (EG) algorithm has been directly applied to aggregate the predictions, which not only requires tuning the learning rate, but also yields a regret bound depending on the loss incurred by the worst model. Our adaptive hedge algorithm is parameter-free and guarantees a regret bound depending on the time series sequence. Table 1 provides a comparison of the online learning algorithms applied to the learning of the ARIMA models. In addition to the theoretical analysis, we also demonstrate the performance of the proposed algorithm using both synthetic and real-world datasets.

Table 1. Algorithms for online learning of ARIMA.

Problem	Algorithm	Reference	Tuning-Free	Loss Function	Regret Dependence
OL for ARIMA	OGD	[6–9]	✗	any	largest gradient norm
OL for ARIMA	ONS	[6–9]	✗	exp-concave	largest gradient norm
PF-OCO	Coin Betting	[14,15]	✔	normalized gradient	gradient vectors
PF-OCO	FreeRex	[16]	✔	any	largest gradient norm
PF-OCO	SF-MD	[17]	✗	any	gradient vectors
PF-OCO	SOLO-FTRL	[17]	✔	any	largest gradient norm
OL for ARIMA	Algorithm 1	This Paper	✔	Lipschitz	data sequence
OL for ARIMA	Algorithm 2	This Paper	✔	squared error	data sequence
OMS for ARIMA	EG	[12,13]	✗	bounded	loss of the worst model
OMS for ARIMA	Algorithm 3	This Paper	✔	local Lipschitz	data sequence

For non-Lipschitz-continuous loss functions, the gradient norm can be unbounded. These algorithms with performance depending on the gradient norm can fail without making further assumptions on the data generation. For OGD, the learning rate and the diameter of the decision set need to be tuned in practice. ONS has an additional hyperparameter controlling the numerical stability. Applying SF-MD to ARIMA, the diameter of the model parameter has to be tuned. To obtain optimal performance, the learning rate of EG has to be tuned.

The rest of the paper is organized as follows. Section 2 reviews the existing work on the subject. The notation, learning model, and formal description of the problem are introduced in Section 3. Next, we present and analyze our algorithms in Section 4. Section 5 demonstrates the empirical performance of the proposed methods. Finally, we conclude our work with some future research directions in Section 6.

Algorithm 1 ARIMA-AdaFTRL.

Input: $L_1 > 0$
Initialize $\theta_{1,i}$ arbitrarily, $\eta_{1,i} = 0$, $G_{i,0} = 0$ for $i = 1, \ldots, m$
for $t = 1$ to T **do**
    for $i = 1$ to m **do**
        $G_{i,t} = \max\{G_{i,t-1}, \|\nabla^d X_{t-i}\|_2\}$
        $\eta_{i,t} = \|\theta_{i,1}\|_F + \sqrt{\sum_{s=1}^{t-1} \|g_{i,s}\|_F^2 + (L_t G_{i,t})^2}$
        if $\eta_{i,t} \neq 0$ **then**
            $\gamma_{i,t} = \frac{\theta_{i,t}}{\eta_{i,t}}$
        else
            $\gamma_{i,t} = 0$
        end if
    end for
    Play $\tilde{X}_t(\gamma_t)$
    Observe X_t and $h_t \in \partial l_t(\tilde{X}_t(\gamma_t))$
    $L_{t+1} = \max\{L_t, \|g_t\|_2\}$
    for $i = 1$ to m **do**
        $g_{i,t} = g_t \nabla^d X_{t-i}^\top$
        $\theta_{i,t+1} = \theta_{i,t} - g_{i,t}$
    end for
end for

Algorithm 2 ARIMA-AdaFTRL-Poly.

Input: $G_0 > 0$
Initialize θ_1 arbitrarily, $G_1 = \max\{G_0, \|\nabla^d X_0\|_2, \ldots, \|\nabla^d X_{-m+1}\|_2\}$
for $t = 1$ to T **do**
    $\eta_t = \|\theta_1\|_F + \sqrt{\sum_{s=1}^{t-1} \|\nabla^d X_s x_s^\top\|_F^2 + (G_t \|x_t\|_2)^2}$
    $\lambda_t = \sqrt{\sum_{s=1}^{t} \|x_s\|_2^4}$
    if $\|\theta_t\|_F \neq 0$ **then**
        Select $c \geq 0$ satisfying $\lambda_t c^3 + \eta_t c = \|\theta_t\|_F$
        $\gamma_t = \frac{c \theta_t}{\|\theta_t\|_F}$
    else
        $\gamma_t = 0$
    end if
    Play $\tilde{X}_t(\gamma_t)$
    Observe X_t and $g_t = \gamma_t x_t - \nabla^d X_t$
    $G_{t+1} = \max\{G_t, \|\nabla^d X_t\|_2\}$
    $\theta_{t+1} = \theta_t - g_t x_t^\top$
end for

Algorithm 3 ARIMA-AO-Hedge.

Input: predictor $\mathcal{A}_1, \ldots, \mathcal{A}_K, d$
Initialize $\theta_{k,1} = 0$, $\eta_1 = 0$ for $i = 1, \ldots, K$
for $t = 1$ to T **do**
    Get prediction $\tilde{X}_t^i$ from $\mathcal{A}_k$ for $i = 1, \ldots, K$
    Set $Y_t = \sum_{i=0}^{d-1} \nabla^i X_{t-1}$
    Set $h_{i,t} = l(Y_t, \tilde{X}_t^i)$ for $i = 1, \ldots, K$
    if $\eta_1 = 0$ **then**
        Set $w_{i,t} = 1$ for some $i \in \arg\max_{j \in \{1, \ldots, K\}} h_{j,t}$
    else
        Set $w_{i,t} = \frac{\exp(\eta_t^{-1}(\theta_{i,t} - h_{i,t}))}{\sum_{i=1}^{K} \exp(\eta_t^{-1}(\theta_{i,t} - h_{i,t}))}$ for $i = 1, \ldots, K$
    end if
    Predict $\tilde{X}_t = \sum_{i=1}^{K} w_{i,t} \tilde{X}_t^i$
    Observe X_t, update $\mathcal{A}_i$, and set $z_{i,t} = l(X_t, \tilde{X}_t^i)$ for $i = 1, \ldots, K$
    $\theta_{t+1} = \theta_t - z_t$
    $\eta_{t+1} = \sqrt{\frac{1}{2 \log K} \sum_{s=1}^{t} \|h_t - z_t\|_\infty^2}$
end for

2. Related Work

An ARIMA model can be fitted using statistical methods such as recursive least square and maximum likelihood estimation, which are not only based on strong assumptions such as the Gaussian distributed noise terms [18], linear dependencies [19], and data generated by a stationary process [20], but also require solution of non-convex optimization problems [21]. Although these assumptions can be relaxed by considering non-Gaussian noise [22,23], non-stationary processes [24], or a convex relaxation [21], the pre-trained models still cannot deal with concept drift [7]. Moreover, retraining is time consuming and memory intensive, especially for large-scale datasets. The idea of applying regret minimization techniques to autoregressive moving average (ARMA) prediction was first introduced in [6]. The authors propose online algorithms incrementally producing predictions close to the values generated by the best ARMA model. This idea was extended to ARIMA(p,q,d) models in [7] by learning the AR(m) model of the higher-order differencing of the time series. Further extensions to multiple time series can be found in [8,9], while the problem of predicting time series with missing data was addressed in [25].

In order to obtain accurate predictions, the lag of the AR model and the order of differencing have to be tuned, which has been well studied in the offline setting. In some textbooks [20,26,27], Akaike's Information Criterion (AIC) and the Bayesian Information Criterion (BIC) are recommended for this task. Both require prior knowledge and strong assumptions about the variance of the noise [20], and are time and space consuming as they require numerical simulation such as cross-validation on previously collected datasets. Nevertheless, given a properly selected lag m and order d, online convex optimization techniques such as online Newton step (ONS) or online gradient descent (OGD) can be applied to fitting the model in the regret minimization framework [6–9]. However, both algorithms introduce additional hyperparameters to control the learning rate and numerical stability.

The idea of selecting hyperparameters for online time series prediction was proposed in [12,13]. Regarding the online AR predictor with different lags as experts, the authors aggregate over predictors by applying a multiplicative weights algorithm for prediction with expert advice. The proposed algorithm is not optimal for time series prediction, since the regret bound of the chosen algorithm depends on the largest loss incurred by the experts [28]. Furthermore, each individual expert still requires that the parameters are taken from a compact decision set, the diameter of which needs to be tuned in practice. A series of recent works on parameter-free online learning have provided possibilities of achieving sublinear regret without prior information on the decision set. In [14], the unconstrained online learning problem is modeled as a betting game, based on which a parameter-free algorithm is developed. The algorithm was further extended in [15], so a better regret bound can be achieved for strongly convex loss functions. However, the coin betting algorithm requires that the gradient vectors are normalized, which is unrealistic for unbounded time series and the squared error loss. In [16,17], the authors introduced parameter-free algorithms without requiring normalized gradient vectors. Unfortunately, the regret upper bounds of the proposed algorithms depend on the norm of the gradient vectors, which could be extremely large in our setting.

The main idea of the current work is based on the combination of the adaptive FTRL framework [10] and the idea of handling relative Lipschitz continuous functions [11], which makes it possible to devise an online algorithm with a data-dependent regret upper bound. To aggregate the results, an adaptive optimistic algorithm is proposed, such that the overall regret depends on the data sequence instead of the worst-case loss.

3. Preliminary and Learning Model

Let X_t denote the value observed at time t of a time series. We assume that X_t is taken from a finite dimensional real vector space $\mathbb{X}$ with norm $\|\cdot\|$. We denote by $\mathcal{L}(\mathbb{X}, \mathbb{X})$ the vector space of bounded linear operators from $\mathbb{X}$ to $\mathbb{X}$ and $\|\alpha\|_{op} = \sup_{x \in \mathbb{X}, x \neq 0} \frac{\|\alpha x\|}{\|x\|}$ the corresponding operator norm. An AR(p) model is given by

$$X_t = \sum_{i=1}^{p} \alpha_i X_{t-i} + \epsilon_t,$$

where $\alpha_i \in \mathcal{L}(\mathbb{X}, \mathbb{X})$ is a linear operator and $\epsilon_t \in \mathbb{X}$ is an error term. The ARMA(p, q) model extends the AR(p) model by adding a moving average (MA) component as follows:

$$X_t = \sum_{i=1}^{p} \alpha_i X_{t-i} + \sum_{i=1}^{q} \beta_i \epsilon_{t-i} + \epsilon_t,$$

where $\epsilon_t \in \mathbb{X}$ is the error term and $\beta_i \in \mathcal{L}(\mathbb{X}, \mathbb{X})$. We define the d-th order differencing of the time series as $\nabla^d X_t = \nabla^{d-1} X_t - \nabla^{d-1} X_{t-1}$ for $d \geq 1$ and $\nabla^0 X_t = X_t$. The ARIMA(p, q, d) model assumes that the d-th order differencing of the time series follows an ARMA(p, q) model. In this section, this general setting suffices for introducing the learning model. In the following sections, we fix the basis of $\mathbb{X}$ to obtain implementable algorithms, for which different kinds of norms and inner products for vectors and matrices are needed. We provide a table of required notation in Appendix C.

In this paper, we consider the setting of online learning, which can be described as an iterative game between a player and an adversary. In each round t of the game, the player makes a prediction $\tilde{X}_t$. Next, the adversary chooses some X_t and reveals it to the player, who then suffers the loss $l(X_t, \tilde{X}_t)$ for some convex loss function $l : \mathbb{X} \times \mathbb{X} \to \mathbb{R}$. The ultimate goal is to design a strategy for the player to minimize the cumulative loss $\sum_{t=1}^{T} l(X_t, \tilde{X}_t)$ of T rounds. For simplicity, we define

$$l_t : \mathbb{X} \to \mathbb{R}, X \mapsto l(X_t, X).$$

In classical textbooks about time series analysis, the signal is assumed to be generated by a model, based on which the predictions are made. In this paper, we make no assumptions on the data generation. Therefore, minimizing the cumulative loss is generally impossible. An achievable objective is to keep a possibly small regret of not having chosen some ARIMA(p, q, d) model to generate the prediction $\tilde{X}_t$. Formally, we denote by $\tilde{X}_t(\alpha, \beta)$ the prediction using the ARIMA(p, q, d) model parameterized by α and β, given by (in this paper, we do not directly address the problem of the cointegration, where the third term should be applied to a low-rank linear operator):

$$\tilde{X}_t(\alpha, \beta) = \sum_{i=1}^{p} \alpha_i \nabla^d X_{t-i} + \sum_{i=1}^{q} \beta_i \epsilon_{t-i} + \sum_{i=0}^{d-1} \nabla^i X_{t-1}. \tag{1}$$

The cumulative regret of T rounds is then given by

$$R_T(\alpha, \beta) = \sum_{t=1}^{T} l_t(\tilde{X}_t) - \sum_{t=1}^{T} l_t(\tilde{X}_t(\alpha, \beta)).$$

The goal of this paper is to design a strategy for the player such that the cumulative regret grows sublinearly in T. In the ideal case, in which the data are actually generated by an ARIMA process, the prediction generated by the player yields a small loss. Otherwise, the predictions are always close to those produced by the best ARIMA model, independent of the data generation. Following the adversarial setting in [6], we allow the sequences $\{X_t\}$, $\{\epsilon_t\}$ and the parameters α, β to be selected by the adversary. Without any restrictions on the model, this is no different than the impossible task of minimizing the cumulative loss, since ϵ_{t-1} can always be selected such that $X_t = \tilde{X}_t(\alpha, \beta)$ holds for all t. Therefore, we make the following assumptions throughout this paper:

Assumption 1. $X_t = \epsilon_t + \tilde{X}_t(\alpha, \beta)$, and there is some $R > 0$ such that $\|\epsilon_t\| \leq R$ for all $t = 1, \ldots T$.

Assumption 2. The coefficients β_i satisfy $\sum_{i=1}^{q} \|\beta_i\|_{\text{op}} \leq 1 - \epsilon$ for some $\epsilon > 0$.

Since we are interested in competing against predictions generated by ARIMA models, we assume that ϵ_t is selected as if X_t is generated by the ARIMA process. Furthermore, we assume the norm $\|\epsilon_t\|$ is upper bounded within T iterations. Assumption 2 is a sufficient condition for the MA component to be invertible, which prevents it from going to infinity as $t \to \infty$ [27].

Our work is based on the fact that we can compete against an ARIMA(p, q, d) model by taking predictions from an AR(m) model of the d-th order differencing for large enough m, which is shown in the following lemma, the proof of which can be found in Appendix A.

Lemma 1. Let $\{X_t\}$, $\{\epsilon_t\}$, α, and β be as assumed in Assumptions 1 and 2. Then there is some $\gamma \in \mathcal{L}(\mathbb{X}, \mathbb{X})^m$ with $m \geq \frac{q \log T}{\log \frac{1}{1-\epsilon}} + p$ such that

$$\|\nabla^d \tilde{X}_t(\gamma) - \nabla^d \tilde{X}_t(\alpha, \beta)\| \leq (1-\epsilon)^{\frac{t}{q}} R + \frac{2R}{T}$$

holds for all $t = 1 \ldots T$, where we define $\nabla^d \tilde{X}_t(\gamma) = \sum_{i=1}^{m} \gamma_i \nabla^d X_{t-i}$.

As can be seen from the lemma, a prediction $\tilde{X}_t(\gamma)$ generated by the process

$$\tilde{X}_t(\gamma) = \sum_{i=1}^{m} \gamma_i \nabla^d X_{t-i} + \sum_{i=0}^{d-1} \nabla^i X_{t-1}$$

is close to the prediction $\tilde{X}_t(\alpha, \beta)$ generated by the ARIMA process. In the previous works [6,7], the loss function l_t is assumed to be Lipschitz continuous to control the difference of loss incurred by the approximation. In general, this does not hold for squared error. However, from Assumption 1 and Lemma 1, it follows that both $\tilde{X}_t(\alpha, \beta)$ and $\tilde{X}_t(\gamma)$

lie in a compact set around X_t with a bounded diameter. Given the convexity of l, which is local Lipschitz continuous in the compact convex domain, we obtain a similar property:

$$l(X_t, \tilde{X}_t(\gamma)) - l(X_t, \tilde{X}_t(\alpha, \beta)) \leq L(X_t)\|\nabla^d \tilde{X}_t(\gamma) - \nabla^d \tilde{X}_t(\alpha, \beta)\|,$$

where $L(X_t)$ is some constant depending on X_t. For squared error, it is easy to verify that the Lipschitz constant depends on $\|\nabla^d X_t\|$, the boundedness of which can be reasonably assumed. To avoid extraneous details, we simply add the third assumption:

Assumption 3. *Define set $\mathcal{X}_t = \{X \in \mathbb{X} | \|X - X_t\| \leq 4R\}$. There is a compact convex set $\mathcal{X} \supseteq \bigcup_{t=1}^{T} \mathcal{X}_t$, such that l_t is L-Lipschitz continuous in $\mathcal{X}$ for $t = 1, \ldots T$.*

The next corollary shows that the losses incurred by the ARIMA and its approximation are close, which allows us to take predictions from the approximation.

Corollary 1. *Let $\{X_t\}$, $\{\epsilon_t\}$, α, β, and l be as assumed in Assumptions 1–3. Then there is some $\gamma \in \mathcal{L}(\mathbb{X}, \mathbb{X})^m$ with $m \geq \frac{q \log T}{\log \frac{1}{1-\epsilon}} + p$, such that*

$$\sum_{t=1}^{T} l_t(\tilde{X}_t(\gamma)) - l_t(\tilde{X}_t(\alpha, \beta)) \leq LR\left(\frac{1}{1-(1-\epsilon)^{\frac{1}{q}}} + 2\right)$$

holds for all $t = 1 \ldots T$.

Proof. It follows from Assumption 1 and Lemma 1 that $\tilde{X}_t(\gamma), \tilde{X}_t(\alpha, \beta) \in \mathcal{X}$ holds for all $t = 1, \ldots T$. Together with Assumption 3, we obtain

$$\sum_{t=1}^{T} (l_t(\tilde{X}_t(\gamma)) - l_t(\tilde{X}_t(\alpha, \beta))) \leq L \sum_{t=1}^{T} \|\tilde{X}_t(\gamma) - \tilde{X}_t(\alpha, \beta)\|.$$

Applying Lemma 1, we obtain the claimed result. □

4. Algorithms and Analysis

From Corollary 1, it follows clearly that an ARIMA(p, q, d) model can be approximated by an integrated AR model with large enough m. However, neither the order of differencing d nor the lag m is known. To circumvent tuning them using a previously collected dataset, we propose a framework with a two-level hierarchical construction, which is described in Algorithm 4.

Algorithm 4 Two-level framework.

Input: K instances of the slave algorithm $\mathcal{A}_1, \ldots, \mathcal{A}_K$. An instance of master algorithm $\mathcal{M}$.
for $t = 1$ to T do
   Get $\tilde{X}_t^i$ from each $\mathcal{A}_i$
   Get $w_t \in \Delta^K$ from $\mathcal{M}$ ▷ Δ^K is the standard K-simplex
   Integrate the prediction: $\tilde{X}_t = \sum_{i=1}^{K} w_t^i \tilde{X}_t^i$
   Observe X_t
   Define $z_t \in \mathbb{R}^K$ with $z_{i,t} = l_t(\tilde{X}_t^i)$
   Update $\mathcal{A}_i$ using $z_{i,t}$ for $i = 1, \ldots, K$
   Update $\mathcal{M}$ using z_t
end for

The idea is to maintain a master algorithm $\mathcal{M}$ and a set of slave algorithms $\{\mathcal{A}_m | m = 1, \ldots, K\}$. At each step t, the master algorithm receives predictions $\tilde{X}_t^k$ from $\mathcal{A}_k$ for $k = 1, \ldots, K$. Then it comes up with a convex combination $\tilde{X}_t = \sum_{i=1}^{K} w_t^i \tilde{X}_t^i$ for some $w_t \in \Delta$ in the simplex. Next, it observes X_t and computes the loss $l_t(X_t^k(\gamma))$ for each slave

$\mathcal{A}_k$, which is then used to update $\mathcal{A}_k$ and w_{t+1}. Let $\{\tilde{X}_t^k\}$ be the sequence generated by some slave k. We define the regret of not having chosen the prediction generated by slave k as

$$R_T(k) = \sum_{t=1}^T l_t(\sum_{i=1}^K w_t^i \tilde{X}_t^i) - \sum_{t=1}^T l_t(\tilde{X}_t^k),$$

and the regret of the slave k

$$R_T(\mathcal{A}_k) = \sum_{t=1}^T l_t(\tilde{X}_t^k) - \sum_{t=1}^T l_t(\tilde{X}_t(\gamma_k)),$$

where $\tilde{X}_t(\gamma_k)$ is the prediction generated by an integrated AR model parameterized by γ_k. Let $\mathcal{A}_k$ be some slave. Then the regret of this two-level framework can obviously be decomposed as

$$R_T(\alpha, \beta) = R_T(k) + R_T(\mathcal{A}_k) + \underbrace{\sum_{t=1}^T l_t(\tilde{X}_t(\gamma_k)) - \sum_{t=1}^T l_t(\tilde{X}_t(\alpha, \beta))}_{\text{Corollary 1}}.$$

For γ_k, α, and β satisfying the condition in Corollary 1 (this is not a condition of having a correct algorithm—with more slaves, there are more α, β satisfying the condition; we increase the freedom of the model by increasing the number of slaves), the marked term above is upper bounded by a constant, that is,

$$\sum_{t=1}^T l_t(\tilde{X}_t(\gamma_k)) - \sum_{t=1}^T l_t(\tilde{X}_t(\alpha, \beta)) \in \mathcal{O}(1).$$

If the regret of the master and the slaves grow sublinearly in T, we can achieve an overall sublinear regret upper bound, which is formally described in the following corollary.

Corollary 2. Let $\mathcal{A}_i$ be an online learning algorithm against an $AR(m_i)$ model parameterized by γ^i for $i = 1, \ldots, K$. For any ARIMA model parameterized by α and β, if there is a $k \in \{1, \ldots, K\}$ such that $\tilde{X}_t(\gamma^k)$, $\tilde{X}_t(\alpha, \beta)$ and $\{X_t\}$ satisfy Assumptions 1–3, then running Algorithm 4 with $\mathcal{M}$ and $\mathcal{A}_1, \ldots, \mathcal{A}_K$ guarantees

$$\sum_{t=1}^T (l_t(\tilde{X}_t) - l_t(\tilde{X}_t(\alpha, \beta))) \leq \mathcal{R}_T(k) + \mathcal{R}_T(\mathcal{A}_k) + \mathcal{O}(1).$$

Next, we design and analyze parameter-free algorithms for the slaves and the master.

4.1. Parameter-Free Online Learning Algorithms

4.1.1. Algorithms for Lipschitz Loss

Given fixed m and d, an integrated $AR(m)$ model can be treated as an ordinary linear regression model. In each iteration t, we select $\gamma_t = (\gamma_{1,t}, \ldots, \gamma_{m,t}) \in \mathcal{L}(\mathbb{X}, \mathbb{X})^m$ and make prediction

$$\tilde{X}_t(\gamma_t) = \sum_{i=1}^m \gamma_{i,t} \nabla^d X_{t-i} + \sum_{i=0}^{d-1} \nabla^i X_{t-1}.$$

Since l_t is convex, there is some subdifferential $g_t \in \partial l_t(\tilde{X}_t(\gamma_t))$ such that

$$l_t(\tilde{X}_t(\gamma_t)) - l_t(\tilde{X}_t(\gamma)) \leq g_t(\sum_{i=1}^m (\gamma_{i,t} - \gamma_i) \nabla^d X_{t-i}),$$

for all $\gamma \in \mathcal{L}(\mathbb{X}, \mathbb{X})^m$. Define $g_{i,t} : \mathcal{L}(\mathbb{X}, \mathbb{X}) \to \mathbb{R}, v \mapsto g_t(v \nabla^d X_{t-i})$. The regret can be further upper bounded by

$$\sum_{t=1}^T l_t(\tilde{X}_t(\gamma_t)) - l_t(\tilde{X}_t(\gamma)) \leq \sum_{t=1}^T \sum_{i=1}^m g_{i,t}(\gamma_{i,t} - \gamma_i). \qquad (2)$$

Thus, we can cast the online linear regression problem to an online linear optimization problem. Unlike the previous work, we focus on the unconstrained setting, where γ_t is not picked from a compact decision set. In this setting, we can apply an FTRL algorithm with an adaptive regularizer. To obtain an efficient implementation, we fix a basis for both $\mathbb{X}$ and $\mathbb{X}_*$. Now we can assume $\mathbb{X} = \mathbb{X}_* = \mathbb{R}^n$ and work with the matrix representation of $\gamma \in \mathcal{L}(\mathbb{X}, \mathbb{X})$. It is easy to verify that (2) can be rewritten as

$$\sum_{t=1}^T l_t(\tilde{X}_t(\gamma_t)) - l_t(\tilde{X}_t(\gamma)) \leq \sum_{t=1}^T \sum_{i=1}^m \langle g_t \nabla^d X_{t-i}^\top, \gamma_{i,t} - \gamma_i \rangle_F,$$

where $\langle A, B \rangle_F = \mathrm{tr}(A^\top B)$ is the Frobenius inner product. It is well known that the Frobenius inner product can be considered as a dot product of vectorized matrices, with which we obtain a simple first-order (the computational complexity per iteration depends linearly on the dimension of the parameter, i.e., $\mathcal{O}(n^2 m)$) algorithm described in Algorithm 1.

The cumulative regret of Algorithm 1 can be upper bounded using the following theorem.

Theorem 1. *Let $\{X_t\}$ be any sequence of vectors taken from $\mathbb{X}$. Algorithm 1 guarantees*

$$\sum_{t=1}^T l_t(\tilde{X}_t(\gamma_t)) - l_t(\tilde{X}_t(\gamma))$$
$$\leq \sum_{i=1}^m \left(\frac{\|\gamma_i\|_F^2 L_{T+1}}{2} + L_{T+1} + \frac{L_{T+1}^2}{L_1} \right) \sqrt{\sum_{t=1}^T \|\nabla^d X_{t-i}\|_2^2}$$
$$+ \sum_{i=1}^m \frac{(L_{T+1} G_{i,T+1} + \|\theta_{i,1}\|_F) \|\gamma_i\|_F^2 + \|\theta_{i,1}\|_F}{2}.$$

For an L-Lipschitz loss function l_t, in which L_{T+1} is upper bounded by L, we obtain a sublinear regret upper bound depending on the sequence of d-th order differencing $\{\nabla^d X_t\}$. In case L is known, we can set $L_0 = L$, otherwise picking L_0 arbitrarily from a reasonable range (e.g., $L_0 = 1$) would not have a devastating impact on the performance of the algorithms.

4.1.2. Algorithms for Squared Errors

For the commonly used squared error given by

$$l_t(\tilde{X}_t(\gamma_t)) = \frac{1}{2} \|\tilde{X}_t(\gamma_t) - X_t\|_2^2,$$

it can be verified that g_t can be represented as a vector

$$g_t = \sum_{i=1}^m \gamma_{i,t} \nabla^d X_{t-i} - \nabla^d X_t$$

for all t. Existing algorithms, which have a regret upper bound depending on $\|g_t\|_2$, could fail since $\|g_t\|_2$ can be set arbitrarily large due to the adversarially selected data sequence $X_1, \ldots, X_t$. To design a parameter-free algorithm for the squared error, we equip FTRL with a time-varying polynomial regularizer described in Algorithm 2.

Define
$$x_t = \begin{pmatrix} \nabla^d X_{t-1} \\ \vdots \\ \nabla^d X_{t-m} \end{pmatrix}$$

and consider the matrix representation $\gamma_t = (\gamma_{1,t} \cdots \gamma_{m,t})$. Then we have $g_t = \gamma_t x_t - \nabla^d X_t$, and the upper bound of the regret can be rewritten as

$$\sum_{t=1}^{T} l_t(\tilde{X}_t(\gamma_t)) - l_t(\tilde{X}_t(\gamma)) \leq \sum_{t=1}^{T} \langle (\gamma_t x_t - \nabla^d X_t) x_t^\top, \gamma_t - \gamma \rangle_F.$$

The idea of Algorithm 2 is to run the FTRL algorithm with a polynomial regularizer

$$\frac{\lambda_t}{4} \|\gamma\|_F^4 + \frac{\eta_t}{2} \|\gamma\|_F^2,$$

for increasing sequences $\{\lambda_t\}$ and $\{\eta_t\}$, which leads to updating rule given by

$$\gamma_t = \arg\max_{\gamma \in \mathcal{L}(\mathbb{X}, \mathbb{X})^m} \langle \theta_t, \gamma \rangle_F - \frac{\lambda_t}{4} \|\gamma\|_F^4 - \frac{\eta_t}{2} \|\gamma\|_F^2 = \frac{c\theta_t}{\|\theta_t\|_F},$$

for c satisfying $\lambda_t c^3 + \eta_t c = \|\theta_t\|_F$. Since we have $\lambda_t \geq 0$ and $\eta_t > 0$ for $\theta_1 \neq 0$, c exists and has a closed-form expression. The computational complexity per iteration has a linear dependency on the dimension of $\mathcal{L}(\mathbb{X}, \mathbb{X})^m$. The following theorem provides a regret upper bound of Algorithm 2.

Theorem 2. *Let $\{X_t\}$ be any sequence of vectors taken from $\mathbb{X}$ and*

$$l_t(\tilde{X}_t(\gamma)) = \frac{1}{2} \|X_t - \tilde{X}_t(\gamma)\|_2^2 = \frac{1}{2} \|\nabla^d X_t - \nabla^d \tilde{X}_t(\gamma)\|_2^2$$

be the squared error. We define $x_t = (\nabla^d X_{t-1} \cdots \nabla^d X_{t-m})^\top$ and $\gamma = (\gamma_1 \cdots \gamma_m)$, the matrix representation of $\gamma_1, \ldots \gamma_m \in \mathcal{L}(\mathbb{X}, \mathbb{X})$. Then, Algorithm 2 guarantees

$$\sum_{t=1}^{T} (l_t(\tilde{X}_t(\gamma_t)) - l_t(\tilde{X}_t(\gamma))) \leq \frac{(\sqrt{m} G_{T+1}^2 + \|\theta_1\|_F) \|\gamma\|_F^2}{2}$$

$$+ \|\theta_1\|_F + (1 + \frac{\|\gamma\|_F^4}{4}) \sqrt{\sum_{t=1}^{T} \|x_t\|_2^4}$$

$$+ (1 + \frac{G_{T+1}}{G_0} + \frac{\|\gamma\|_F^2}{2}) \sqrt{\sum_{t=1}^{T} \|\nabla^d X_t x_t^\top\|_F^2}$$

for all $\gamma \in \mathcal{L}(\mathbb{X}, \mathbb{X})^m$.

For squared error, Algorithm 2 does not require a compact decision set and ensures a sublinear regret bound depending on the data sequence. Similar to Algorithm 1, one can set G_0 according to the prior knowledge about the bounds of the time series. Alternatively, we can simply set $G_0 = 1$ to obtain a reasonable performance.

4.2. Online Model Selection Using Master Algorithms

The straightforward choice of the master algorithm would be the exponentiated gradient algorithm for prediction with expert advice. However, this algorithm requires tuning of the learning rate and losses bounded by a small quantity, which can not be assumed for our case. The AdaHedge algorithm [29] solves these problems. However, it yields a worst-case regret bound depending on the largest loss observed, which could be much worse compared to a data-dependent regret bound.

Our idea is based on the adaptive optimistic follow the regularized leader (AO-FTRL) framework [10]. Given a sequence of hints $\{h_t\}$ and loss vectors $\{z_t\}$, AO-FTRL guarantees a regret bound related to $\sum_{t=1}^{T} \|z_t - h_t\|_t^2$ for some time-varying norm $\|\cdot\|_t$. In our case, where the loss incurred by a slave is given by $l(X_t, \tilde{X}_t^k)$ at iteration t, we simply choose $h_{k,t} = l(\sum_{i=0}^{d-1} \nabla^i X_{t-1}, \tilde{X}_t^k)$. If l is L-Lipschitz in its first argument, then we have $|z_{k,t} - h_{k,t}| \leq L\|\nabla^d X_t\|$, which leads to a data-dependent regret. The obtained algorithm is described in Algorithm 3. Its regret is upper bounded by the following theorem, the proof of which is provided in Appendix B.

Theorem 3. *Let $\{\tilde{X}_t\}$, $\{\tilde{X}_t^k\}$, $\{z_t\}$, $\{h_t\}$, and $\{w_t\}$ be as generated in Algorithm 3. Assume l is L-Lipschitz in its first argument and convex in its second argument. Then for any sequence $\{X_t\}$ and slave algorithm $\mathcal{A}_k$, we have*

$$\mathcal{R}_T(k) \leq (\sqrt{2 \log K} + \sqrt{\frac{8}{\log K}})\sqrt{\sum_{t=1}^{T} L^2 \|\nabla^d X_t\|_2^2}.$$

By Corollary 2, combining Algorithm 3 with Algorithms 1 or 2 guarantees a data-dependent regret upper bound sublinear in T. Note that there is an input parameter d for Algorithm 3, which can be adjusted according to the prior knowledge of the dataset such that $\|\nabla^d X_t\|_2^2$ can be bounded by a small quantity. In case no prior knowledge can be obtained, we can set d to the maximal order of differencing used in the slave algorithms. Arguably, the Lipschitz continuity is not a reasonable assumption for squared error with unbounded domain. With a bounded $\|\nabla^d X_t\|_2^2$, we can assume that the loss function is locally Lipschitz, but with a Lipschitz constant depending on the prediction. In the next section, we show the performance of Algorithm 3 in combination with Algorithms 1 and 2 in different experimental settings.

5. Experiments and Results

In this section, we carry out experiments on both synthetic and real-world data to show that the proposed algorithms can generate promising predictions without tuning hyperparameters.

5.1. Experiment Settings

The synthetic data was generated randomly. We run 20 trials for each synthetic experiment and average the results. For numerical stability, we scale the real-world data down so that the values are between 0 and 10. Note that the range of the data are not assumed or used in the algorithms.

Setting 1: Sanity Check

For a sanity check, we generate a stationary 10-dimensional ARIMA$(5,2,1)$ process using randomly drawn coefficients.

Setting 2: Time-Varying Parameters

Aimed at demonstrating the effectiveness of the proposed algorithm in the non-stationary case, we generate the non-stationary 10-dimensional ARIMA$(5,2,1)$ process using time-varying parameters. We draw α_1, α_2, and β_1, β_2 randomly and independent, and generate data at iteration t with the ARIMA$(5,2,1)$ model parameterized by $\alpha_t = \frac{t}{10^4}\alpha_1 + (1 - \frac{t}{10^4})\alpha_2$ and $\beta_t = \frac{t}{10^4}\beta_1 + (1 - \frac{t}{10^4})\beta_2$.

Setting 3: Time-Varying Models

To get more adversarially selected time series values, we generate the first half of the values using a stationary 10-dimensional ARIMA(5, 2, 1) model and the second half of the values using a stationary 10-dimensional ARIMA(5, 2, 0) model. The model parameters are drawn randomly.

Stock Data: Time Series with Trend

Following the experiments in [8], we collect the daily stock prices of seven technology companies from Yahoo Finance together with the S&P 500 index for over twenty years, which has an obvious increasing trend and is believed to exhibit integration.

Google Flu Data: Time Series with Seasonality

We collect estimates of influenza activity of the northern hemisphere countries, which has an obvious seasonal pattern. In the experiment, we examine the performance of the algorithms for handling regular and predictable changes that occur over a fixed period.

Electricity Demand: Trend and Seasonality

In this setting, we collect monthly load, gross electricity production, net electricity consumption, and gross demand in Turkey from 1976 to 2010. The dataset contains both trend and seasonality.

5.2. Experiments for the Slave Algorithms

We first fix $d = 1$ and $m = 16$ and compare our slave algorithms with ONS and OGD from [9] for squared error $l_t(\tilde{X}_t) = \frac{1}{2}\|X_t - \tilde{X}_t\|_2^2$ and Euclidean distance $l_t(\tilde{X}_t) = \|X_t - \tilde{X}_t\|_2$. ONS and OGD stack and vectorize the parameter matrices, and incrementally update the vectorized parameter respectively using the following rules

$$w_{t+1} = \Pi_{\mathcal{W}}(w_t - \eta(\sum_{s=1}^{t} g_t g_t^\top + \lambda I)^{-1} g_t)$$

and

$$w_{t+1} = \Pi_{\mathcal{W}}(w_t - \eta g_t),$$

where g_t is the vectorized gradient at step t, $\mathcal{W}$ is the decision set satisfying $\sup_{u \in \mathcal{W}} \|u\|_2 \leq c$, and the operator $\Pi_{\mathcal{W}}(v)$ projects v into $\mathcal{W}$. We select a list of candidate values for each hyperparameter, evaluate their performance on the whole dataset, and select the configuration with the best performance for comparison. Since the synthetic data are generated randomly, we average the results over 20 trials for stability. The corresponding results are shown in Figures 1–6 (to amplify the differences of the algorithms, we use log plots for the y-axis for all settings; for the synthetic datasets, we also use log plot for the x-axis, so that the behavior of the algorithms in the first 1000 steps can be better observed). To show the impact of the hyperparameters on the performance of the baseline algorithm, we also plot their performance using sub-optimal configurations. Note that since the error term ϵ_t cannot be predicted, an ideal predictor would suffer an average error rate of at least $\|\epsilon_t\|_2^2$ and $\|\epsilon_t\|_2$ for the two kinds of loss function. This is known for the synthetic datasets and plotted in the figures.

In all settings, both AdaFTRL and AdaFTRL-Poly have a performance on par with well-tuned OGD and ONS, which can have extremely bad performance using sub-optimal hyperparameter configurations. In the experiments using synthetic datasets, AdaFTRL suffers large loss at the beginning while generating accurate predictions after 1000 iterations. The relative performances of the proposed algorithms after the first 1000 iterations compared to the best tuned baseline algorithms are plotted in Appendix D. AdaFTRL-Poly has more stable performance compared to AdaFTRL. In the experiment with Google Flu data, all algorithms suffer huge losses around iteration 300 due to an abrupt change in the dataset. OGD and ONS with sub-optimal hyperparameter configurations, despite good performance for the first half of the data, generate very inaccurate predictions after the abrupt change in the dataset. This could lead to a catastrophic failure in practice, when certain patterns do not appear in the dataset collected for hyperparameter tuning. Our algorithms are more robust against this change and perform similarly to OGD and ONS with optimal hyperparameter configurations.

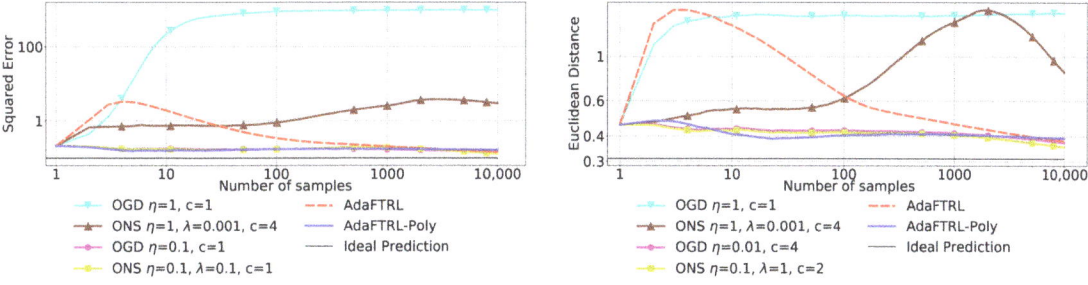

Figure 1. Results for setting 1 (sanity check), using a stationary ARIMA(5,2,1) model.

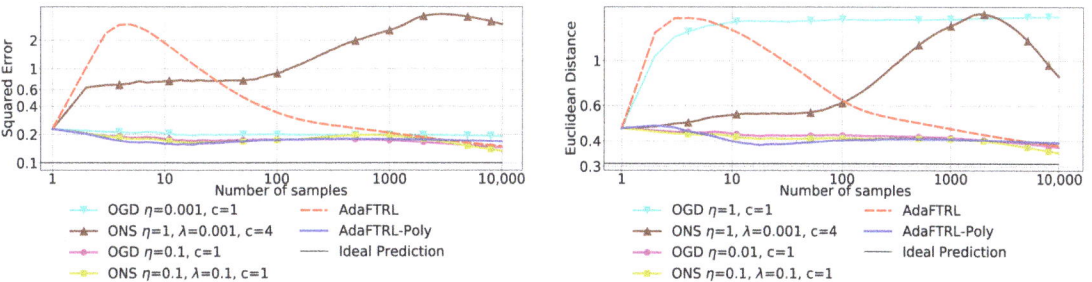

Figure 2. Results for setting 2 (time-varying parameters), using a non-stationary ARIMA(5,2,1) model.

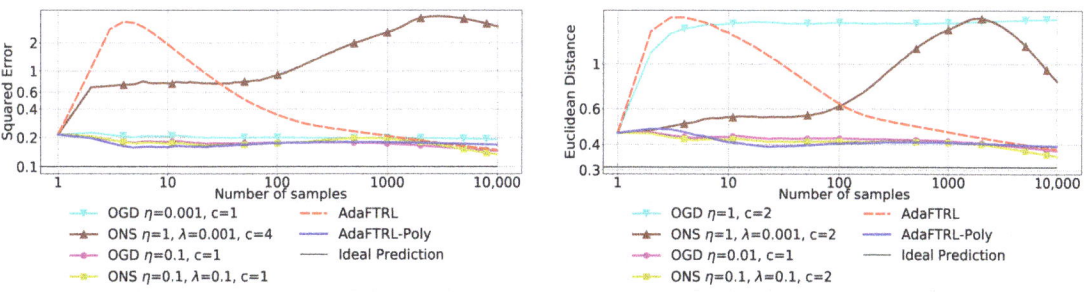

Figure 3. Results for setting 3 (time-varying models), using a combination of stationary ARIMA(5,2,1) and ARIMA(5,2,0) models.

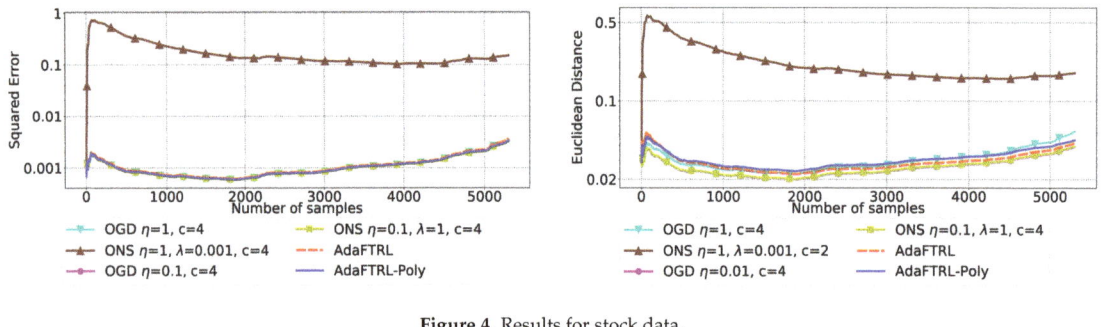

Figure 4. Results for stock data.

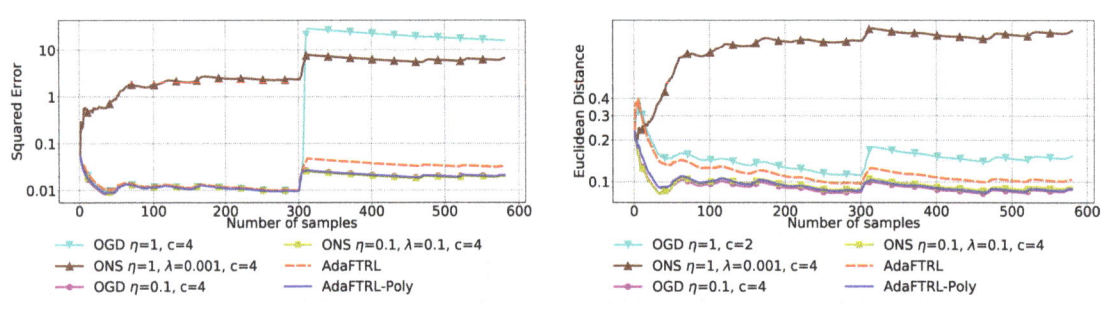

Figure 5. Results for Google Flu data.

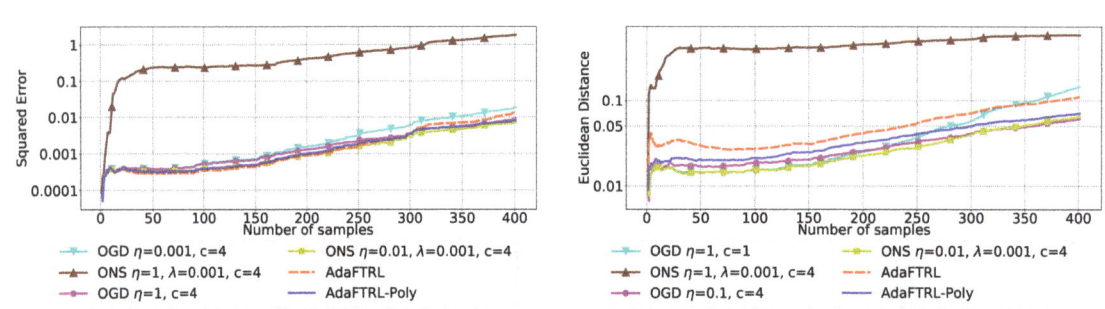

Figure 6. Results for electricity demand data.

5.3. Experiments for Online Model Selection

The performance of the two-level framework and Algorithm 3 for online model selection is demonstrated in Figures 7–12. We simultaneously maintain 96 AR(m) models of d-th-order differencing for $m = 1, \ldots 32$ and $d = 0, \ldots 2$, which are updated by Algorithms 1 and 2 for squared error and Euclidean distance, respectively. The predictions generated by the AR models are aggregated using Algorithm 3 and the aggregation algorithm (AA) introduced in [13] with learning rate set to $\sqrt{T}$. We compare the average losses incurred by the aggregated predictions with those incurred by the best AR model. To show the impact of m and d, we also plot the average loss of some other sub-optimal AR models.

In all settings, AO-Hedge outperforms AA, although the differences are very slight in some of the experiments. We would like to stress again that the choice of the hyperparameters has a great impact on the performance of the AR model. In settings 1–3, the AR model with 0-th-order differencing has the best performance, although the data are generated using $d = 1$, which suggests that the prior knowledge about the data generation may not be helpful for the model selection in all cases. The experimental results also show that AO-Hedge has a performance similar to the best AR model.

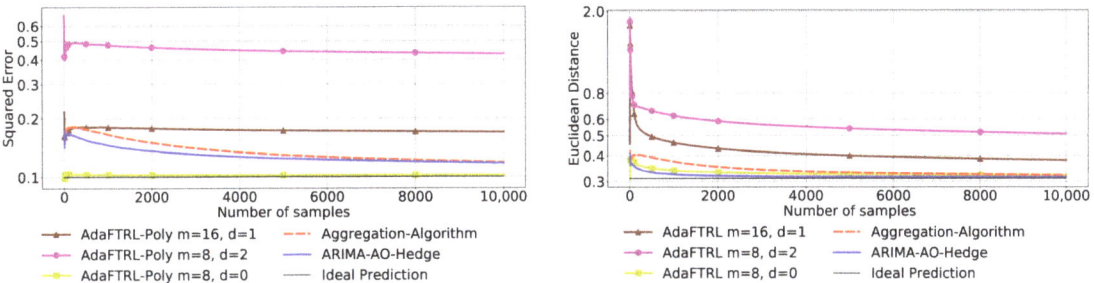

Figure 7. Model selection in setting 1.

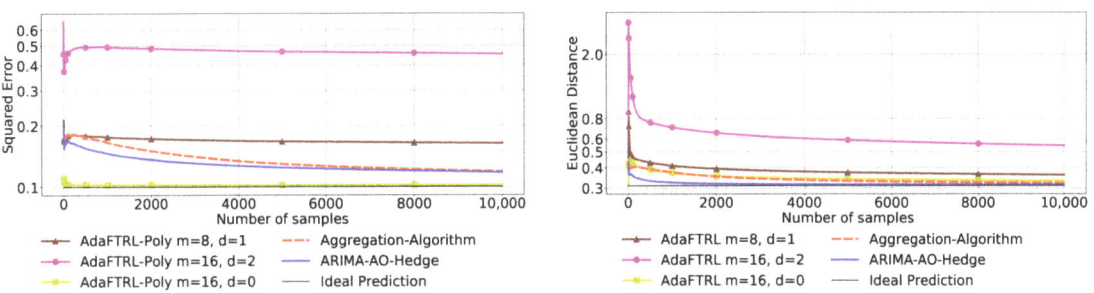

Figure 8. Model selection in setting 2.

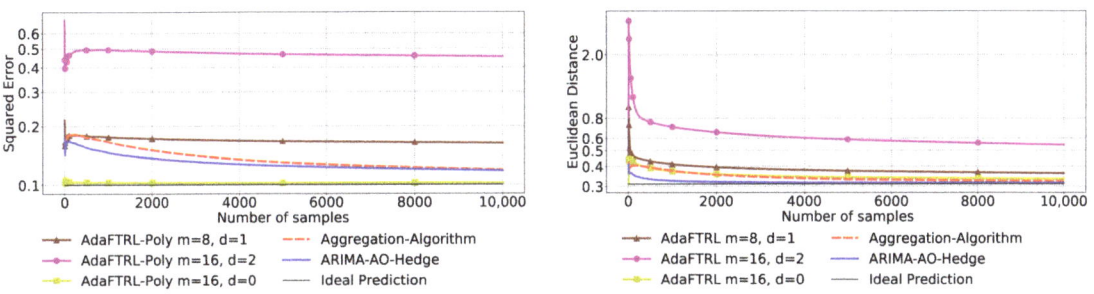

Figure 9. Model selection in setting 3.

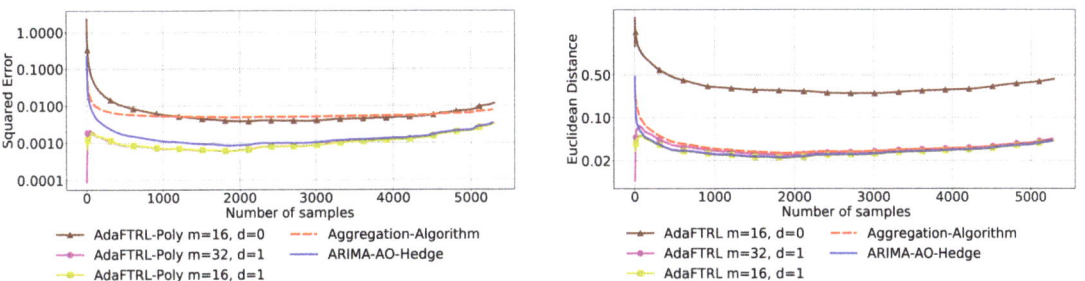

Figure 10. Model selection for stock data.

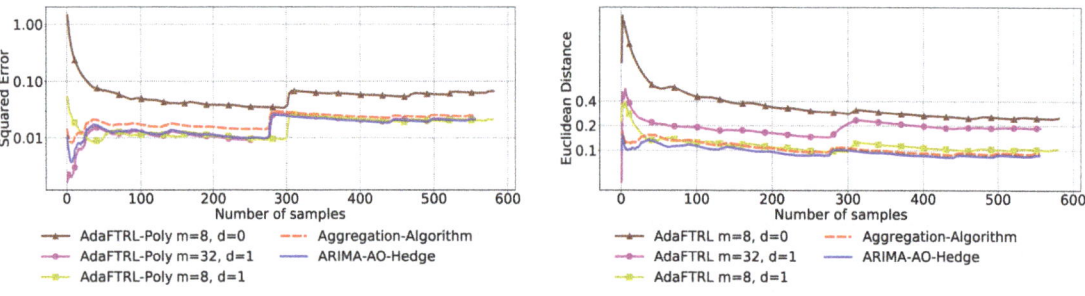

Figure 11. Model selection for Google Flu.

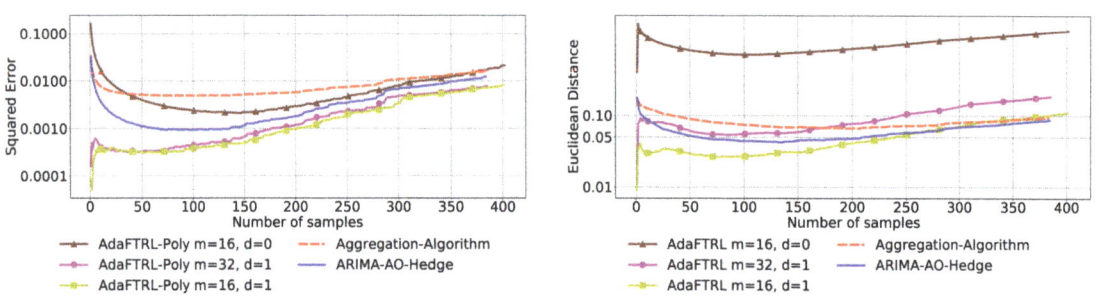

Figure 12. Model Selection for electricity demand.

6. Conclusions

We proposed algorithms for fitting ARIMA models in an online manner without requiring prior knowledge or tuning hyperparameters. We showed that the cumulative regret of our method grows sublinearly with the number of iterations and depends on the values of the time series. The comparison study on both synthetic and real-world datasets suggests that the proposed algorithms have a performance on par with the well-tuned state-of-the-art algorithms.

There are still several remaining issues that we want to address in future research. Firstly, it would be interesting to also develop a parameter-free algorithm for the cointegrated vector ARMA model. Secondly, we believe that the strong assumption on the β coefficient can be relaxed for multi-dimensional time series by generalizing Lemma 2 in [7]. Furthermore, we are also interested in applying online learning to other time series models such as the (generalized) ARCH model [30]. Finally, the proposed algorithms need to be empirically analyzed using more real-world datasets and loss functions, and compared with more recent predictive models such as recurrent neural networks and the models combining neural networks and ARIMA models [31].

Author Contributions: Conceptualization, W.S.; methodology, W.S. and L.F.R.; validation, W.S., L.F.R., and F.S.; formal analysis, W.S.; investigation, W.S. and L.F.R.; writing—original draft preparation, W.S. and L.F.R.; writing—review and editing, W.S., L.F.R., F.S., and S.A.; visualization, L.F.R.; supervision, F.S. and S.A. All authors have read and agreed to the published version of the manuscript.

Funding: We acknowledge support by the German Research Foundation and the Open Access Publication Fund of TU Berlin.

Institutional Review Board Statement: Not applicable

Informed Consent Statement: Not applicable

Data Availability Statement: The source code for generating the synthetic data set, the implementation of the algorithms, and the detailed information about our experiments are available on GitHub: https://github.com/OnlinePredictorTS/AOLForTimeSeries (accessed on March 2021). The stock data are collected from https://finance.yahoo.com/ (accessed on March 2021). The Google Flu data are available in https://github.com/datalit/googleflutrends/ (accessed on March 2021). The detailed information about the electricity demand can be found in [32].

Conflicts of Interest: The authors declare no conflicts of interest.

Appendix A

We prove Lemma 1 in this section. Consider the ARIMA model given by

$$\nabla^d X_t(\alpha,\beta) = \sum_{i=1}^{p} \alpha_i \nabla^d X_{t-i} + \sum_{i=1}^{q} \beta_i \epsilon_{t-i} + \epsilon_t$$

with $\nabla^d X_t(\alpha,\beta) = \nabla^d X_t$ for $t \leq 0$. Let

$$X_t(\alpha,\beta) = \nabla^d X_t(\alpha,\beta) + \sum_{i=0}^{d-1} \nabla^i X_{t-1}$$

be the t-th value generated by the ARIMA process. To prove Lemma 1, we generalize the proof provided in [6]. To remove the MA component, we first recursively define a growing process of the d-th-order differencing

$$\nabla^d X_t^\infty(\alpha,\beta) = \sum_{i=1}^{p} \alpha_i \nabla^d X_{t-i} + \sum_{i=1}^{q} \beta_i (\nabla^d X_{t-i} - \nabla^d X_{t-i}^\infty(\alpha,\beta))$$

with $\nabla^d X_t^\infty(\alpha,\beta) = \nabla^d X_t$ for $t \leq 0$. Let

$$X_t^\infty(\alpha,\beta) = \nabla^d X_t^\infty(\alpha,\beta) + \sum_{i=0}^{d-1} \nabla^i X_{t-1}$$

be the t-th value generated by this process.

The next lemma shows that it approximates an ARIMA(p,q,d) process.

Lemma A1. *For any α, β, and $\{\epsilon_t\}$ satisfying Assumptions 1 and 2, we have, for $t = 1, \ldots, T$,*

$$\|X_t^\infty(\alpha,\beta) - \tilde{X}_t(\alpha,\beta)\| \leq (1-\epsilon)^{\frac{t}{q}} R.$$

Proof. First of all, we have

$$X_t^\infty(\alpha,\beta) - \tilde{X}_t(\alpha,\beta) = \nabla^d X_t^\infty(\alpha,\beta) - \nabla^d \tilde{X}_t(\alpha,\beta)$$

$$= \sum_{i=1}^{q} \beta_i (\nabla^d X_{t-i} - \nabla^d X_{t-i}^\infty(\alpha,\beta) - \epsilon_{t-i})$$

for $t \geq 0$. Define $Y_t = \nabla^d X_t - \nabla^d X_t^\infty(\alpha,\beta) - \epsilon_t$. W.l.o.g. we can assume $\|\epsilon_t\| \leq R$ for $t \leq 0$. Next, we prove by induction on t that $\|Y_\tau\| \leq (1-\epsilon)^{\frac{\tau}{q}} R$ holds for all $\tau \leq t$. For the induction basis, we have

$$\|Y_\tau\| = \|-\epsilon_t\| \leq R$$

for all $\tau \leq 0$. We assume the claim holds for some t, then we have

$$\begin{aligned}
\|Y_{t+1}\| &= \|\nabla^d X_{t+1} - \nabla^d X_{t+1}^\infty(\alpha,\beta) - \epsilon_{t+1}\| \\
&= \|\nabla^d X_{t+1} - \sum_{i=1}^p \alpha_i \nabla^d X_{t+1-i} - \sum_{i=1}^q \beta_i \epsilon_{t+1-i} - \epsilon_{t+1}\| + \|\sum_{i=1}^q \beta_i Y_{t+1-i}\| \\
&= \sum_{i=1}^q \|Y_{t+1-i}\| \|\beta_i\|_{\text{op}} \\
&\leq (1-\epsilon)^{\frac{t+1-q}{q}} R \sum_{i=1}^q \|\beta_i\|_{\text{op}} \\
&\leq (1-\epsilon)^{\frac{t+1}{q}} R,
\end{aligned}$$

which concludes the induction. Finally, we have

$$\begin{aligned}
\|X_t^\infty(\alpha,\beta) - \tilde{X}_t(\alpha,\beta)\| &= \|\sum_{i=1}^q \beta_i(\nabla^d X_{t-i}(\alpha,\beta) - \nabla^d X_{t-i}^\infty(\alpha,\beta) - \epsilon_{t-i})\| \\
&\leq \sum_{i=1}^q \|\beta_i\|_{\text{op}} \|Y_{t-i}\| \\
&\leq (1-\epsilon)(1-\epsilon)^{\frac{t-q}{q}} R \\
&= (1-\epsilon)^{\frac{t}{q}} R,
\end{aligned}$$

which is the claimed result. □

Next, we recursively define the following process:

$$\nabla^d X_t^m(\alpha,\beta) = \sum_{i=1}^p \alpha_i \nabla^d X_{t-i} + \sum_{i=1}^q \beta_i (\nabla^d X_{t-i} - \nabla^d X_{t-i}^{m-i}(\alpha,\beta)), \tag{A1}$$

where $\nabla^d X_t^m(\alpha,\beta) = \nabla^d X_t$ for $m \leq 0$. Let $\{X_t^m(\alpha,\beta)\}$ be the sequence generated as follows:

$$X_t^m(\alpha,\beta) = \nabla^d X_t^m(\alpha,\beta) + \sum_{i=0}^{d-1} \nabla^i X_{t-1}. \tag{A2}$$

We show in the next lemma that it is close to $\{X_t^\infty(\alpha,\beta)\}$.

Lemma A2. *For any α, β, $\{l_t\}$, and $\{\epsilon_t\}$ satisfying A1–A2, we have*

$$\|X_t^m(\alpha,\beta) - X_t^\infty(\alpha,\beta)\| \leq \frac{2R}{T},$$

for $m = \frac{q \log T}{\log \frac{1}{1-\epsilon}}$.

Proof. Define $Z_t^m = \nabla^d X_t^m(\alpha,\beta) - \nabla^d X_t^\infty(\alpha,\beta)$. We prove by induction on m that

$$\|Z_t^{\tilde{m}}\| \leq (1-\epsilon)^{\frac{\tilde{m}}{q}} 2R$$

holds for all $t = 1, \ldots, T$ and $0 \leq \tilde{m} \leq m$. For $m = 0$, we have for $t = 1, \ldots, T$

$$\begin{aligned}
\|Z_t^0\| &= \|\nabla^d X_t^0(\alpha,\beta) - \nabla^d X_t^\infty(\alpha,\beta)\| \\
&= \|\nabla^d X_t - \nabla^d X_t^\infty(\alpha,\beta)\|.
\end{aligned}$$

By the definition of the stochastic process $\{\nabla^d X^\infty(\alpha,\beta)\}$, we have

$$-\nabla^d X_t + \nabla^d X_t^\infty(\alpha,\beta)$$
$$= -\nabla^d X_t + \sum_{i=1}^p \alpha_i \nabla^d X_{t-i} + \sum_{i=1}^q \beta_i(\nabla^d X_{t-i}(\alpha,\beta) - \nabla^d X_{t-i}^\infty(\alpha,\beta))$$
$$= -\nabla^d X_t + \sum_{i=1}^p \alpha_i \nabla^d X_{t-i} + \sum_{i=1}^q \beta_i \epsilon_{t-i} + \sum_{i=1}^q \beta_i(\nabla^d X_{t-i}(\alpha,\beta) - \nabla^d X_{t-i}^\infty(\alpha,\beta) - \epsilon_{t-i})$$
$$= \nabla^d \tilde{X}_t(\alpha,\beta) - \nabla^d X_t + \sum_{i=1}^q \beta_i(\nabla^d X_{t-i}(\alpha,\beta) - \nabla^d X_{t-i}^\infty(\alpha,\beta) - \epsilon_{t-i})$$
$$= \nabla^d \tilde{X}_t(\alpha,\beta) - \nabla^d X_t + \sum_{i=1}^q \beta_i Y_{t-i},$$

where Y_{t-i} is defined as in the proof of Lemma A1. From the assumption, we have $\|\nabla^d \tilde{X}_t(\alpha,\beta) - \nabla^d X_t\| = \|\epsilon_t\| \leq R$, and, as we have proved in Lemma A1, $\|Y_t\| \leq R$ holds. Therefore, we obtain $\|Z_t^0\| \leq 2R$, which is the induction basis. Next, assume the claim holds for all $0, \ldots, m-1$. Then we have

$$\|Z_t^m\| = \|\sum_{i=1}^q \beta^i(\nabla^d X_{t-i} - \nabla^d X_{t-i}^{m-i}(\alpha,\beta) - \nabla^d X_{t-i} + \nabla^d X_{t-i}^\infty(\alpha,\beta))\|$$
$$\leq \|\sum_{i=1}^q \beta_i(\nabla^d X_{t-i}^\infty(\alpha,\beta) - \nabla^d X_{t-i}^{m-i}(\alpha,\beta))\|$$
$$\leq \sum_{i=1}^m \|\beta_i(\nabla^d X_{t-i}^\infty(\alpha,\beta) - \nabla^d X_{t-i}^{m-i}(\alpha,\beta))\|$$
$$+ \sum_{i=m+1}^q \|\beta_i(\nabla^d X_{t-i}^\infty(\alpha,\beta) - \nabla^d X_{t-i})\|$$

From the induction hypothesis, we have

$$\|\nabla^d X_{t-i}^\infty(\alpha,\beta) - \nabla^d X_{t-i}^{m-i}(\alpha,\beta)\| \leq (1-\epsilon)^{\frac{m-i}{q}} 2R.$$

From the proof of the induction basis, we have

$$\sum_{i=m+1}^q \|\beta_i(\nabla^d X_{t-i}^\infty(\alpha,\beta) - \nabla^d X_{t-i})\| \leq 2R \sum_{i=m+1}^q \|\beta_i\|_{\text{op}}.$$

Therefore, $\|Z_t^m\|$ can be further bounded using

$$\|Z_t^m\| \leq 2R \sum_{i=1}^m \|\beta^i\|_{\text{op}} (1-\epsilon)^{\frac{m-i}{q}} + 2R \sum_{i=m+1}^q \|\beta^i\|_{\text{op}}$$
$$\leq 2R \sum_{i=1}^m \|\beta^i\|_{\text{op}} (1-\epsilon)^{\frac{m-i}{q}} + 2R \sum_{i=m+1}^q \|\beta^i\|_{\text{op}} (1-\epsilon)^{\frac{m-i}{q}}$$
$$\leq (1-\epsilon)^{\frac{m-q}{q}} 2R \sum_{i=1}^q \|\beta^i\|_{\text{op}}$$
$$\leq (1-\epsilon)^{\frac{m}{q}} 2R.$$

Choosing $m \geq \frac{q \log T}{\log \frac{1}{1-\epsilon}} = q \log_{1-\epsilon}(T)^{-1}$, we have

$$\|X_t^m(\alpha,\beta) - X_t^\infty(\alpha,\beta)\| \leq \frac{2R}{T},$$

which is the claimed result. □

This process of the d-th-order differencing is actually an integrated AR($m+p$) process with order d, which is shown in the following lemma.

Lemma A3. *For any data sequence $\{X_t^m(\alpha, \beta)\}$ generated by a process of the d-th-order differencing given by (A1) and (A2) there is a $\gamma \in \mathcal{L}(\mathbb{X}, \mathbb{X})^{m+p}$ such that*

$$\sum_{i=1}^{m+p} \gamma_i \nabla^d X_{t-i} + \sum_{i=0}^{d-1} \nabla^i X_{t-1} = X_t^m(\alpha, \beta)$$

holds for all t.

Proof. Let $\{\nabla^d X_t^m(\alpha, \beta)\}$ be the sequence generated by (A1). We prove by induction on m that for all $\tilde{m} \leq m$ there is a $\gamma \in \mathcal{L}(\mathbb{X}, \mathbb{X})^{\tilde{m}+p}$ such that

$$\nabla^d X_t^{\tilde{m}}(\alpha, \beta) = \sum_{i=1}^{\tilde{m}+p} \gamma_i \nabla^d X_{t-i}$$

holds for all α and β. The induction basis follows directly from the definition that

$$\nabla^d X_t^0(\alpha, \beta) = \sum_{i=1}^{p} \alpha_i \nabla^d X_{t-i}.$$

Assume that the claim holds for some m. Let α_i be the zero linear functional for $i > p$ and β_i be the zero linear functional for $i > q$. Then we have

$$\nabla^d X_t^{m+1}(\alpha, \beta)$$
$$= \sum_{i=1}^{p} \alpha_i \nabla^d X_{t-i} + \sum_{i=1}^{q} \beta_i (\nabla^d X_{t-i} - \nabla^d X_{t-i}^{m+1-i}(\alpha, \beta))$$
$$= \sum_{i=1}^{p} \alpha_i \nabla^d X_{t-i} + \sum_{i=1}^{m+1} \beta_i \nabla^d X_{t-i} - \sum_{i=1}^{m+1} \beta_i \nabla^d X_{t-i}^{m+1-i}(\alpha, \beta)$$
$$= \sum_{i=1}^{p} \alpha_i \nabla^d X_{t-i} + \sum_{i=1}^{m+1} \beta_i \nabla^d X_{t-i} - \sum_{i=1}^{m+1} \beta_i \sum_{j=1}^{m+1-i+p} \gamma_j^{m+1-i} \nabla^d X_{t-i-j}$$
$$= \sum_{i=1}^{p} \alpha_i \nabla^d X_{t-i} + \sum_{i=1}^{m+1} \beta_i \nabla^d X_{t-i} - \sum_{i=1}^{m+p+1} (\sum_{j=1}^{m+1} \beta_j \sum_{k=1}^{i-j} \gamma_k^{m+1-j}) \nabla^d X_{t-i},$$

where the second equality follows from the fact that $\beta_i(\nabla^d X_{t-i} - \nabla^d X_{t-i}^{m+1-i}(\alpha, \beta)) = 0$ for $i > m+1$, the third line uses the induction hypothesis and the last line is obtained by rearranging and setting $\sum_{i=m}^{n} a_i = 0$ for $m > n$. The induction step is obtained by setting

$$\gamma_i^{m+1} = \alpha_i + \beta_i - \sum_{j=1}^{m+1} \beta_j \sum_{k=1}^{i-j} \gamma_k^{m+1-j}$$

for $i = 1, \ldots, m + p + 1$, and the claimed result follows. □

Finally, we prove Lemma 1 by combining the results.

Proof of Lemma 1. From Lemmas A1, A2, and A3, there is some $\gamma \in \mathcal{L}(\mathbb{X}, \mathbb{X})^m$ with $m \geq \frac{q \log T}{\log \frac{1}{1-\epsilon}} + p$ such that

$$\|\nabla^d X_t(\gamma) - \nabla^d \tilde{X}_t(\alpha,\beta)\|$$
$$=\|\nabla^d X_t^m(\gamma) - \nabla^d \tilde{X}_t(\alpha,\beta)\|$$
$$\leq \|\nabla^d X_t^m(\gamma) - \nabla^d X_t^\infty(\alpha,\beta)\| + \|\nabla^d X_t^\infty(\gamma) - \nabla^d \tilde{X}_t(\alpha,\beta)\|$$
$$\leq (1-\epsilon)^{\frac{t}{q}} R + \frac{2R}{T},$$

which is the claimed result. □

Appendix B

In this section, we prove the theorems in Section 4. The required notation is summarized in Appendix C. We apply some important properties of convex functions and their convex conjugate defined on a general vector space, which can be found in [17]. The proposed algorithms are instances of the adaptive optimistic follow the regularized leader (AO-FTRL) [10], which is described in Algorithm A1.

Algorithm A1 AO-FTRL.

Input: closed convex set $\mathcal{W} \subseteq \mathbb{X}$
Initialize: θ_1 arbitrary
for $t = 1$ to T do
  Get hint h_t
  $w_t = \nabla \psi_t^*(\theta_t - h_t)$
  Observe $g_t \in \mathbb{X}_*$
  $\theta_{t+1} = \theta_t - g_t$
end for

Lemma A4. *We run AO-FTRL with closed convex regularizers $\psi_1, \ldots, \psi_T$ defined on $\mathcal{W} \subseteq \mathbb{X}$ satisfying $\psi_t(w) \leq \psi_{t+1}(w)$s for all $w \in \mathcal{W}$ and $t = 1, \ldots, T$. Then, for all $u \in \mathcal{W}$, we have*

$$\sum_{t=1}^T g_t(w_t - u) \leq \psi_{T+1}(u) + \psi_1^*(\theta_1) + \sum_{t=1}^T \mathcal{B}_{\psi_t^*}(\theta_{t+1}, \theta_t - h_t),$$

where $\mathcal{B}_{\psi_t^*}(\theta_{t+1}, \theta_t - h_t)$ is the Bregman divergence associated with ψ_t^*.

Proof. W.l.o.g. we assume $h_{T+1} = 0$, since it is not involved in the algorithm. Then we have

$$\sum_{t=1}^T (\psi_{t+1}^*(\theta_{t+1} - h_{t+1}) - \psi_t^*(\theta_t - h_t))$$
$$= \psi_{T+1}^*(\theta_{T+1} - h_{T+1}) - (\theta_1 - h_1)w_1 + \psi_1(w_1)$$
$$\geq (\theta_{T+1} - h_{T+1})u - \psi_{T+1}(u) + h_1 w_1 - \theta_1 w_1 + \psi_1(w_1)$$
$$\geq \theta_{T+1} u - \psi_{T+1}(u) + h_1 w_1 - \sup_{w \in \mathcal{W}}(\theta_1 w_1 - \psi_1(w_1))$$
$$= -\sum_{t=1}^T g_t u - \psi_{T+1}(u) + h_1 w_1 - \psi_1^*(\theta_1).$$

Furthermore, we have

$$\psi_{t+1}^*(\theta_{t+1} - h_{t+1}) - \psi_t^*(\theta_t - h_t)$$
$$= \psi_{t+1}^*(\theta_{t+1} - h_{t+1}) - \psi_t^*(\theta_{t+1}) + \psi_t^*(\theta_{t+1}) - \psi_t^*(\theta_t - h_t)$$
$$\leq (\theta_{t+1} - h_{t+1})w_{t+1} - \psi_{t+1}(w_{t+1}) - \theta_{t+1}w_{t+1} + \psi_t(w_{t+1}) + \psi_t^*(\theta_{t+1}) - \psi_t^*(\theta_t - h_t)$$
$$\leq \psi_t^*(\theta_{t+1}) - \psi_t^*(\theta_t - h_t) - h_{t+1}w_{t+1}$$

Combining the inequalities above, rearranging and adding $\sum_{t=1}^{T}\langle g_t, w_t\rangle$ to both sides, we obtain

$$\sum_{t=1}^{T} g_t(w_t - u)$$
$$\leq \psi_{T+1}(u) + \psi_1^*(\theta_1) + \sum_{t=1}^{T}(\psi_t^*(\theta_{t+1}) - \psi_t^*(\theta_t - h_t) + g_t w_t - h_t w_t)$$
$$= \psi_{T+1}(u) + \psi_1^*(\theta_1) + \sum_{t=1}^{T}(\psi_t^*(\theta_{t+1}) - \psi_t^*(\theta_t - h_t) - (\theta_{t+1} - \theta_t + h_t)\nabla \psi_t^*(\theta_t - h_t))$$
$$= \psi_{T+1}(u) + \psi_1^*(\theta_1) + \sum_{t=1}^{T} \mathcal{B}_{\psi_t^*}(\theta_{t+1}, \theta_t - h_t),$$

which is the claimed result. □

Proof of Theorem 1. First of all, since we have

$$\sum_{t=1}^{T} l_t(\tilde{X}_t(\gamma_t)) - l_t(\tilde{X}_t(\gamma)) \leq \sum_{t=1}^{T}\sum_{i=1}^{m} g_{i,t}(\gamma_{i,t} - \gamma_i)$$
$$= \sum_{i=1}^{m}(\sum_{t=1}^{T} g_{i,t}(\gamma_{i,t} - \gamma_i)),$$

the overall regret can be considered as the sum of the regrets $\sum_{t=1}^{T} g_{i,t}(\gamma_{i,t} - \gamma_i)$. Next, we analyse the regret of each $i = 1, \ldots m$. Define $\psi_{i,t}(\gamma_i) = \frac{\eta_{i,t}}{2}\|\gamma_i\|_F^2$. It is easy to verify $\gamma_{i,t} \in \partial \psi_{i,t}^*(\theta_{i,t})$ for $t = 1, \ldots, T$. Applying Lemma A4 with $h_t = 0$, we obtain

$$\sum_{t=1}^{T} g_{i,t}(\gamma_{i,t} - \gamma_i) \leq \psi_{i,T+1}(\gamma_i) + \psi_{i,1}^*(\theta_{i,1}) + \sum_{t=1}^{T}\mathcal{B}_{\psi_{i,t}^*}(\theta_{i,t+1}, \theta_{i,t}).$$

From the updating rule of $G_{i,t}$, we have $g_{i,t} = 0$ for $G_{i,t} = 0$. Let t_0 be the smallest index such that $G_{i,t_0} > 0$. Then we have

$$\sum_{t=1}^{T}\mathcal{B}_{\psi_{i,t}^*}(\theta_{i,t+1}, \theta_{i,t}) = \sum_{t=t_0}^{T}\mathcal{B}_{\psi_{i,t}^*}(\theta_{i,t+1}, \theta_{i,t}).$$

For $G_{i,t} > 0$, $\psi_{i,t}$ is $\eta_{i,t}$-strongly convex with respect to $\|\cdot\|_F$. From the duality of strong convexity and strong smoothness (see Proposition 2 in [17]), we have

$$\sum_{t=t_0}^{T}\mathcal{B}_{\psi_{i,t}^*}(\theta_{i,t+1}, \theta_{i,t}) \leq \sum_{t=t_0}^{T}\frac{1}{2\eta_{i,t}}\|g_{i,t}\|_F^2 = \sum_{t=t_0}^{T}\frac{\|g_{i,t}\|_F^2}{2\sqrt{\sum_{s=1}^{t-1}\|g_{i,s}\|_F^2 + (L_t G_{i,t})^2}}.$$

From the definition of Frobenius norm, we have

$$\|g_{i,t}\|_F^2 = \|h_t \nabla^d X_{t-i}^\top\|_F^2 = \|h_t\|_2^2 \|\nabla^d X_{t-i}\|_2^2 \leq \frac{\|h_t\|_2^2}{L_t^2} L_t^2 G_{i,t}^2.$$

Then, we obtain

$$\sum_{t=t_0}^{T} \frac{\|g_{i,t}\|_F^2}{2\sqrt{\sum_{s=1}^{t-1}\|g_{i,s}\|_F^2 + (L_t G_{i,t})^2}} \leq \sum_{t=t_0}^{T} \frac{\max\{1, \frac{\|h_t\|_2}{L_t}\}\|g_{i,t}\|_F^2}{2\sqrt{\sum_{s=1}^{t}\|g_{i,s}\|_F^2}}$$

$$\leq \max\{1, \frac{\|h_1\|_2}{L_1}, \ldots, \frac{\|h_T\|_2}{L_T}\} \sqrt{\sum_{t=1}^{T}\|g_{i,t}\|_F^2}$$

$$\leq (1 + \frac{L_{T+1}}{L_1}) \sqrt{\sum_{t=1}^{T}\|g_{i,t}\|_F^2}$$

$$\leq (L_{T+1} + \frac{L_{T+1}^2}{L_1}) \sqrt{\sum_{t=1}^{T}\|\nabla^d X_{t-i}\|_2^2},$$

where the second inequality uses Lemma 4 in [17] and the last inequality follows from the fact that $\|g_{i,t}\|_F \leq L_t \|\nabla^d X_{t-i}\|_2 \leq L_{T+1}\|\nabla^d X_{t-i}\|_2$. Furthermore, we have

$$\psi_{i,T+1}(\gamma_i) \leq \frac{\|\gamma_i\|_F^2}{2} \sqrt{\sum_{t=1}^{T}\|g_{i,t}\|_F^2} + \frac{L_{T+1} G_{i,T+1}\|\gamma_i\|_F^2}{2}$$

$$\leq \frac{\|\gamma_i\|_F^2 L_{T+1}}{2} \sqrt{\sum_{t=1}^{T}\|\nabla^d X_{t-i}\|_2^2} + \frac{L_{T+1} G_{i,T+1}\|\gamma_i\|_F^2}{2},$$

and $\psi_{i,1}^*(\theta_{i,1}) \leq \frac{\|\theta_{i,1}\|_F}{2}$. Adding up from 1 to m, we have

$$\sum_{t=1}^{T} l_t(\tilde{X}_t(\gamma_t)) - l_t(\tilde{X}_t(\gamma))$$

$$\leq \sum_{i=1}^{m} (\frac{\|\gamma_i\|_F^2 L_{T+1}}{2} + L_{T+1} + \frac{L_{T+1}^2}{L_1}) \sqrt{\sum_{t=1}^{T}\|\nabla^d X_{t-i}\|_2^2}$$

$$+ \sum_{i=1}^{m} \frac{L_{T+1} G_{i,T+1}\|\gamma_i\|_F^2 + \|\theta_{i,1}\|_F}{2}$$

□

Proof of Theorem 2. Define $\psi_t(\gamma) = \frac{\lambda_t \|\gamma\|^4}{4} + \frac{\lambda_t \|\gamma\|^2}{2}$. First of all, it is easy to verify that $\gamma_t \in \partial \psi_t^*(\theta_t)$. Applying Lemma A4 with $h_t = 0$, we have

$$\sum_{t=1}^{T} \langle g_t x_t^\top, \gamma_t - \gamma \rangle_F \leq \psi_{T+1}(\gamma) + \psi_1^*(\theta_1) + \sum_{t=1}^{T} \mathcal{B}_{\psi_t^*}(\theta_{t+1}, \theta_t). \tag{A3}$$

Define $v_t \in \partial \psi_{t+1}^*(\theta_t)$. Then we have

$$\begin{aligned}
\mathcal{B}_{\psi_t^*}(\theta_{t+1}, \theta_t) &= \psi_t^*(\theta_{t+1}) - \psi_t^*(\theta_t) - \langle \gamma_t, \theta_{t+1} - \theta_t \rangle_F \\
&= \langle \theta_{t+1}, v_t \rangle_F - \psi_t(v_t) - \langle \theta_t, \gamma_t \rangle_F + \psi_t(\gamma_t) - \langle \gamma_t, \theta_{t+1} - \theta_t \rangle_F \\
&= \langle \theta_{t+1}, v_t \rangle_F - \psi_t(v_t) + \psi_t(\gamma_t) - \langle \gamma_t, \theta_{t+1} \rangle_F \\
&= \langle \theta_{t+1}, v_t - \gamma_t \rangle_F - \psi_t(v_t) + \psi_t(\gamma_t) \\
&= \langle g_t x_t^\top, \gamma_t - v_t \rangle_F - \psi_t(v_t) + \psi_t(\gamma_t) + \langle \theta_t, v_t - \gamma_t \rangle_F \\
&= \langle g_t x_t^\top, \gamma_t - v_t \rangle_F - \mathcal{B}_{\psi_t}(v_t, \gamma_t) \\
&= \langle \gamma_t x_t x_t^\top, \gamma_t - v_t \rangle_F + \langle -\nabla^d X_t x_t^\top, \gamma_t - v_t \rangle_F - \mathcal{B}_{\psi_t}(v_t, \gamma_t) \\
&= \langle \gamma_t x_t x_t^\top, \gamma_t - v_t \rangle_F - \mathcal{B}_{\tilde{\psi}_t}(v_t, \gamma_t) \\
&\quad + \langle -\nabla^d X_t x_t^\top, \gamma_t - v_t \rangle_F - \mathcal{B}_{\check{\psi}_t}(v_t, \gamma_t),
\end{aligned} \tag{A4}$$

23

where we define $\tilde{\psi}_t(\gamma) = \frac{\lambda_t}{4}\|\gamma\|_F^4$ and $\bar{\psi}_t(\gamma) = \frac{\eta_t}{2}\|\gamma\|_F^2$. From the properties of the Frobenius norm, we have

$$\langle \gamma_t x_t x_t^\top, \gamma_t - v_t\rangle_F \leq \|\gamma_t x_t x_t^\top\|_F \|\gamma_t - v_t\|_F$$
$$\leq \|x_t\|_2^2 \|\gamma_t\|_F \|\gamma_t - v_t\|_F$$

Following the idea of [33], we can upper bound $\|\gamma_t\|_F^2 \|\gamma_t - v_t\|_F^2$ as follows:

$$\frac{\lambda_t}{2}\|\gamma_t\|_F^2 \|\gamma_t - v_t\|_F^2$$
$$= \frac{\lambda_t}{2}\|\gamma_t\|_F^2 (\|\gamma_t\|_F^2 + \|v_t\|_F^2 - 2\langle \gamma_t, v_t\rangle_F)$$
$$\leq \frac{\lambda_t}{4}(\|\gamma_t\|_F^4 + \|v_t\|_F^4 - 2\|\gamma_t\|_F^2\|v_t\|_F^2) + \frac{\lambda_t}{2}\|\gamma_t\|_F^2(\|\gamma_t\|_F^2 + \|v_t\|_F^2 - 2\langle \gamma_t, v_t\rangle_F)$$
$$= \frac{\lambda_t}{4}\|v_t\|_F^4 + \frac{3\lambda_t}{4}\|\gamma_t\|_F^4 - \lambda_t\|\gamma_t\|_F^2 \langle \gamma_t, v_t\rangle_F$$
$$= \frac{\lambda_t}{4}\|v_t\|_F^4 - \frac{\lambda_t}{4}\|\gamma_t\|_F^4 + \lambda_t\|\gamma_t\|_F^2\langle \gamma_t, \gamma_t\rangle_F - \lambda_t\|\gamma_t\|_F^2\langle \gamma_t, v_t\rangle_F$$
$$= \frac{\lambda_t}{4}\|v_t\|_F^4 - \frac{\lambda_t}{4}\|\gamma_t\|_F^4 - \lambda_t\|\gamma_t\|_F^2\langle \gamma_t, v_t - \gamma_t\rangle_F$$
$$= \mathcal{B}_{\tilde{\psi}_t}(v_t, \gamma_t)$$

Thus, for $\lambda_t \neq 0$, we have

$$\langle \gamma_t x_t x_t^\top, \gamma_t - v_t\rangle_F - \mathcal{B}_{\tilde{\psi}_t}(v_t, \gamma_t) \leq 2\sqrt{\frac{\|x_t\|_2^4}{2\lambda_t}\mathcal{B}_{\tilde{\psi}_t}(v_t, \gamma_t)} - \mathcal{B}_{\tilde{\psi}_t}(v_t, \gamma_t)$$
$$\leq \frac{\|x_t\|_2^4}{2\lambda_t},$$

where the second inequality uses the fact that $2ab - b^2 \leq a^2$. Let t_0 be the smallest index such that $\lambda_{t_0} > 0$. Then we have

$$\sum_{t=1}^T (\langle \gamma_t x_t x_t^\top, \gamma_t - v_t\rangle_F - \mathcal{B}_{\tilde{\psi}_t}(v_t, \gamma_t))$$
$$\leq \sum_{t=t_0}^T \frac{\|x_t\|_2^4}{2\lambda_t}$$
$$= \sum_{t=t_0}^T \frac{\|x_t\|_2^4}{2\sqrt{\sum_{s=1}^t \|x_t\|_2^4}} \quad \text{(A5)}$$
$$\leq \sqrt{\sum_{t=1}^T \|x_t\|_2^4},$$

where the last inequality uses Lemma 4 in [17]. Similarly, let t_1 be the smallest index such that $\eta_{t_0} > 0$. Then we obtain the upper bound

$$\sum_{t=1}^{T}(\langle -\nabla^d X_t x_t^\top, \gamma_t - v_t\rangle_F - \mathcal{B}_{\bar{\psi}_t}(v_t, \gamma_t))$$

$$\leq \sum_{t=1}^{T}(\|\nabla^d X_t x_t^\top\|_F \|\gamma_t - v_t\|_F - \mathcal{B}_{\bar{\psi}_t}(v_t, \gamma_t))$$

$$\leq \sum_{t=t_1}^{T}(\sqrt{\frac{2\|\nabla^d X_t x_t^\top\|_F^2}{\eta_t} \mathcal{B}_{\bar{\psi}_t}(v_t, \gamma_t)} - \mathcal{B}_{\bar{\psi}_t}(v_t, \gamma_t))$$

$$\leq \sum_{t=t_1}^{T}(2\sqrt{\frac{\|\nabla^d X_t x_t^\top\|_F^2}{2\eta_t} \mathcal{B}_{\bar{\psi}_t}(v_t, \gamma_t)} - \mathcal{B}_{\bar{\psi}_t}(v_t, \gamma_t))$$

$$\leq \sum_{t=t_1}^{T} \frac{\|\nabla^d X_t x_t^\top\|_F^2}{2\eta_t} \tag{A6}$$

$$= \sum_{t=t_1}^{T} \frac{\|\nabla^d X_t x_t^\top\|_F^2}{2\sqrt{\sum_{s=1}^{t-1}\|\nabla^d X_s x_s^\top\|_F^2 + L_t^2 \|x_t\|_2^2}}$$

$$\leq \max\{1, \frac{\|\nabla^d X_1 x_1^\top\|_F}{G_1}, \ldots, \frac{\|\nabla^d X_T x_T^\top\|_F}{G_T}\} \sum_{t=t_1}^{T} \frac{\|\nabla^d X_t x_t^\top\|_F^2}{2\sqrt{\sum_{s=1}^{t}\|\nabla^d X_s x_s^\top\|_F^2}}$$

$$\leq \max\{1, \frac{\|\nabla^d X_1 x_1^\top\|_F}{G_1}, \ldots, \frac{\|\nabla^d X_T x_T^\top\|_F}{G_T}\} \sqrt{\sum_{t=1}^{T}\|\nabla^d X_t x_t^\top\|_F^2}$$

$$\leq (1 + \frac{G_{T+1}}{G_1}) \sqrt{\sum_{t=1}^{T}\|\nabla^d X_t x_t^\top\|_F^2}$$

Combining (A3)–(A6), we obtain

$$\sum_{t=1}^{T}\langle g_t x_t^\top, \gamma_t - \gamma\rangle_F \leq \frac{(\sqrt{m}G_{T+1}^2 + \|\theta_1\|_F)\|\gamma\|_F^2}{2} + \psi_1^*(\theta_1) + (1 + \frac{\|\gamma\|_F^4}{4})\sqrt{\sum_{t=1}^{T}\|x_t\|_2^4}$$

$$+ (1 + \frac{G_{T+1}}{G_1} + \frac{\|\gamma\|_F^2}{2})\sqrt{\sum_{t=1}^{T}\|\nabla^d X_t x_t^\top\|_F^2}.$$

For $\theta_1 \neq 0$, it is easy to verify that $\psi_1^*(\theta_1) \leq \langle w_1, \theta_1\rangle_F \leq \frac{\|\theta_1\|_F^2}{\eta_1} \leq \|\theta_1\|_F$. By putting this in the inequality above, we obtain the claimed result. □

Proof of Theorem 3.
Proof. Define

$$\psi_t : \Delta \to \mathbb{R}, w \mapsto \eta_t \sum_{k \in I_w}^{K} w_k \log w_k + \eta_t \log K,$$

where $I_w = \{i = 1, \ldots, k | w_i \neq 0\}$. It can be verified that $w_t \in \partial \psi_t^*(\theta_t)$. Applying Lemma A4, we obtain

$$\sum_{t=1}^{T} z_t^\top (w_t - u) \leq \psi_{T+1}(u) + \psi_1^*(\theta_1) + \sum_{t=1}^{T} \mathcal{B}_{\psi_t^*}(\theta_{t+1}, \theta_t - h_t).$$

From the definition of ψ_t, it follows that $\psi_{T+1}(u) \leq \sqrt{\frac{\log K}{2} \sum_{t=1}^{T}\|z_t - h_t\|_\infty^2}$ and $\psi_1^*(\theta_1) = 0$ hold. Define $v_t \in \partial \psi_t^*(\theta_{t+1})$. Next, we bound the third term as follows:

$$\mathcal{B}_{\psi_t^*}(\theta_{t+1}, \theta_t - h_t)$$
$$=\psi_t^*(\theta_{t+1}) - \psi_t^*(\theta_t - h_t) - (h_t - z_t)^\top w_t$$
$$=\theta_{t+1}^\top v_t - \psi_t(v_t) - (\theta_t - h_t)^\top w_t + \psi_t(w_t) - (h_t - z_t)^\top w_t$$
$$=(h_t - z_t)^\top (v_t - w_t) - (\psi_t(v_t) - \psi_t(w_t) - (\theta_t - h_t)^\top (v_t - w_t))$$
$$=(h_t - z_t)^\top (v_t - w_t) - \mathcal{B}_{\psi_t}(v_t, w_t)$$
$$=(h_t - z_t)^\top (v_t - w_t) - \eta_{t+1}\|v_t - w_t\|_1^2 + \eta_{t+1}\|v_t - w_t\|_1^2 - \mathcal{B}_{\psi_t}(v_t, w_t)$$
$$\leq (h_t - z_t)^\top (v_t - w_t) - \eta_{t+1}\|v_t - w_t\|_1^2 + (\eta_{t+1} - \eta_t)\|v_t - w_t\|_1^2$$
$$\leq \|h_t - z_t\|_\infty \|v_t - w_t\|_1 - \eta_{t+1}\|v_t - w_t\|_1^2 + 4(\eta_{t+1} - \eta_t)$$
$$\leq \frac{\|h_t - z_t\|_\infty^2}{4\eta_{t+1}} + 4(\eta_{t+1} - \eta_t),$$

where the first inequality uses the fact that ψ_t is $2\eta_t$ strongly convex w.r.t. $\|\cdot\|_1$. Adding up from 1 to T, we have

$$\sum_{t=1}^T \mathcal{B}_{\psi_t^*}(\theta_{t+1}, \theta_t - h_t) \leq \sum_{t=1}^T \left(\frac{\|h_t - z_t\|_\infty^2}{4\eta_{t+1}} + 4(\eta_{t+1} - \eta_t) \right)$$
$$\leq \sqrt{\frac{\log K}{2} \sum_{t=1}^T \|h_t - z_t\|_\infty^2} + 4\eta_{T+1}$$
$$\leq \sqrt{\frac{\log K}{2} \sum_{t=1}^T \|h_t - z_t\|_\infty^2} + \sqrt{\frac{8}{\log K} \sum_{t=1}^T \|h_t - z_t\|_\infty^2}.$$

Combining the inequalities, we obtain

$$\sum_{t=1}^T l(X_t, \sum_{i=1}^K w_{i,t} \tilde{X}_t^i) - \sum_{t=1}^T l(X_t, \tilde{X}_t^k)$$
$$\leq \sum_{t=1}^T \sum_{i=1}^K w_{i,t} l(X_t, \tilde{X}_t^i) - \sum_{t=1}^T l(X_t, \tilde{X}_t^k)$$
$$= \sum_{t=1}^T w_t^\top z_t - \sum_{t=1}^T l(X_t, \tilde{X}_t^k)$$
$$\leq \left(\sqrt{2\log K} + \sqrt{\frac{8}{\log K}}\right) \sqrt{\sum_{t=1}^T \|h_t - z_t\|_\infty^2},$$

where the first inequality follows from Jensen's inequality. Furthermore, if l is L-Lipschitz in its first argument, then we have

$$\|h_t - z_t\|_\infty = \max_{i \in \{1,\ldots,K\}} |z_{i,t} - h_{i,t}| \leq L \|\nabla^d X_t\|_2.$$

Finally, we obtain the regret upper bound

$$\sum_{t=1}^T l(X_t, \sum_{i=1}^K w_{i,t} \tilde{X}_t^i) - \sum_{t=1}^T l(X_t, \tilde{X}_t^k) \leq \left(\sqrt{2\log K} + \sqrt{\frac{8}{\log K}}\right) \sqrt{\sum_{t=1}^T L^2 \|\nabla^d X_t\|_2^2},$$

which is the claimed result. □

Appendix C

We summarize the main notations used throughout the article in Table A1.

Table A1. Nomenclature.

$(\mathbb{X}, \|\cdot\|)$	finite dimensional norm space
$(\mathbb{X}_*, \|\cdot\|_*)$	the dual space with dual norm of $(\mathbb{X}, \|\cdot\|)$
$\mathcal{L}(\mathbb{X}, \mathbb{X})$	vector space of bounded linear operators
$\|\alpha\|_{op} = \sup_{x \in \mathbb{X}, x \neq 0} \frac{\|\alpha x\|}{\|x\|}$	the operator norm of $\alpha \in \mathcal{L}(\mathbb{X}, \mathbb{X})$
$\|x\|_2 = \sqrt{\sum_{i=1}^d x_i^2}$	2 norm for $x \in \mathbb{R}^d$
$\|x\|_1 = \sum_{i=1}^d |x_i|$	1 norm for $x \in \mathbb{R}^d$
$\|x\|_\infty = \max\{|x_1|, \ldots, |x_d|\}$	max norm for $x \in \mathbb{R}^d$
$\langle A, B \rangle_F = \text{tr}(A^\top B)$	Frobenius inner product
$\|A\|_F = \sqrt{\langle A, A \rangle_F}$	Frobenius norm
$\Delta^d : \{x \in \mathbb{R}^d | \sum_{i=1}^d x_i = 1, x_i \geq 0\}$	standard d-simplex
$\psi : \mathcal{W} \to \mathbb{R}$	closed convex function
$\partial \psi(w) = \{g \in \mathbb{X}_* | \forall v \in \mathcal{W}. \psi(v) - \psi(w) \geq g(v - w)\}$	the set of subdifferential of ψ at w
$\psi^* : \mathbb{X}_* \to \mathbb{R}, \theta \mapsto \sup_{w \in \mathcal{W}} \theta w - \psi(w)$	convex conjugate of ψ
$\mathcal{B}_\psi(u, v) = \psi(u) - \psi(v) - g(u - v)$, where $g \in \partial \psi(u)$	the Bregman divergence

Appendix D

For the synthetic data, the relative performance of the proposed algorithms after the first 1000 iterations are plotted in Figures A1–A3. For each setting, we calculate the average loss after the first 1000 iterations and plot the difference of the proposed algorithms compared to the average loss incurred by the best baseline algorithm.

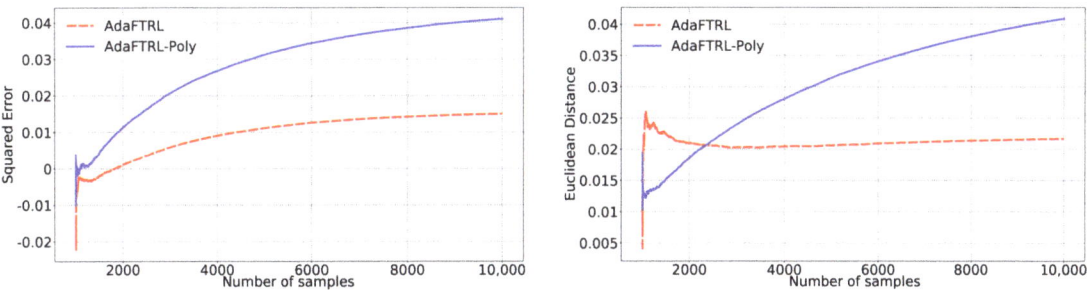

Figure A1. Relative performance for setting 1.

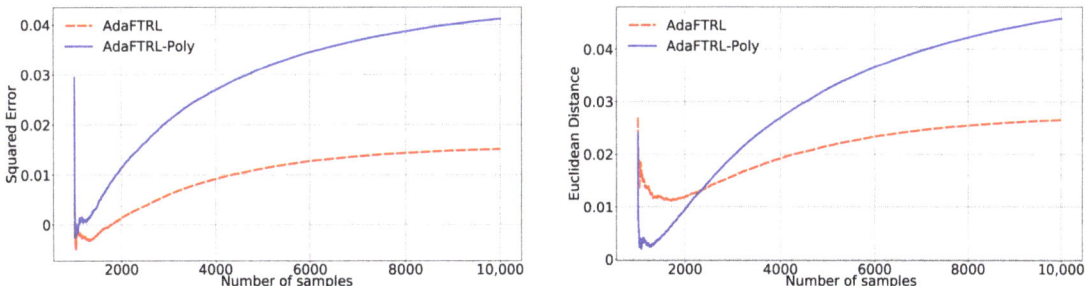

Figure A2. Relative performance for setting 2.

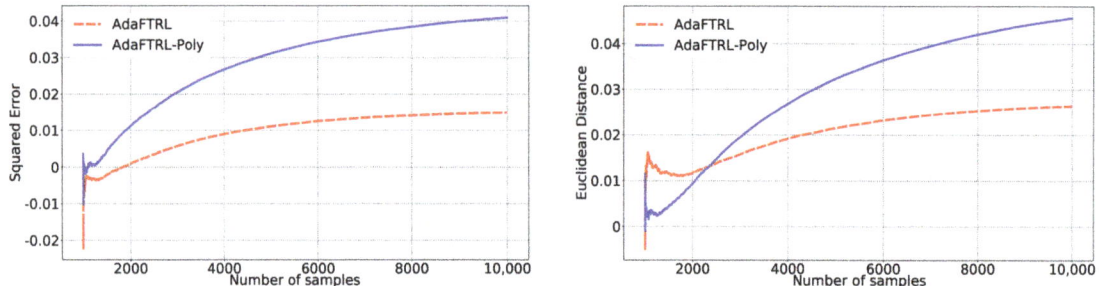

Figure A3. Relative performance for setting 3.

Similarly, we plot the relative performance for the real-world data over the time horizon in Figures A4–A6.

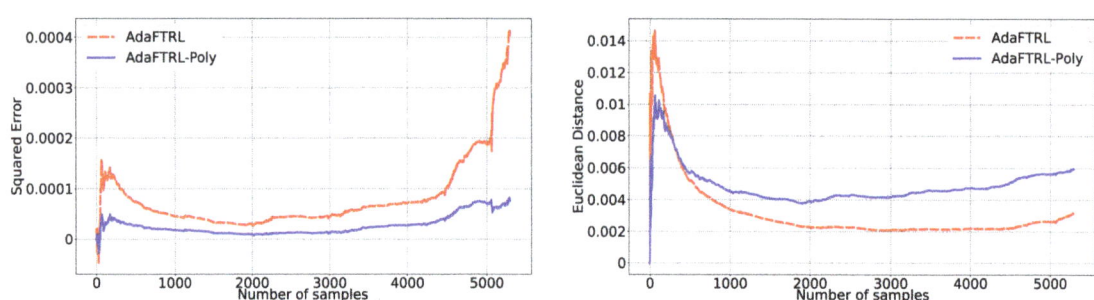

Figure A4. Relative performance for stock data.

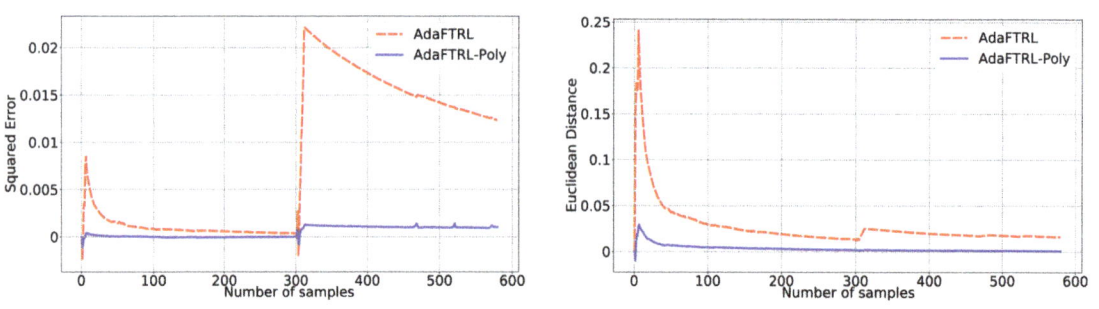

Figure A5. Relative performance for Google Flu.

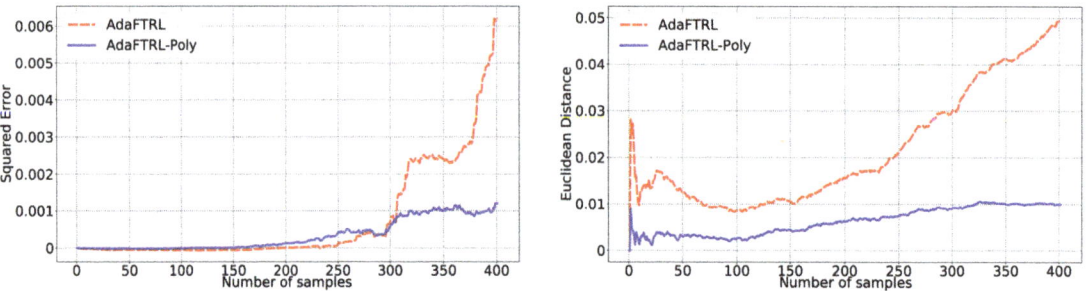

Figure A6. Relative Performance for electricity demand.

References

1. Shumway, R.; Stoffer, D. *Time Series Analysis and Its Applications: With R Examples*; Springer Texts in Statistics; Springer: New York, NY, USA, 2010.
2. Chujai, P.; Kerdprasop, N.; Kerdprasop, K. Time series analysis of household electric consumption with ARIMA and ARMA models. In Proceedings of the International MultiConference of Engineers and Computer Scientists, Hong Kong, China, 13–15 March 2013; Volume 1, pp. 295–300.
3. Ghofrani, M.; Arabali, A.; Etezadi-Amoli, M.; Fadali, M.S. Smart scheduling and cost-benefit analysis of grid-enabled electric vehicles for wind power integration. *IEEE Trans. Smart Grid* **2014**, *5*, 2306–2313. [CrossRef]
4. Rounaghi, M.M.; Zadeh, F.N. Investigation of market efficiency and financial stability between S&P 500 and London stock exchange: Monthly and yearly forecasting of time series stock returns using ARMA model. *Phys. A Stat. Mech. Its Appl.* **2016**, *456*, 10–21.
5. Zhu, B.; Chevallier, J. Carbon price forecasting with a hybrid Arima and least squares support vector machines methodology. In *Pricing and Forecasting Carbon Markets*; Springer: Berlin/Heidelberg, Germany, 2017; pp. 87–107.
6. Anava, O.; Hazan, E.; Mannor, S.; Shamir, O. Online learning for time series prediction. In Proceedings of the Conference on Learning Theory, Princeton, NJ, USA, 23–26 June 2013; pp. 172–184.
7. Liu, C.; Hoi, S.C.; Zhao, P.; Sun, J. Online ARIMA algorithms for time series prediction. In Proceedings of the Thirtieth AAAI Conference on Artificial Intelligence, Phoenix, AZ, USA, 12–17 February 2016; pp. 1867–1873.
8. Xie, C.; Bijral, A.; Ferres, J.L. Nonstop: A nonstationary online prediction method for time series. *IEEE Signal Process. Lett.* **2018**, *25*, 1545–1549. [CrossRef]
9. Yang, H.; Pan, Z.; Tao, Q.; Qiu, J. Online learning for vector autoregressive moving-average time series prediction. *Neurocomputing* **2018**, *315*, 9–17. [CrossRef]
10. Joulani, P.; György, A.; Szepesvári, C. A modular analysis of adaptive (non-) convex optimization: Optimism, composite objectives, variance reduction, and variational bounds. *Theor. Comput. Sci.* **2020**, *808*, 108–138. [CrossRef]
11. Zhou, Y.; Sanches Portella, V.; Schmidt, M.; Harvey, N. Regret Bounds without Lipschitz Continuity: Online Learning with Relative-Lipschitz Losses. *Adv. Neural Inf. Process. Syst.* **2020**, *33*, 15823–15833.
12. Jamil, W.; Bouchachia, A. Model selection in online learning for times series forecasting. In *UK Workshop on Computational Intelligence*; Springer: Berlin/Heidelberg, Germany, 2018; pp. 83–95.
13. Jamil, W.; Kalnishkan, Y.; Bouchachia, H. Aggregation Algorithm vs. Average For Time Series Prediction. In Proceedings of the ECML PKDD 2016 Workshop on Large-Scale Learning from Data Streams in Evolving Environments, Riva del Garda, Italy, 23 September 2016; pp. 1–14.
14. Orabona, F.; Pál, D. Coin betting and parameter-free online learning. In Proceedings of the 30th International Conference on Neural Information Processing Systems, Barcelona, Spain, 4–9 December 2016; pp. 577–585.
15. Cutkosky, A.; Orabona, F. Black-box reductions for parameter-free online learning in banach spaces. In Proceedings of the Conference on Learning Theory, Stockholm, Sweden, 6–9 July 2018; pp. 1493–1529.
16. Cutkosky, A.; Boahen, K. Online learning without prior information. In Proceedings of the Conference on Learning Theory, Amsterdam, The Netherlands, 7–10 July 2017; pp. 643–677.
17. Orabona, F.; Pál, D. Scale-free online learning. *Theor. Comput. Sci.* **2018**, *716*, 50–69. [CrossRef]
18. Hamilton, J.D. *Time Series Analysis*; Princeton University Press: Princeton, NJ, USA, 1994; Volume 2.
19. Box, G.E.; Jenkins, G.M.; Reinsel, G.C.; Ljung, G.M. *Time Series Analysis: Forecasting and Control*; John Wiley & Sons: Hoboken, NJ, USA, 2015.
20. Brockwell, P.J.; Davis, R.A. *Time Series: Theory and Methods*; Springer Science & Business Media: Berlin/Heidelberg, Germany, 2013.
21. Georgiou, T.T.; Lindquist, A. A convex optimization approach to ARMA modeling. *IEEE Trans. Autom. Control* **2008**, *53*, 1108–1119. [CrossRef]
22. Lii, K.S. Identification and estimation of non-Gaussian ARMA processes. *IEEE Trans. Acoust. Speech Signal Process.* **1990**, *38*, 1266–1276. [CrossRef]
23. Huang, S.J.; Shih, K.R. Short-term load forecasting via ARMA model identification including non-Gaussian process considerations. *IEEE Trans. Power Syst.* **2003**, *18*, 673–679. [CrossRef]
24. Ding, F.; Shi, Y.; Chen, T. Performance analysis of estimation algorithms of nonstationary ARMA processes. *IEEE Trans. Signal Process.* **2006**, *54*, 1041–1053. [CrossRef]
25. Yang, H.; Pan, Z.; Tao, Q. Online Learning for Time Series Prediction of AR Model with Missing Data. *Neural Process. Lett.* **2019**, *50*, 2247–2263. [CrossRef]
26. Ding, J.; Noshad, M.; Tarokh, V. Order selection of autoregressive processes using bridge criterion. In Proceedings of the 2015 IEEE International Conference on Data Mining Workshop (ICDMW), Atlantic City, NJ, USA, 14–17 November 2015; pp. 615–622.
27. Lütkepohl, H. *New Introduction to Multiple Time Series Analysis*; Springer Science & Business Media: Berlin/Heidelberg, Germany, 2005.
28. Steinhardt, J.; Liang, P. Adaptivity and optimism: An improved exponentiated gradient algorithm. In Proceedings of the International Conference on Machine Learning, PMLR, Bejing, China, 22–24 June 2014; pp. 1593–1601.
29. De Rooij, S.; Van Erven, T.; Grünwald, P.D.; Koolen, W.M. Follow the leader if you can, hedge if you must. *J. Mach. Learn. Res.* **2014**, *15*, 1281–1316.

30. Bollerslev, T. Generalized autoregressive conditional heteroskedasticity. *J. Econom.* **1986**, *31*, 307–327. [CrossRef]
31. Deng, Y.; Fan, H.; Wu, S. A hybrid ARIMA-LSTM model optimized by BP in the forecast of outpatient visits. *J. Ambient. Intell. Humaniz. Comput.* **2020**. [CrossRef]
32. Tutun, S.; Chou, C.A.; Canıyılmaz, E. A new forecasting framework for volatile behavior in net electricity consumption: A case study in Turkey. *Energy* **2015**, *93*, 2406–2422. [CrossRef]
33. Lu, H. "Relative Continuity" for Non-Lipschitz Nonsmooth Convex Optimization Using Stochastic (or Deterministic) Mirror Descent. *Informs J. Optim.* **2019**, *1*, 288–303. [CrossRef]

Article

AutoNowP: An Approach Using Deep Autoencoders for Precipitation Nowcasting Based on Weather Radar Reflectivity Prediction

Gabriela Czibula [1,*,†], Andrei Mihai [1,†], Alexandra-Ioana Albu [1,†], Istvan-Gergely Czibula [1,†], Sorin Burcea [2] and Abdelkader Mezghani [3]

- [1] Department of Computer Science, Babeş-Bolyai University, 400084 Cluj-Napoca, Romania; mihai.andrei@ubbcluj.ro (A.M.); alexandra.albu@ubbcluj.ro (A.-I.A.); istvan.czibula@ubbcluj.ro (I.-G.C.)
- [2] Romanian National Meteorological Administration, 013686 Bucharest, Romania; sorin.burcea@meteoromania.ro
- [3] Meteorologisk Instittut, 0371 Oslo, Norway; abdelkader.mezghani@met.no
- * Correspondence: gabriela.czibula@ubbcluj.ro or gabis@cs.ubbcluj.ro; Tel.: +40-264-405327
- † These authors contributed equally to this work.

Citation: Czibula, G.; Mihai, A.; Albu, A.-I.; Czibula, I.G.; Burcea, S.; Mezghani, A. *AutoNowP*: An Approach Using Deep Autoencoders for Precipitation Nowcasting Based on Weather Radar Reflectivity Prediction. *Mathematics* **2021**, *9*, 1653. https://doi.org/10.3390/math9141653

Academic Editor: Freddy Gabbay

Received: 31 May 2021
Accepted: 30 June 2021
Published: 14 July 2021

Publisher's Note: MDPI stays neutral with regard to jurisdictional claims in published maps and institutional affiliations.

Copyright: © 2021 by the authors. Licensee MDPI, Basel, Switzerland. This article is an open access article distributed under the terms and conditions of the Creative Commons Attribution (CC BY) license (https://creativecommons.org/licenses/by/4.0/).

Abstract: Short-term quantitative precipitation forecast is a challenging topic in meteorology, as the number of severe meteorological phenomena is increasing in most regions of the world. Weather radar data is of utmost importance to meteorologists for issuing short-term weather forecast and warnings of severe weather phenomena. We are proposing *AutoNowP*, a binary classification model intended for precipitation nowcasting based on weather radar reflectivity prediction. Specifically, *AutoNowP* uses two convolutional autoencoders, being trained on radar data collected on both stratiform and convective weather conditions for learning to predict whether the radar reflectivity values will be above or below a certain threshold. *AutoNowP* is intended to be a proof of concept that autoencoders are useful in distinguishing between convective and stratiform precipitation. Real radar data provided by the Romanian National Meteorological Administration and the Norwegian Meteorological Institute is used for evaluating the effectiveness of *AutoNowP*. Results showed that *AutoNowP* surpassed other binary classifiers used in the supervised learning literature in terms of probability of detection and negative predictive value, highlighting its predictive performance.

Keywords: precipitation nowcasting; deep learning; autoencoders; radar data

1. Introduction

Forecast of severe weather phenomena, including the quantitative precipitation forecast (QPF), represents a challenging topic in meteorology. Due to the increase in the number of heavy rainfall events in most regions of the world, population safety could be affected and significant damage may occur. The short-term weather forecasting is known as nowcasting and is of particular interest as it has an important role in risks management and crisis control. The problem of weather nowcasting is a complex and difficult one, due to its high dependence on numerous environmental conditions. Precipitation nowcasting represents a challenging and actual research topic, referring to producing predictions of rainfall intensities over a certain region in the near future, and playing an important role in daily life [1].

At global scale, flood threat is increasing because of climate change impact of heavy precipitation, as for instance the total urban area being exposed to flood has dramatically increased in Europe over the past century. Also, various socioeconomic sectors are impacted by climate change induced hazards, such as extreme rainfall, which amplify both the intensity and probability of floods [2]. Research on the exposure of flood hazard, using climate models simulations, showed that the climate change presents the potential to actively change the human, assets, and urban areas exposure to flood hazard, but nevertheless considerable uncertainty in the magnitude of the climate change impact in different regions around the globe exists [3].

Nowadays, integrating crowdsourced observations into research studies can contribute to reducing the risk and the costs related to extreme events. Citizens around the world have, currently, at their disposal a great number of sources of information and amazing possibilities to report and to study meteorological phenomena. Hence, these volunteers who collect, report and/or process the data they observe are citizen scientists. They are active not only in the field of meteorology, but also in sciences as astronomy, archeology, natural history and others [4]. Their contribution to science can have a practical effect, especially by increasing the awareness and perception on climate change related risks, thus helping in mitigating the effects.

Although significant progress has been made recently on nowcasting systems in general, and precipitation nowcasting in particular, the challenges remain as, for instance, severe convective storms are localized, occurring on a small spatial area (i.e., mesoscale) and having an overall short lifecycle. Due to its high spatiotemporal resolution, radar data is used both in the so-called expert nowcasting systems and in the less complex forms that involve processing the radar data solely [5,6]. These systems blend radar data and other observations with numerical weather prediction (NWP) models to generate forecasts up to 6 h [7]. Although NWP significantly improves the precipitation nowcasting, there are still issues to be resolved, like the predictability of precipitation systems, the improvement of rapid update NWP, and the need for improvement of mesoscale observation networks [8].

Some of the most used radar products in weather nowcasting are reflectivity (R) and Doppler radial velocity (V). For instance, operational meteorologists are mainly using the values of reflectivity and radial velocity to monitor the spatiotemporal evolution of precipitating clouds, while operational radar algorithms use the reflectivity for rainfall estimation and storm tracking and classification: R values above a certain threshold (e.g., 35 dBZ [5,9]) indicate possible convective storms occurrence associated with heavy rainfall. Estimating the values of the radar products based on their historical values is important for QPF. NWP models [10] represent the main techniques for QPF, but there are still errors in rainfall forecasting due to difficulties in modelling cloud dynamics and microphysics [11].

Deep learning methods [12–14] are believed to have the potential to overcome the limitations of NWP methods through modeling patterns in large amounts of historical meteorological data. Deep learning methods offer data-driven solutions for the nowcasting problem, by learning dependencies between radar measurements at consecutive time steps [15]. A central characteristic of deep neural networks is represented by their ability to learn abstract representations of the input data through stacking multiple layers and thus forming deep architectures. Autoencoders (AEs) are a type of neural network that can be trained to learn low dimensional representations that capture the relevant characteristics of the input data [16]. AEs are trained to learn data representations by reconstructing their inputs. They are built of two components, an encoder that maps the input to a latent representation and a decoder that uses this representation to reconstruct the input. Typically, the dimensionality of the latent representation is chosen to be smaller than the input space dimensionality, thus obtaining a so-called undercomplete autoencoder. Autoencoders can be trained using gradient descent methods to minimize the error between the input data and the predicted reconstruction [16]. Convolutional autoencoders (ConvAEs) are able to capture spatial patterns in the input data by using convolutions as their building blocks. Convolutional encoder-decoder architectures have been extensively used in various computer vision tasks and they are the typical choice for modeling the spatial characteristics of meteorological measurements gathered along geographical locations [15,17,18].

The contribution of the paper is threefold. First, we aim at introducing a supervised classifier *AutoNowP* that uses two convolutional autoencoders for distinguishing between convective and stratiform rainfall based on radar reflectivity prediction. *AutoNowP* is based on training two ConvAEs trained on radar data collected on both stratiform and convective weather conditions. After the training step, *AutoNowP* will learn to predict whether the radar reflectivity values will be higher than a certain threshold, and thus indicating if a convective storm is likely to happen. *AutoNowP* is intended to be a proof of concept that AEs applied on radar data are useful in distinguishing between convective and stratiform rainfall. Secondly, the effectiveness of *AutoNowP* is empirically proven on two case studies consisting of real radar data collected from the Romanian National Meteorological Administration (NMA) and the Norwegian Meteorological Institute (MET). The obtained results are compared to the results of recent similar approaches in the field of precipitation nowcasting. As an additional goal we aim at analyzing the relevance of the obtained results from a meteorological perspective, as a proof of concept that autoencoders are able to capture relevant meteorological knowledge. To the best of our knowledge, an approach similar to *AutoNowP* has not been proposed in the nowcasting literature so far.

To summarize, the research conducted in the paper is oriented toward answering the following research questions:

RQ1 How to use an ensemble of ConvAEs to supervisedly discriminate between severe and normal rainfall conditions, considering the encoded relationships between radar products values corresponding to both normal and severe weather events?

RQ2 What is the performance of *AutoNowP* introduced for answering RQ1 on real radar data collected from Romania and Norway and how does it compare to similar related work?

The rest of the paper is organized as follows. A literature review on recent deep learning methods for precipitation nowcasting is presented in Section 2. Section 3 introduces our binary classification model *AutoNowP* for predicting if the radar reflectivity values are above or below a specific threshold. The performed experiments and the obtained results are described in Section 4, while a discussion on the results and a comparison to related approaches is provided in Section 5. Section 6 presents the conclusions of our research and highlights directions for future work.

2. Literature Review on Machine-Learning-Based Precipitation Nowcasting

A lot of work has been carried out lately in the field of machine-learning-based precipitation nowcasting. We are reviewing, in the following, several recent approaches in the field.

Shi et al. [19] have approached precipitation nowcasting by introducing an extension of a long short-term memory (LSTM) network, named ConvLSTM, suitable for handling spatiotemporal data by preserving due to the convolutional structure of the spatiotemporal features. Their architecture is composed of two networks, a ConvLSTM encoder and a ConvLSTM decoder. As precipitation nowcasting performance indicators, a Rainfall Mean Squared Error (Rainfall-MSE) of 1.420, a Critical Success Index (CSI) of 0.577, a False Alarm Rate (FAR) of 0.195 and a Probability of Detection (POD) of 0.660 have been obtained.

Heye et al. [20] investigated a precipitation nowcasting approach based on a 3D ConvLSTM architecture. In their experiments, a vanilla sequence-to-sequence model achieved better performance than a model using attention layers. Overall, the CSI varied between 0.40 and 0.43, the FAR ranged from 0.28 to 0.31, and the POD fluctuated between 0.46 and 0.51.

A method for precipitation nowcasting, combining the advantages of convolutional gated recurrent networks (ConvGRU) and adversarial training was introduced by Tian et al. [21]. The method aimed at improving the sharpness of the predicted precipitation maps by means of adversarial training. The system is composed of a generator network, represented by the ConvGRU, which learns to generate realistically looking precipitation maps and a discriminator represented by a convolutional neural network that is trained to distinguish between predicted ground truth maps. Their method achieved better performance in terms of probability of detection than an optical flow algorithm and the original ConvGRU. Han et al. [22] used 3D convolutions to build a neural network for convective storm nowcasting. The task was formulated as a binary classification problem and their multisource approach achieved a CSI of 0.44, FAR of 0.45, and POD of 0.69 for 30 min forecasts, outperforming a Support-Vector Machine using hand-crafted features.

The MetNet model [15] has been introduced by Sønderby et al. using both radar and satellite data for precipitation forecasting with a lead time of up to 8 h. The model incorporates three components—a feature extractor formed of a succession of downsampling convolutional layers, a ConvLSTM component used for modeling dependencies on the past time steps and an attention module composed of several axial self-attention blocks that aim to capture relationships among geographic locations situated far away in the map. By including the forecasted time in the data given as input and thus conditioning the entire model on it, predictions for multiple time steps can be obtained in parallel. The loss function was computed only for points on good quality maps from the data set in order to account for possible noisy or incorrect labels. MetNet outperformed the persistence model, an optical flow-based algorithm, as well as the High-Resolution Rapid Refresh (HRRR) for forecasts up to 8 h in the future. By performing ablation studies, they pointed out that using a large spatial context leads to better performance than using a smaller context on long-term predictions. However, reducing the temporal context up to 30 min did not decrease the model's performance. Moreover, the authors pointed out that radar data plays a more important role in the overall model performance for short-term predictions than for long-term ones. These results can be explained by the fact that long-term predictions need to take into account a larger spatial context that cannot be typically captured by radar, thus highlighting the importance of incorporating satellite data for this type of predictions.

The model proposed by Franch et al. [23] aimed to improve the performance of nowcasting systems on extreme events prediction by training an ensemble of Trajectory Gated Recurrent Units (TrajGRUs), each optimized by over-weighting the objective for a specific precipitation threshold. In addition to the ensemble components, a model stacking strategy that consists of training an additional model using the outputs of the ensemble components is employed. Moreover, their approach enhances the radar data with orographic features. The proposed model achieved overall better performance than several TrajGRU baselines and two models obtained by using only part of the components—an ensemble model without orographic features, and a single model trained with orographic features.

Chen et al. [1] improved upon the training of ConvLSTMs by introducing a multi-sigmoid loss function tailored for the precipitation nowcasting task and incorporating residual connections in the recurrent architecture. Additionally, the group normalization mechanism proved to be beneficial for the model's performance. The model was trained on radar images and predictions were evaluated for lead times of up to one hour.

The Small Attention-Unet (SmaAt-Unet) [17] precipitation nowcasting model introduced by Trebing et al. is a modified U-Net architecture, in which traditional convolutions have been replaced by depthwise separable convolutions and convolutional block attention modules have been added to the encoder. The proposed approach achieved an overall comparable performance to the original U-Net, while using a quarter of the number of parameters. The nowcasting is done for up to 30 min in the future using 1 h of past radar data, sampled at a frequency of 5 min. Similarly to other U-Net-based methods, different time stamps are concatenated channelwise and given as input to the network. Patterns across the channel dimension are captured by the attention modules. As precipitation nowcasting performance indicators, a CSI of 0.647, a FAR of 0.270, and an *F-score* of 0.768 have been obtained.

An approach for weather forecasting using ConvLSTMs and attention was introduced in [18]. Their proposed method was tested on the ECMWF (European Centre for Medium-Range Weather Forecasts) Reanalysis v5 (ERA5) data set, which contains several weather measurements such as temperature, geopotential, humidity and vertical velocity at a time resolution of one hour. The approach was shown to outperform other methods such as Simple Moving Average, U-Net, and ConvLSTM, achieving MSE values between 1.32 and 2.47.

Jeong et al. [24] alternatively proposed a weighted broadcasting strategy for ConvLSTMs, which is based on the idea of overweighting the last time stamp in the input sequence. Their approach reached generally better performance than the baseline ConvLSTM architecture, with CSI values ranging between 0.0108 and 0.5031, FAR between 0.2960 and 0.5653, POD values in the range 0.0110-0.6403 and Heidke skill score (HSS) between 0.01 and 0.3.

A deep learning approach for precipitation estimation from reflectivity values was introduced by Yo et al. [25]. The proposed approach was compared to an operational precipitation estimation method used by the Central Weather Bureau in Taiwan and was shown to slightly outperform it, especially in predicting extreme meteorological events. However, the improvement was not statistically significant, the proposed method obtaining an average POD of 0.8 and FAR of 0.0134.

3. Methodology

With the goal of answering research question **RQ1**, this section introduces our binary classification model proposal, *AutoNowP*, that consists of two ConvAEs, trained on radar data collected from rainfall conditions with different classes of severity, for recognizing severe phenomena. More specifically, *AutoNowP* is trained for learning to predict whether the radar reflectivity values will be above or below a specific threshold. The ConvAE models are used due to their ability to preserve the structure of the input data and to detect underlying structural relationships within the data.

AutoNowP is aimed to empirically demonstrate that autoencoders are able to learn, by self-supervision, features that are relevant for distinguishing structural relationships in radar data collected in both stratiform and convective weather conditions. The model is designed to classify if a radar product Rp is below or above a threshold τ. In the experiments we will use two radar products, the reflectivity at the first elevation level (R01) and the composite reflectivity, and different values for the threshold τ (e.g., 5, 20, 35 dBZ). *AutoNowP* consists of three stages depicted in Figure 1: data representation and preprocessing, training, and testing (evaluation). These stages will be further detailed.

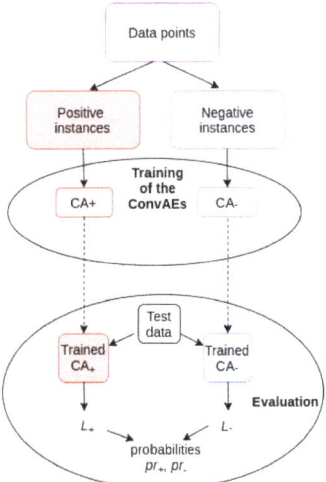

Figure 1. Overview of *AutoNowP*.

3.1. Data Representation and Preprocessing

The raw radar data used in our experiments is converted into two-dimensional arrays, with a grid cell representing a geographical location. A cell in the matrix stores the value of a specific radar product at a given time stamp. A sequence of such matrices is available for a given day, each matrix storing the values for a specific radar product p at a time moment t. We assume that np radar products are available and thus, the radar data at a time moment t may be visualized as a data grid with np channels.

In our previous works [11,26] we highlighted that similar values for the radar products in a specific location l at a time t are encoded in similar neighborhoods of the location l at time $t-1$. For a specific location l at time t, a d^2–dimensional vector containing the values of a radar product Rp from the sub-grid of diameter d centered on l (at time $t-1$) will be assigned. The d^2—dimensional instance will be labeled with the of Rp for the location l at time t [11]. A sample data grid containing the values for the product R01 at time t is shown in Figure 2a, while Figure 2b depicts the data grid at time $t-1$.

15	5	10	10	15
10	25	15	5	10
20	0	**10**	10	10
15	10	25	15	5
15	0	15	5	0

20	30	10	15	10
15	15	10	20	5
20	10	**15**	20	25
10	5	10	10	20
15	5	10	15	20

(a) The data matrix at time stamp t. In red is the value of R01 at location $l = (3,3)$.

(b) The data grid at time stamp $t-1$. In blue is the neighborhood of the location $l = (3,3)$ of diameter $d = 3$.

Figure 2. Sample data grids at time stamp t and $t-1$ highlighting an instance sample at location $l = (3,3)$ and a diameter $d = 3$ for the neighborhood.

For the example from Figure 2, the instance corresponding to the location (3,3) at time t is the vector (15,10,20,10,15, 20,5,10,10) and is labeled with 10 (the value of R01 at location (3,3) and time t).

Consequently, considering a specific diameter d for the neighborhood, a data set R is built from the instances (d^2—dimensional points) associated to each location from the data grid and all available time moments [11]. The radar data set R will be divided in two classes: the positive class (denoted as "+") composed by the instances having the label (i.e., values for the radar product Rp at a certain time t) higher than a threshold τ, while the negative class (denoted as "−") contains the instances having the label lower or equal to the threshold τ. The data set representing the positive class is denoted by R_+, while R_- denotes the set of instances belonging to the negative class. We note that the dimensionality of R_- is significantly larger than the cardinality of R_+, as the number of severe weather events is often small.

Both data sets are then normalized so that the value Rp of a radar product is transformed to be in the $[0,1]$ range. For normalization purposes, we use the classic min/max normalization formula:

$$Rp'(l,t) = \frac{Rp(l,t) - Rp_{min}}{Rp_{max} - Rp_{min}},$$

where:

- $Rp(l,t)$ is the value of Rp at time t and location l;
- $Rp'(l,t)$ is the normalized value of Rp at time t and location l;
- Rp_{min} is the minimum value in the domain of Rp;
- Rp_{max} is the maximum value in the domain of Rp.

It should be noted that we are using the minimum and maximum values from a radar product's domain to ensure that both R_+ and R_- data sets are normalized in the same way (i.e., the same value in different data sets is mapped to the same normalized value), as the positive data set may have different minimum and maximums than the negative data set. *AutoNowP* is trained and tested on the normalized data.

3.2. AutoNowP Classification Model

Considering the notations from Section 3.1, the classification problem is formalized as the approximation of two target functions (i.e., one target function for each class) $t_c : \mathcal{R}_+ \cup \mathcal{R}_- \to [0,1]$ ($\forall c \in \{+,-\}$) that express the probability of instances from $\mathcal{R}_+ \cup \mathcal{R}_-$ to belong to the "+" or "−" classes. Thus, the learning goal of *AutoNowP* will be to approximate the functions t_+ and t_-. *AutoNowP* consists of two ConvAEs, one for the "+" class (CA_+) and one for the "−" class (CA_-). For training an autoencoder CA_c ($c \in \{+,-\}$) 47% from the data set R_c (i.e., 70% from the data not used for testing) will be used for training, 20% for the model validation and the rest of 33% from R_c will be further used for testing, using a 3-fold cross-validation testing methodology.

3.2.1. Training

As previously stated, *AutoNowP* classifier will be trained to predict, based on the radar products values from the neighborhood of a geographical location at time $t-1$, whether the value of a radar product Rp at time t will be higher than a threshold τ. For instance, if Rp is chosen as R01 and τ as 35 dBZ, then *AutoNowP* will be trained to predict if, in a certain geographical location or area, a convective storm is likely to occur (i.e., if the value of R01 will be higher than 35 dBZ in that geographical location).

AutoNowP is trained to recognize both normal and severe weather events, and thus it will learn to predict if a certain instance is likely to indicate stormy or normal weather. Each of the two autoencoders CA_+ and CA_- will be self-supervisedly trained on the data set of positive and negative instances, respectively ($\mathcal{R}_+$ and $\mathcal{R}_-$).

The prediction is based on estimating the probabilities (denoted by p_+ and p_-) that a high-dimensional instance corresponding to a particular geographic location (as described in Section 3.1) belongs to the positive and negative classes. The method for computing these probabilities will be detailed in Section 3.2.2.

Autoencoders Architecture

The current study uses convolutional undercomplete AEs to learn meaningful lower-dimensional representations for radar data. The autoencoders were implemented in Python, using the Keras framework with Tensorflow backend. Both autoencoders (CA_+ and CA_-) have the same architecture. The input data of the AEs is the 2D grid of the neighborhood of diameter d for one location (as exemplified in Figure 2b)—i.e., the 2D grid representing the values of an instance from $\mathcal{R}_+ \cup \mathcal{R}_-$. As we have to choose a different diameter d for our experiments on different data sets (see Section 4.1), we made the architecture so that it minimally changes with d: while the number, type and hyper-parameters of each layer of the network remain the same, the number of neurons on each layer changes, proportionally, depending on d.

Even if the architecture of the autoencoder may be adapted to the diameter d of the neighborhood (i.e., the dimensionality d^2 of the input data), the value of d may influence the performance of *AutoNowP* model. Intuitively, high values for d will make the AEs to harder distinguish between the positive and negative instances. This may happen since, hypothetically speaking, it would be possible that two neighboring points at time t (one positive and one negative) have a large number of identical neighbors at time $t-1$ (i.e., the data instances representing the two locations are similar) and thus the AEs are unable to distinguish between them. On the other hand, a small number of neighbors for a data point (i.e, small values for d) is not enough for *AutoNowP* classifier to discriminate between the input instances. For determining the most appropriate value for the diameter d, a grid search was performed for selecting the value d that provides the best performance for *AutoNowP*.

In the following, we will present the architecture of the autoencoders and the hyper-parameters used, without mentioning the number of neurons, so that the following description is valid for the *AutoNowP* model in general, regardless of the specific experiment. Figure 3 illustrates the architecture of the autoencoder (as mentioned above, both autoencoders, CA_- and CA_+, have the same architecture). This is a Convolutional Autoencoder, thus the main layers are the Conv2D—2-dimensional convolution layers—represented in yellow in the figure. These layers reduce the data grid input in three steps, leading to an encoding layer (the blue layer in the figure). From the encoding, the autoencoder needs to recreate the input, thus the inverse of the Conv2D is needed: Conv2DTranspose (the orange layers). Using the Conv2DTranspose layers we apply the reverse transformation so that it recreates the data grid as it was before the convolutions. When using convolutions, we need to reduce the size of the image, and this works best if the size of the image is even. However, our input layer has always an odd size: since the input represents the neighborhood of one point, having that point in the center, for a given radius r, the size will be $(2r+1, 2r+1)$—i.e., we take r neighbors from all sides of the center; for example, r neighbors on the right with r neighbors on the left plus the center itself results in an $2r+1$ length. Since the input is always odd in size, we need to adjust it so that we can perform the convolutions. For this, we use a ZeroPadding2D layer: after the Input layer (first gray layer), we pad the margins of the data grid with zeros until it reaches the desired size, using the ZeroPadding2D layer (the red layer). Afterwards, the convolutions can occur. The transpose convolutions will recreate the data grid as it was before the convolutions—that is, after padding—so it is not the same size as the input. Since it is an autoencoder, we want to match the output to the input, thus, we need to adjust the transpose convolutions output so that the final size of the autoencoder output fits the size of its input. To readjust the size, we use a Cropping2D layer, which will also be the output layer of the autoencoder (the second gray layer represented in the figure).

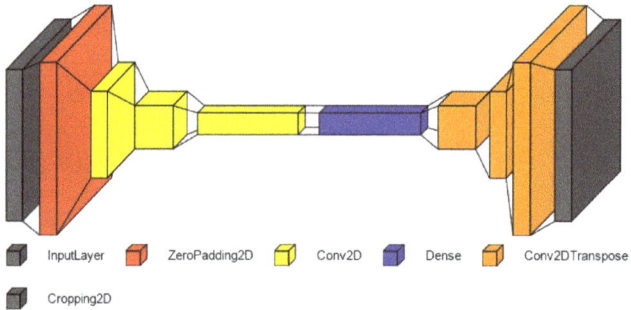

Figure 3. Architecture of a Convolutional Autoencoder (CA_c).

As with other neural networks, while the architecture is the principal element of the network, there are other metaparameters that need to be tuned that change the network's behavior. One of these is the number of neurons on each hidden layer, but as we mentioned above, this number may differ among the experiments if the input size changes; however, while the absolute number changes, the proportion of neurons on the hidden layers are preserved. Then, we have the activation used for the layers: for all convolutional layers, transpose convolutional layers and the dense layers, except for the last transpose convolutional layer, we use the SELU activation function (Scaled Exponential Linear Unit [27]). For the last transpose convolutional layer, we used the sigmoid activation function, so that the output of the autoencoder is between 0 and 1, as is the input. For all convolutional layers and transpose convolutional layers, we used a kernel size of 4 and 2 strides.

The training configuration was the following: we used a batch size of 1024 and we trained each autoencoder for 500 epochs in the case of the NMA data set and for 200 epochs for the MET data set; the Adam optimizer [28] was used with learning rates of 0.01 and 0.001 respectively for the NMA and MET data sets and *epsilon* of 0.00001.

Loss Functions

As explained in Section 3.1, the high-dimensional input instance x may be visualized as a data grid, i.e., the neighborhood around the location of the value we want to predict. The autoencoders learn to encode and decode each instance, the output of the autoencoder being the reconstruction of the instance. The loss functions represent the difference between the original instances and their reconstruction; lower values for the loss indicate better reconstructions (i.e., closer to the input), with a loss equal to 0 meaning no difference. The loss is based on a modified mean squared error (MSE), to assign a priority to the values greater than the threshold τ relative to the other values. More specifically, we wanted to be able to make the autoencoders prioritize values in the neighborhood that are either greater or lower or equal to the given threshold τ. We also wanted to be able to change this prioritization between CA_- and CA_+ (i.e., CA_- is trained to prioritize negative points, while CA_+ is trained by over-weighting positive points in the neighborhood) and between experiments, so we introduced a parameter, α, that controls this prioritization. We split the computation of MSE in two parts: computing the MSE for values greater than τ (Formula (1)) and computing the MSE for values lesser or equal to τ (Formula (2)). The final loss value (Formula (3)) is expressed as a linear combination between the two separately computed MSEs; we use the α parameter to decide how to prioritize the values greater than τ relative to the values less or equal to τ. The exact way to compute the loss function $L(x, x')$ for a given instance $x \in \mathcal{R}_+ \cup \mathcal{R}_-$ is given by Formulae (1)–(3):

$$MSE_{greater}(x, x') = \frac{1}{d^2} \sum_{\substack{1 \leq i \leq d^2 \\ x_i > \tau}} (x_i - x'_i)^2 \qquad (1)$$

$$MSE_{lesser}(x, x') = \frac{1}{d^2} \sum_{\substack{1 \leq i \leq d^2 \\ x_i \leq \tau}} (x_i - x'_i)^2 \qquad (2)$$

$$L(x, x') = \alpha \cdot MSE_{greater}(x, x') + (1 - \alpha) \cdot MSE_{lesser}(x, x') \qquad (3)$$

where:
- d is the diameter of the neighborhood used for characterizing the input instances x (see Section 4.1);
- $x \in \mathcal{R}_+ \cup \mathcal{R}_-$ is the d^2-dimensional instance for which we compute the loss;
- x' is the autoencoder output for instance x (the reconstruction of x);
- τ is the chosen threshold that differentiates between positive and negative class;
- α is the parameter that we introduced for the loss;
- x_i and x'_i denote the ith component from x and x' respectively.

3.2.2. Classification Using *AutoNowP*

After *AutoNowP* has been trained as described in Section 3.2.1, when an unseen query instance q has to be classified, the probabilities $p_+(q)$ (that q belongs to the positive class) and $p_-(q)$ (that q belongs to the negative class) are computed. As shown above, a query instance q is a high-dimensional vector (Section 3.1) consisting of radar products values from the neighborhood of a specific geographical location l at time t. *AutoNowP* will classify q as "+" (i.e., the value of the radar product Rp at time $t+1$ is likely to be higher than the threshold τ) iff $p_+(q) \geq p_-(q)$, i.e., $p_+(q) \geq 0.5$.

The underlying idea behind deciding that a query instance q is likely to belong to the "+" class (i.e., $p_+(q) \geq p_-(q)$) is the following. We started from the assumption that an AE is able to encode the structure of the class of instances it was trained on well and with the intention to further reconstruct data similar to the training data. In addition, the AE will be unable to reconstruct, through its learned latent space representation, the instances that are dissimilar to the training data (i.e., likely to belong to another class than the class on which the AE was trained on). Thus, if for a certain instance q the MSE between q and the reconstruction of q by CA_+ is less than the MSE between q and the reconstruction of q by CA_-, then it is likely that the query instance belongs to the "+" class, as it is more similar to the information encoded for the positive class.

Definition 1. *Let us denote by $MSE_c(\hat{q}, q)$ the MSE between q and the reconstruction ($\hat{q}$) of q by the autoencoder CA_c ($c \in \{+, -\}$) and by τ the threshold considered. The probabilities $p_+(q)$ and $p_-(q)$ are computed as given in Formulae (4) and (5).*

$$p_+(q) = 0.5 + \frac{MSE_-(\hat{q}, q) - MSE_+(\hat{q}, q)}{2 \cdot (MSE_-(\hat{q}, q) + MSE_+(\hat{q}, q))} \qquad (4)$$

$$p_-(q) = 1 - p_+(q). \qquad (5)$$

From Formula (4) we observe that $0 \leq p_+(q) \leq 1$ and that if $MSE_+(\hat{q}, q) \leq MSE_-(\hat{q}, q)$, then $pr_+(q) \geq 0.5$, meaning that q is classified by $AutoNowP$ as being positive. Much more, we note that:

- if $MSE_+(\hat{q}, q) = 0$ (and consequently $MSE_-(\hat{q}, q) \neq 0$) it follows that $p_+(q) = 1$;
- $p_+(q)$ increases as $MSE_+(\hat{q}, q)$ decreases;
- if $MSE_+(\hat{q}, q) > MSE_-(\hat{q}, q)$, then $pr_+(q) < 0.5$, meaning that q is classified by $AutoNowP$ as being negative.

After the probabilities $p_+(q)$ and $p_-(q)$ were computed from the training data, the classification $c(q)$ of q is computed as shown in Formula (6).

$$c(q) = \begin{cases} + & \text{if } pr_+(q) \geq 0.5 \\ - & \text{otherwise}. \end{cases} \qquad (6)$$

3.3. Testing

After $AutoNowP$ was trained as described in Section 3.2.1, it is evaluated on 33% of the instances from each data set R_+ and R_- that were unseen during the training stage. The classification of a query instance q is made as described in Section 3.2.2.

For evaluating the performance of $AutoNowP$ on a testing data set, the confusion matrix is computed [29], composed by the number of true positives—TP, true negatives—TN, false positives—FP, and false negatives—FN. Then, based on the values from the confusion matrix, evaluation measures used for assessing the performance of supervised classifiers and weather predictors are employed:

1. Critical success index (CSI) computed as $CSI = \frac{TP}{TP+FN+FP}$ is used for convective storms nowcasting based on radar data [30].
2. True skill statistic (TSS), $TSS = \frac{TP \cdot TN - FP \cdot FN}{(TP+FN) \cdot (FP+TN)}$.
3. Probability of detection (POD), also known as sensitivity or recall, is the true positive rate (TPRate), $POD = \frac{TP}{TP+FN}$.
4. Precision for the positive class, also known as positive predictive value (PPV), $PV = \frac{TP}{TP+FP}$.
5. Precision for the negative class, also known as negative predictive value (NPV), $NPV = \frac{TN}{TN+FN}$.
6. Specificity ($Spec$), also known as true negative rate (TNRate), $Spec = \frac{TN}{TN+FP}$.
7. Area Under the ROC Curve (AUC). The AUC measure is recommended in case of imbalanced data and is computed as the average between the true positive rate and the true negative rate, $AUC = \frac{Spec+POD}{2}$.
8. Area Under the Precision–Recall Curve ($AUPRC$), computed as the average between the precision and recall values, $AUPRC = \frac{Precision+Recall}{2}$.

All these measures take values in the [0, 1] range, with higher values indicating better predictors, excepting FAR that should be minimized for a better performance.

A three-fold cross-validation testing methodology is then applied. The value for each of the performance measures previously described are averaged over the three runs. The mean values are computed together with their 95% confidence intervals (CI) [31].

4. Data and Experiments

In this section, we answer research question **RQ2** by describing the experiments conducted for evaluating the performance of *AutoNowP* and analyzing the obtained experimental results.

4.1. Data Sets

For assessing the performance of *AutoNowP*, experiments were conducted on real radar data provided by the Romanian National Meteorological Administration (NMA) and the Norwegian Meteorological Institute (MET).

4.1.1. NMA Radar Data Set

The NMA radar data set was collected over central Romania by a single polarization S-band Weather Surveillance Radar—98 Doppler (WSR-98D) located near the village of Bobohalma. The radar completes a full volume scan every 6 min, gathering data about the location, intensity and movement direction, and speed of atmospheric cloud systems. Volume scan data is collected by employing a scan strategy consisting in 9 elevation angles, the raw data being afterwards processed to compute a large variety of radar products. For *AutoNowP* experiments, we used the base Reflectivity product (R) sampled at the lowest elevation angle (R01), being expressed in decibels relative to the reflectivity factor Z (dBZ). Using the so-called Z-R relationships, the base reflectivity is used to derive the rainfall rate, and further, the radar estimated precipitation accumulation over a given area and time interval.

The radar data set used herein contains the quality controlled (cleaned) values of the raw R01 product. The cleaning is needed, as during the radar scans, both meteorological and nonmeteorological targets can be detected. Various clutter sources (e.g., terrain, buildings), biological targets (e.g., insects, birds) and external electromagnetic sources (e.g., sun) can impact the data quality within the volume scan, and although the signal processing can effectively mitigate the effects of this data contamination, additional processing is required to identify and remove the residual nonmeteorological echoes. Herein, the quality control algorithm is applied in a two-way process, by firstly detecting and removing the contaminated radar data, and secondly tuning the key variables to mitigate the effects of the first step on good data. The method used to clean and filter the reflectivity data is based on the three-dimensional structure of the measured data, in terms of computing horizontal and vertical data quality parameters. The computation algorithm is executed on radar data projected on a polar grid to not alter the measurements and to remain at the level of data recording, and it is built considering various key quality issues like ground clutter echoes and external electromagnetic interferences. First, the radar data is passed through a noise filter to remove the isolated ground clutter reflectivity bins, and then the algorithm performs the identification and removal of echoes generated by external signals and calculates the horizontal texture and the vertical gradient of reflectivity. The outputs of these steps (i.e., sub-algorithms) are finally used to reconstruct the quality-controlled reflectivity field.

Within AutoNowP, the NMA radar data was processed by selecting a value of 7 for the diameter d of the neighborhood (introduced in Section 3.1), representing about 7 km on the physical map, and this distance commonly determines small gradients of the meteorological parameters [30]. The value 7 for d provided the best performance for *AutoNowP*.

4.1.2. MET Radar Data Set

The MET radar data set used in our experiments consists of composite reflectivity values gathered from the MET Norway Thredds Data Server [32].

The reflectivity product, available at [33] was derived from the raw reflectivity values by considering the best radar scan out of all considered elevations. Thus, it is a composite product, obtained by applying an interpolation scheme that weights radar volume sources differently based on their quality flags and various properties that may influence the measurement. The considered properties include ground or sea clutter, ships or airplanes, beam blockage, RLAN, sun flare, height above CAPPI level (typically 1000 m msl), range, and azimuth displacement. The measurements used in our experiments were collected by the radar at a time resolution of 7.5 min.

The dimension d of the neighborhood data grid was set to 15 for the MET experiment, since this dimensionality provided the best performance for $AutoNowP$.

Table 1 describes the data sets used as our case studies. The second column in the table indicates the radar product Rp of interest. The next three columns contain the number of instances from the data sets (both "+" and "−") and the percentage of positive and negative instances obtained using a threshold of 10 dBZ. The last column illustrates the entropy of each data set. The entropy is used for measuring the imbalancement of each data set [34]: lower entropy values indicate a higher degree of imbalancement.

Table 1. Description of the data sets.

Data Set	Product of Interest (Rp)	# Instances	% of "+" Instances	% of "−" Instances	Entropy
NMA	R01	9003688	3.44%	96.56%	0.216
MET	Composite reflectivity	6607836	31.97%	68.03%	0.904

From Table 1 we can see that the NMA data set is severely imbalanced: only 3.44% of the instances belong to the positive class, leading to a negative to positive ratio of about 28:1. Another element that highlights the high degree of data imbalancement is the entropy; where an entropy value of 1 reflects a perfectly balanced data set, the NMA data set entropy of 0.216 reflects a data set with low diversity, heavily weighted in favor of one class to the detriment of the other. The MET data set, on the other hand, showed a higher proportion of positive samples for this choice of threshold, as reflected by a higher entropy. In this setting, the negative to positive ratio is approximately 2:1.

The two-dimensional PCA [35] projections of the instances from both NMA and MET data sets from Figure 4 highlight the difficulty of the classification task. For both data sets, there is a low degree of separation between the class of negative instances (blue colored) and the class of positive instances (red colored).

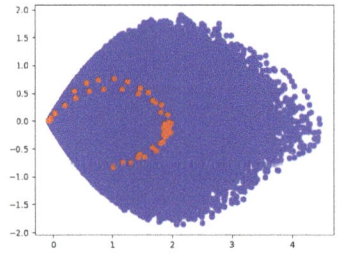

(a) 2D PCA plot for the NMA data set.   (b) 2D PCA plot for the MET data set.

Figure 4. 2D PCA visualization of the NMA data set (**a**) and MET data set (**b**).

The NMA data sets used in our experiments are publicly available at [36], while the MET data is publicly available at [37].

4.2. Results

This section presents the experimental results obtained by applying *AutoNowP* classifier on the data sets described in Section 4.1. For the ConvAEs, the implementation from the Keras deep learning API [38] using the Tensorflow framework was employed.

The experiments were performed on a workstation laptop, with an Intel i9-10980HK CPU, 32 GB RAM and Nvidia RTX 2080 Super for GPU acceleration; and on a Google cloud instance with 12 vCPUs, 64 GB RAM and access to a Nvidia Tesla V100 for GPU acceleration.

The evaluation measures and the testing methodology described in Section 3.3 were employed. Table 2 depicts the obtained results for both data sets used in our case studies, for various values of the threshold τ. The 95% confidence intervals (CIs) are used for the results.

The thresholds we decided to use were chosen considering both computational and meteorological factors. In the literature, there is no convention on thresholds for R. For example, Han et al. [9,39] chose to use the 35 dBZ threshold while Tran and Song [40] studied their prediction performance using the 5, 20 and 40 dBZ thresholds. Thus, the values 10, 20 and 30 were chosen for τ for the NMA data and 10, 15, 20 for the MET data set. Since the MET data contains few instances whose values are higher than 30 dBZ, *AutoNowP* could not be applied for this threshold. The best values obtained for the evaluation measures are highlighted for both data sets.

Table 2. Experimental results, using 95% CIs.

Data Set	τ	CSI	TSS	POD	PPV	NPV	Spec	AUC	AUPRC
NMA	10	0.615 ± 0.018	0.861 ± 0.012	0.876 ± 0.012	0.674 ± 0.017	0.996 ± 0.001	0.985 ± 0.002	0.931 ± 0.006	0.775 ± 0.013
	20	0.425 ± 0.072	0.471 ± 0.091	0.474 ± 0.092	0.810 ± 0.015	0.989 ± 0.001	0.997 ± 0.001	0.736 ± 0.046	0.642 ± 0.039
	30	0.151 ± 0.046	0.157 ± 0.051	0.157 ± 0.028	0.812 ± 0.031	0.993 ± 0.001	1.000 ± 0.000	0.579 ± 0.014	0.485 ± 0.007
MET	10	0.681 ± 0.014	0.740 ± 0.009	0.872 ± 0.019	0.757 ± 0.027	0.936 ± 0.005	0.867 ± 0.026	0.870 ± 0.005	0.814 ± 0.008
	15	0.566 ± 0.05	0.626 ± 0.09	0.675 ± 0.12	0.793 ± 0.08	0.920 ± 0.03	0.951 ± 0.03	0.813 ± 0.05	0.734 ± 0.029
	20	0.401 ± 0.090	0.500 ± 0.223	0.536 ± 0.269	0.710 ± 0.173	0.947 ± 0.026	0.963 ± 0.046	0.750 ± 0.111	0.623 ± 0.048

As shown in Table 2, the values for most of the evaluation measures decrease as the threshold τ increases. This is normal behavior, as the prediction becomes more difficult for higher values. The precision values (both for the positive and negative classes—*PPV* and *NPV*) and the true negative rate (*Spec*) increase for higher thresholds, denoting that the negative class is easier to predict for high values for τ and the number of false predictions decreases. However, the number of true positives significantly decreases for higher thresholds and this is reflected in the other performance metrics that decrease. High values (around 0.9) were obtained for sensitivity (POD), specificity, and AUC for $\tau = 10$ denoting a good enough performance of $AutoNowP$. In addition, the small values obtained for the 95% CI reveal the stability of the model.

5. Discussion

With the goal of better highlighting the performance of $AutoNowP$, this section discusses the obtained results and then provides a comparison between $AutoNowP$ and similar approaches from the nowcasting literature.

5.1. Analysis of AutoNowP performance

As shown in Table 2, $AutoNowP$ succeeds in recognizing the negative class (high specificity) and detecting the positive class (probability of detection higher than 0.85 for $\tau = 10$). This is a strength of $AutoNowP$, the ability to detect severe phenomena well. However, we observed false predictions, both for the positive and negative classes and these occur mostly close to the decision boundary. The performance of $AutoNowP$ is impacted mainly by a large enough amount of false positive predictions, but most of these errors appear near the edges of radar echoes. In these areas the difference between classes becomes blurred, as the neighborhood contains some high values, not enough to be similar enough to the center of the event, but not few enough to be outside the event. These kinds of neighborhoods are close to both classes, the dissimilarity between them and either class is small. For these kinds of instances, AutoNowP has the most prediction errors. In order to better understand the areas where these instances appear, we have created a visualization in Figure 5. This figure shows the actual R01 values read by the radar in two consecutive time steps, color-coded by the dBZ value at each location. In the figure, there are also white and black regions, which represent the regions where most of the errors made by AutoNowP appear. The aforementioned regions were found by studying the erroneous predictions of the model and discovering the common elements of the neighborhoods that are problematic, both for false negative errors and false positive errors. Then, in Figure 5, we changed a pixel to white or black if its neighborhood is problematic, if it belongs to the false negative problems or, respectively, false positive problems. In short, in the image are represented with black points the locations where the model is highly likely to erroneously predict them as positive and, similarly, with white points where it tends to wrongly predict them as negative. The black and white areas in the image account for more than 98% of AutoNowP's errors.

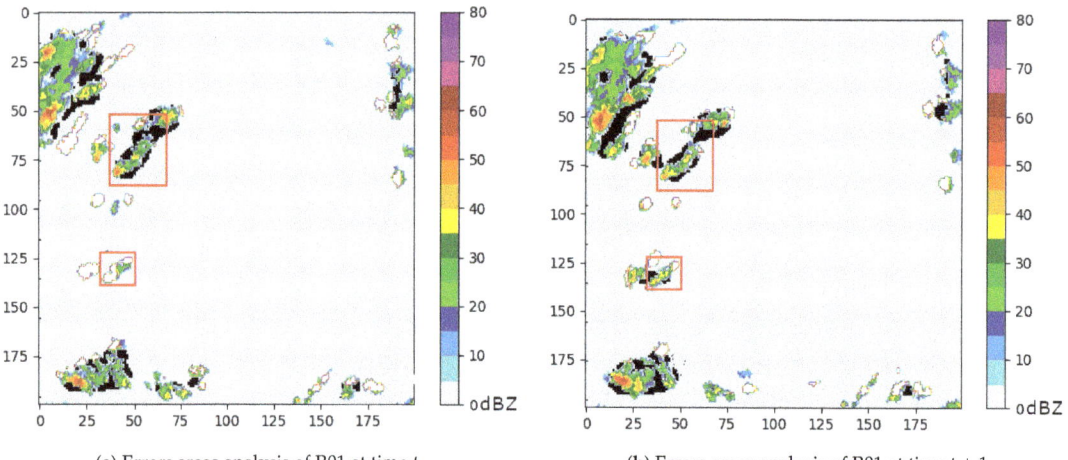

(a) Errors areas analysis of R01 at time t. (b) Errors areas analysis of R01 at time $t + 1$.

Figure 5. Visualization of AutoNowP errors areas analysis for two consecutive time steps. In white are the areas where the model usually predicts false negatives and in black are areas where the model usually predicts false positives.

In Figure 5, it can be observed that most errors appear either at the edges of meteorological events, mostly in case of false positives, or in areas where there are few positive values, in case of false negatives. In case of false positives (in black), the problem areas show a tendency of the model to smooth the predictions out, i.e., to create shapes that are much more uniform. This is not an effect typical for AutoNowP; it is a general problem affecting radar reflectivity prediction models (e.g., the RadRAR model [11]). In Figure 5, a region containing false positives is exemplified in the first highlighted region (the bigger one, around the pixel at (75,50)); it can be seen that the black region surrounds the actual meteorological event, smoothing it out, creating much more homogenous shapes. This tendency is kept from one time step to another, the smoothed shape following closely the real shape.

In case of false negatives (in white), the problems appear generally in areas where there are few positive values, i.e., the neighborhoods of locations contain many zero or close to zero values and few values higher than the threshold. For these kinds of neighborhoods, it is hard to differentiate between classes as they appear both at the start of meteorological events and at the end of meteorological events. The beginning of meteorological events is especially hard to predict, as there is no indication if and where a meteorological event will form; for this reason, the model generally predicts locations with these kinds of neighborhoods as being negative, introducing some false negative errors. In Figure 5, an example area containing a false negative region can be observed in the second highlight (the small one, around the pixel (125,50)). In that highlight, in the first time step (left side) it can be observed that the meteorological event is small, while in the next time step (right side), the region of the meteorological event has more than doubled in size. Since in the first time step the event region is so small, the model has problems predicting the relatively big changes that will happen until the next time step, thus introducing false negative errors, visualized as white regions.

Analyzing the false negative predictions of $AutoNowP$, we also noticed (in both NMA and MET experiments) situations as the one depicted in Figure 6. The figure presents the composite reflectivity for two consecutive radar acquisitions from MET data. The red rectangles highlight a region that illustrates a sample case where $AutoNowP$ provides false predictions.

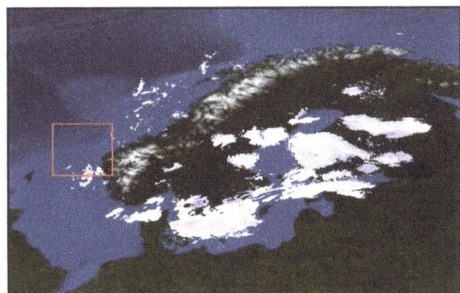 

Figure 6. Actual composite reflectivity values on two consecutive acquisitions (t—(**left**) side image— and $t + 1$—(**right**) side image) from MET data.

From Figure 6 one observes that at time t (left side image) there are no values in the highlighted region for the composite reflectivity, but at $t + 1$ (the next data received from the radar—right side image) high values for composite reflectivity are suddenly detected. Some of the data points inside the rectangle should be classified as positive instances (higher values are displayed in red), but the model fails to predict the correct class (i.e., the positive one) as the input for $AutoNowP$ (the data at t) contained mostly zero-valued data. While these situations are relatively infrequent in real life (the values are usually increasing slowly between consecutive time stamps), they still contribute to a lower prediction accuracy. However, even if $AutoNowP$ is unable to detect the positive instances at time step $t + 1$, in the next step, at time $t + 2$, the model will correctly classify the data points. This is not a limitation of $AutoNowP$, as such unexpected events cannot be detected by a learning model that was trained to predict time $t + 1$ based on time t. A possible solution would be to include more previous time steps in the prediction ($t - 1$, $t - 2$, etc).

In order to assess how the cleaning of the raw radar data impacts the predictive performance of our model, $AutoNowP$ was trained on the uncleaned NMA data as well. A threshold $\tau = 10$ and the methodology introduced in Section 3 were applied for building the $AutoNowP$ classification model on the uncleaned data. Table 3 depicts the obtained results. One observes a significant performance improvement on the cleaned data. For a specific evaluation measure P, the performance improvement is computed as $\frac{P_{cleaned} - P_{uncleaned}}{P_{uncleaned}}$ being shown in the last row of the table.

Table 3. Experimental results obtained applying $AutoNowP$ on the uncleaned NMA data and the improvement achieved on the cleaned data, for all performance measures.

	CSI	TSS	POD	PPV	NPV	Spec	AUC	AUPRC
Value	0.364	0.439	0.463	0.641	0.953	0.976	0.719	0.552
95% CI	± 0.035	± 0.072	± 0.084	± 0.050	± 0.003	± 0.012	± 0.036	± 0.018
Improvement	**69%**	**96%**	**89%**	**5%**	**4%**	**1%**	**29%**	**40%**

Table 3 highlights an average improvement of 42% on the performance measures when using the cleaned data. The highest improvements are observed on TSS (96%), POD (89%) and on CSI (69%), while the lowest improvements are on PPV, NPV and $Spec$ (less than 5%). These variations in the measures occurs because the uncleaned data introduces many false negative errors while marginally introducing true positive errors, thus for measures reliant on false negatives, such as POD, the difference is great while for measures reliant on false positives, such as $Spec$, the difference is small. We can speculate why this happens by analyzing uncleaned data and how it might affect the model: as explained in Section 4.1, the cleaning of the NMA data removes noise and clutter introduced by the interference of nonmeteorological targets during the scan. Effectively, this means that in the uncleaned data there are many locations where there are wrong values, higher than zero instead of zero. Because of this, during training, the model receives many locations labeled as negative where the neighborhood still has a large number of high-valued locations (the erroneous values), thus leading the model to make a false negative prediction (i.e., it will predict "−" even where there were actual meteorological events with a similar pattern as the erroneous training instance).

5.2. Comparison to Related Work

As shown in Section 2, most of the approaches introduced in the literature are for precipitation nowcasting. The existing methods based on radar reflectivity nowcasting were applied to radar data collected from various geographical regions, using various parameters settings, testing methodologies and various thresholds for the radar reflectivity values. The analysis of the recent literature highlighted CSI values ranging from 0.40 [20] to 0.647 [17]; POD values ranging from 0.46 [20] to 0.71 [21]; F-score values ranging from 0.58 [15] to 0.786 [15]. The performance of $AutoNowP$ on both data sets used in our experiments (Table 2) compares favorably with the literature results, considering the magnitude of the evaluation measures for a threshold of 10 (CSI higher than 0.61, POD higher than 0.87, F-score higher than 0.8).

As the literature approaches for nowcasting do not use the same data model as our approach, an exact comparison with these methods cannot be made. For a more exact comparison, we decided to apply four well-known machine learning classifiers on the data sets described in Section 4.1, using $\tau = 10$ and following the testing methodology used for evaluating the performance of *AutoNowP* (the performance measures were computed as shown in Section 3.3 and the testing was repeated 3 times for each training–validation split): logistic regression (LR), linear support vector classifier (linear SVC), decison trees (DT), and nearest centroid classification (NCC). We have selected these classifiers as baseline methods so as to cover a diverse set of methods—linear classifiers, rule-based, and distance-based.

These classifiers were implemented in Python using the scikit-learn [41] machine learning library. The comparative results are depicted in Table 4, with a 95% CIs for the values averaged over the three runs of the classifiers. The best values obtained for each performance metric are highlighted.

Table 4. Comparative results between *AutoNowP* and other classifiers. 95% CIs are used for the results.

Data set	Model	CSI	TSS	POD	PPV	NPV	Spec	AUC	AUPRC
NMA	AutoNowP	0.615 ± 0.018	**0.861** ± 0.012	**0.876** ± 0.012	0.674 ± 0.017	**0.996** ± 0.001	0.985 ± 0.002	**0.931** ± 0.006	0.775 ± 0.013
	LR	0.672 ± 0.012	0.752 ± 0.013	0.757 ± 0.013	0.857 ± 0.005	0.992 ± 0.001	**0.996** ± 0.000	0.876 ± 0.007	0.807 ± 0.008
	Linear SVC	**0.685** ± 0.012	0.778 ± 0.007	0.783 ± 0.007	0.845 ± 0.015	0.992 ± 0.000	0.995 ± 0.000	0.889 ± 0.003	**0.814** ± 0.009
	DT	0.574 ± 0.007	0.725 ± 0.004	0.734 ± 0.006	0.724 ± 0.012	0.991 ± 0.001	0.990 ± 0.002	0.862 ± 0.002	0.729 ± 0.006
	NCC	0.571 ± 0.006	0.793 ± 0.013	0.807 ± 0.013	0.662 ± 0.015	0.993 ± 0.001	0.986 ± 0.001	0.896 ± 0.006	0.735 ± 0.003
MET	AutoNowP	0.681 ± 0.014	0.740 ± 0.009	**0.872** ± 0.019	0.757 ± 0.027	**0.936** ± 0.005	0.867 ± 0.026	0.870 ± 0.005	0.814 ± 0.008
	LR	0.760 ± 0.006	0.796 ± 0.002	0.853 ± 0.001	0.875 ± 0.007	0.932 ± 0.003	**0.943** ± 0.002	0.898 ± 0.001	**0.864** ± 0.004
	Linear SVC	**0.761** ± 0.006	**0.798** ± 0.002	0.858 ± 0.001	0.870 ± 0.007	0.934 ± 0.003	0.940 ± 0.003	**0.899** ± 0.001	**0.864** ± 0.004
	DT	0.670 ± 0.010	0.710 ± 0.004	0.804 ± 0.005	0.801 ± 0.009	0.908 ± 0.003	0.906 ± 0.002	0.855 ± 0.002	0.803 ± 0.007
	NCC	0.681 ± 0.009	0.728 ± 0.005	0.831 ± 0.009	0.791 ± 0.007	0.919 ± 0.001	0.897 ± 0.006	0.864 ± 0.003	0.811 ± 0.007

The comparative results from Table 4 reveal that *AutoNowP* obtained the best results in terms of *POD* and *NPV* for both data sets. In addition, for the NMA data set, our classifier provided the highest *TSS* and *AUC* values. Figures 7 and 8 illustrate the ROC curves for the classifiers from Table 4 on NMA and MET data sets.

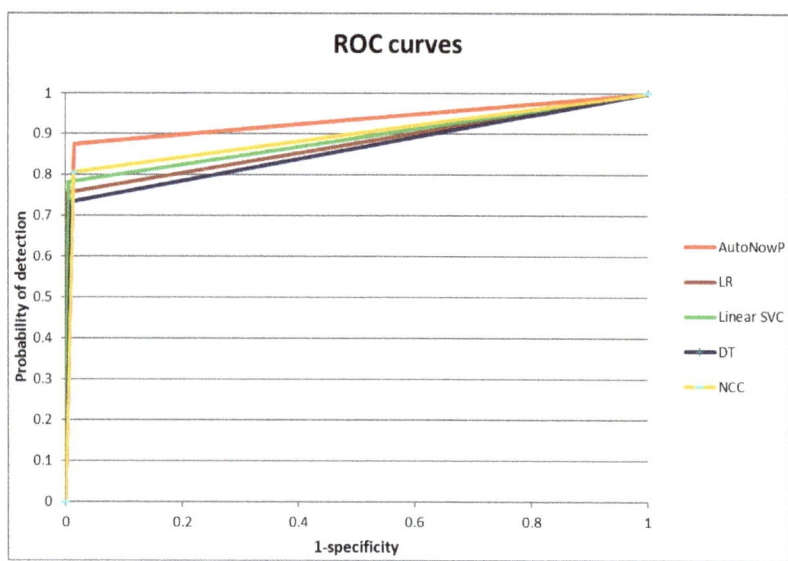

Figure 7. ROC curves for the classifiers from Table 4 on NMA data set.

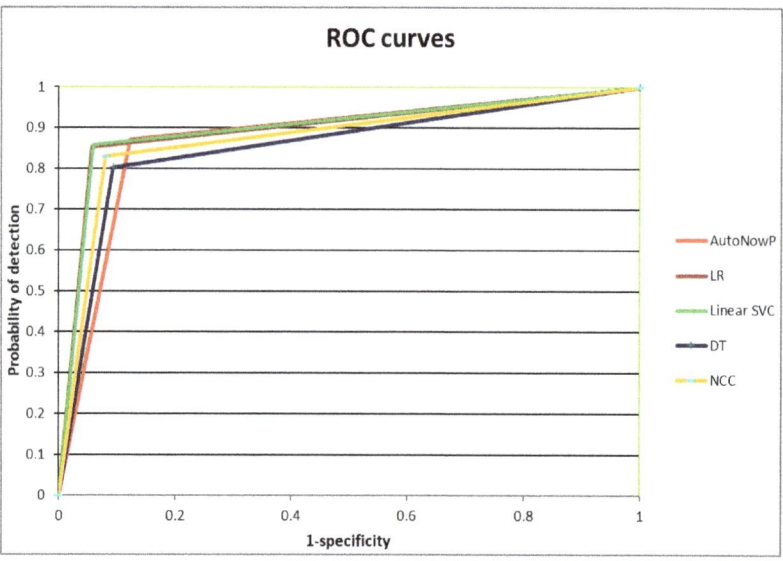

Figure 8. ROC curves for the classifiers from Table 4 on MET data set.

Table 5 summarizes the results of the comparison between *AutoNowP* and the classifiers from Table 4. The table indicates, for both the NMA and MET data sets, the number of comparisons **won** (first row) and **lost** (second row) by *AutoNowP* considering all the evaluation measures and the classifiers from Table 4. More specifically, a comparison between our approach and a classifier c, considering a specific performance measure p, is won by *AutoNowP* if the value for p provided by *AutoNowP* is greater than the one provided by the classifier c. Similarly, the comparison is lost by *AutoNowP* if the value for p provided by *AutoNowP* is lower than the one provided by the classifier c.

Table 5. Summary of the comparison between *AutoNowP* and existing classifiers.

	NMA Data	MET Data	Total
WIN	21	16	37
LOSE	11	16	27
% WIN	66%	50%	58%

The results from Table 5 highlight that *AutoNowP* outperforms similar classifiers in 66% of the cases for the NMA data set and in 50% of the cases for the MET data set out. Overall, out of 64 comparisons, our *AutoNowP* approach wins in 37 cases, i.e in 58% of the cases.

One of the main current limitations of *AutoNowP* is the training data: in order for the model to have a high performance it needs to be trained using large amounts of relevant data. While there are large amounts of historical meteorological data, finding a cohesive set of relevant, high-quality data is not trivial. Due to the large training data set needed, the training process of *AutoNowP* tends to take quite some time, which may hamper the practicality of the model. The data model might be another drawback of the *AutoNowP*, as the way it is currently designed, it might lead to the confounding of the 2 classes in some special cases, as presented in Section 5.1. Nevertheless, these limitations can be addressed, which we plan to do in the future: the long training time can be improved by parallelizing the training process, while the data model can be improved, for example by extending it to contain more than one previous time step.

6. Conclusions and Future Work

The paper introduced *AutoNowP*, a new binary classification model for precipitation nowcasting based on radar reflectivity. *AutoNowP* used two convolutional autoencoders that are trained on radar data collected on both stratiform and convective weather conditions for learning to predict if the value for the radar reflectivity on a specific location will be above or below a certain threshold. *AutoNowP* was introduced in this paper as a proof a concept that autoencoders are helpful in distinguishing between convective and stratiform rainfall. Experiments performed on radar data provided by the Romanian National Meteorological Administration and the Norwegian Meteorological Institute highlighted that the ConvAEs used in *AutoNowP* are able to learn structural characteristics from radar data and thus the lower-dimensional radar data encoded in the ConvAEs latent space is consistent with the meteorological evidence.

The generality of *AutoNowP* classifier has to be noted. Even if it was introduced and evaluated in the context of precipitation nowcasting, it may be extended and applied for other meteorological data sources and binary classification tasks.

AutoNowP is one step toward the end goal of our research: to create machine-learning-based prediction models to be integrated in existing national weather nowcasting systems. The integration of these models aims to improve the Early Warning System frameworks, as the predictions create the possibility of issuing more accurate early warnings. Better early warnings can lead to avoidance of loss and damage due to heavy precipitations, for example in events such as flash floods in densely populated areas [42].

Future work will be conducted in order to extend the data sets used in the experimental evaluation. In addition, we aim to apply *AutoNowP* to other meteorological data sources (such as satellite data) and thus using the model for other nowcasting scenarios.

Author Contributions: Conceptualization, G.C., A.-I.A., A.M. (Andrei Mihai) and I.-G.C.; methodology, G.C., A.-I.A., A.M. (Andrei Mihai) and I.-G.C.; software, A.-I.A., A.M. (Andrei Mihai) and I.-G.C.; validation, G.C., A.-I.A., A.M. (Andrei Mihai) and I.-G.C.; formal analysis, G.C., A.-I.A., A.M. (Andrei Mihai) and I.-G.C.; investigation, G.C., A.-I.A., A.M. (Andrei Mihai) and I.-G.C.; resources, G.C., A.-I.A., A.M. (Andrei Mihai), I.-G.C., S.B. and A.M. (Abdelkader Mezghani); data curation, S.B.; writing—original draft preparation, G.C.; writing—review and editing, G.C., A.-I.A., A.M. (Andrei Mihai), S.B. and A.M. (Abdelkader Mezghani); visualization, G.C., A.-I.A., A.M. (Andrei Mihai) and I.-G.C.; funding acquisition, G.C., A.-I.A., A.M. (Andrei Mihai), I.-G.C., S.B. and A.M. (Abdelkader Mezghani). All authors have read and agreed to the published version of the manuscript.

Funding: The research leading to these results has received funding from the NO Grants 2014–2021, under Project contract No. 26/2020.

Institutional Review Board Statement: Not applicable.

Informed Consent Statement: Not applicable.

Data Availability Statement: The NMA data sets used in our experiments are publicly available at [36], while the MET data is publicly available at [37].

Acknowledgments: The authors would like to thank the editor and the anonymous reviewers for their useful suggestions and comments that helped to improve the paper and the presentation. The research leading to these results has received funding from the NO Grants 2014–2021, under Project contract No. 26/2020.

Conflicts of Interest: The authors declare no conflict of interest. The funders had no role in the design of the study; in the collection, analyses, or interpretation of data; in the writing of the manuscript, or in the decision to publish the results.

References

1. Chen, L.; Cao, Y.; Ma, L.; Zhang, J. A Deep Learning-Based Methodology for Precipitation Nowcasting With Radar. *Earth Space Sci.* **2020**, *7*, e2019EA000812. [CrossRef]
2. Jongman, B. Effective adaptation to rising flood risk. *Nat. Commun.* **2018**, *9*, 1–3. [CrossRef]
3. Arnell, N.; Gosling, S. The impacts of climate change on river flood risk at the global scale. *Clim. Chang.* **2016**, *134*, 387–401. [CrossRef]
4. Silvertown, J. A new dawn for citizen science. *Trends Ecol. Evol.* **2009**, *24*, 467–471. [CrossRef] [PubMed]
5. Dixon, M.; Wiener, G. TITAN: Thunderstorm Identification, Tracking, Analysis, and Nowcasting—A radar-based methodology. *J. Atmos. Ocean. Technol.* **1993**, *10*, 785–797. [CrossRef]
6. Johnson, J.T.; MacKeen, P.L.; Witt, A.; Mitchell, E.D.W.; Stumpf, G.J.; Eilts, M.D.; Thomas, K.W. The Storm Cell Identification and Tracking Algorithm: An Enhanced WSR-88D Algorithm. *Weather Forecast.* **1998**, *13*, 263–276. [CrossRef]
7. Haiden, T.; Kann, A.; Wittmann, C.; Pistotnik, G.; Bica, B.; Gruber, C. The Integrated Nowcasting through Comprehensive Analysis (INCA) System and Its Validation over the Eastern Alpine Region. *Weather Forecast.* **2011**, *26*, 166–183. [CrossRef]
8. Sun, J.; Xue, M.; Wilson, J.W.; Zawadzki, I.; Ballard, S.; onvlee hooiMeyer, J.; Joe, P.; Barker, D.; Li, P.W.; Golding, B.; et al. Use of NWP for Nowcasting Convective Precipitation: Recent Progress and Challenges. *Bull. Am. Meteorol. Soc.* **2014**, *95*, 409–426. [CrossRef]
9. Han, L.; Sun, J.; Zhang, W. Convolutional Neural Network for Convective Storm Nowcasting Using 3D Doppler Weather Radar Data. *arXiv* **2019**, arXiv:1911.06185.
10. Tan, C.; Feng, X.; Long, J.; Geng, L. FORECAST-CLSTM: A New Convolutional LSTM Network for Cloudage Nowcasting. In Proceedings of the 2018 IEEE Visual Communications and Image Processing (VCIP), Taichung, Taiwan, 10–12 December 2018; pp. 1–4.
11. Czibula, G.; Mihai, A.; Czibula, I.G. RadRAR: A relational association rule mining approach for nowcasting based on predicting radar products' values. *Procedia Comput. Sci.* **2020**, *176*, 300–309. [CrossRef]
12. Hao, L.; Kim, J.; Kwon, S.; Ha, I.D. Deep Learning-Based Survival Analysis for High-Dimensional Survival Data. *Mathematics* **2021**, *9*, 1244. [CrossRef]
13. Mousavi, S.M.; Ghasemi, M.; Dehghan Manshadi, M.; Mosavi, A. Deep Learning for Wave Energy Converter Modeling Using Long Short-Term Memory. *Mathematics* **2021**, *9*, 871. [CrossRef]
14. Castorena, C.M.; Abundez, I.M.; Alejo, R.; Granda-Gutiérrez, E.E.; Rendón, E.; Villegas, O. Deep Neural Network for Gender-Based Violence Detection on Twitter Messages. *Mathematics* **2021**, *9*, 807. [CrossRef]
15. Sønderby, C.K.; Espeholt, L.; Heek, J.; Dehghani, M.; Oliver, A.; Salimans, T.; Hickey, J.; Agrawal, S.; Kalchbrenner, N. MetNet: A Neural Weather Model for Precipitation Forecasting. *arXiv* **2020**, arXiv:2003.12140.
16. Alain, G.; Bengio, Y. What regularized auto-encoders learn from the data-generating distribution. *J. Mach. Learn. Res.* **2014**, *15*, 3563–3593.

17. Trebing, K.; Stanczyk, T.; Mehrkanoon, S. SmaAt-UNet: Precipitation nowcasting using a small attention-UNet architecture. *Pattern Recognit. Lett.* **2021**, *145*, 178–186. [CrossRef]
18. Tekin, S.F.; Karaahmetoglu, O.; Ilhan, F.; Balaban, I.; Kozat, S.S. Spatio-temporal Weather Forecasting and Attention Mechanism on Convolutional LSTMs. *arXiv* **2021**, arXiv:2102.00696.
19. Shi, X.; Chen, Z.; Wang, H.; Yeung, D.Y.; Wong, W.k.; Woo, W.c. Convolutional LSTM Network: A ML Approach for Precipitation Nowcasting. In Proceedings of the 28th International Conference on Neural Information Processing Systems, Montreal, QC, Canada, 7–12 December 2015; MIT Press: Cambridge, UK, 2015; Volume 1, pp. 802–810.
20. Heye, A.; Venkatesan, K.; Cain, J. Precipitation Nowcasting: Leveraging Deep Convolutional Recurrent Neural Networks. In Proceedings of the 31st Conference on Neural Information Processing Systems, Long Beach, NY, USA, 4–9 December 2017; pp.1–8.
21. Tian, L.; Li, X.; Ye, Y.; Xie, P.; Li, Y. A Generative Adversarial Gated Recurrent Unit Model for Precipitation Nowcasting. *IEEE Geosci. Remote Sens. Lett.* **2020**, *17*, 601–605. [CrossRef]
22. Han, L.; Sun, J.; Zhang, W. Convolutional Neural Network for Convective Storm Nowcasting Using 3-D Doppler Weather Radar Data. *IEEE Trans. Geosci. Remote Sens.* **2020**, *58*, 1487–1495. [CrossRef]
23. Franch, G.; Nerini, D.; Pendesini, M.; Coviello, L.; Jurman, G.; Furlanello, C. Precipitation Nowcasting with Orographic Enhanced Stacked Generalization: Improving Deep Learning Predictions on Extreme Events. *Atmosphere* **2020**, *11*, 267. [CrossRef]
24. Jeong, C.H.; Kim, W.; Joo, W.; Jang, D.; Yi, M.Y. Enhancing the Encoding-Forecasting Model for Precipitation Nowcasting by Putting High Emphasis on the Latest Data of the Time Step. *Atmosphere* **2021**, *12*, 261. [CrossRef]
25. Yo, T.S.; Su, S.H.; Chu, J.L.; Chang, C.W.; Kuo, H.C. A Deep Learning Approach to Radar-Based QPE. *Earth Space Sci.* **2021**, *8*, e2020EA001340. [CrossRef]
26. Mihai, A.; Czibula, G.; Mihulet, E. Analyzing Meteorological Data Using Unsupervised Learning Techniques. In Proceedings of the ICCP 2019: IEEE 15th International Conference on Intelligent Computer Communication and Processing, Cluj-Napoca, Romania, 5–7 September 2019; IEEE Computer Society: Washington, DC, USA, 2019; pp. 529–536.
27. Klambauer, G.; Unterthiner, T.; Mayr, A.; Hochreiter, S. Self-Normalizing Neural Networks. In Proceedings of the 31st Conference on Neural Information Processing Systems, Long Beach, NY, USA, 4–9 December 2017; pp. 972–981.
28. Kingma, D.P.; Ba, J. Adam: A Method for Stochastic Optimization. *arXiv* **2017**, arXiv:1412.6980.
29. Gu, Q.; Zhu, L.; Cai, Z. Evaluation Measures of the Classification Performance of Imbalanced Data Sets. In *Computational Intelligence and Intelligent Systems*; Springer: Berlin/Heidelberg, Germany, 2009; pp. 461–471.
30. Czibula, G.; Mihai, A.; Mihulet, E. NowDeepN: An Ensemble of Deep Learning Models for Weather Nowcasting Based on Radar Products' Values Prediction. *Appl. Sci.* **2021**, *11*, 125. [CrossRef]
31. Brown, L.; Cat, T.; DasGupta, A. Interval Estimation for a proportion. *Stat. Sci.* **2001**, *16*, 101–133. [CrossRef]
32. MET Norway Thredds Data Server. Available online: https://thredds.met.no/thredds/catalog.html (accessed on 7 May 2021).
33. Composite Reflectivity Product—MET Norway Thredds Data Server. Available online: https://thredds.met.no/thredds/catalog/remotesensing/reflectivity-nordic/catalog.html (accessed on 15 May 2021).
34. Sekerka, R.F. 15—Entropy and Information Theory. In *Thermal Physics*; Sekerka, R.F., Ed.; Elsevier: Amsterdam, The Netherlands, 2015; pp. 247–256.
35. Jolliffe, I.T.; Cadima, J. Principal component analysis: A review and recent developments. *Philos. Trans. R. Soc. A Math. Phys. Eng. Sci.* **2016**, *374*, 20150202. [CrossRef]
36. NMA Data Set. Available online: http://www.cs.ubbcluj.ro/~mihai.andrei/datasets/autonowp/ (accessed on 15 May 2021).
37. MET Data Set. Available online: https://thredds.met.no/thredds/catalog/remotesensing/reflectivity-nordic/2019/05/catalog.html?dataset=remotesensing/reflectivity-nordic/2019/05/yrwms-nordic.mos.pcappi-0-dbz.noclass-clfilter-novpr-clcorr-block.laea-yrwms-1000.20190522.nc (accessed on 15 May 2021).
38. Keras. The Python Deep Learning Library. 2018. Available online: https://keras.io/ (accessed on 15 May 2021).
39. Han, L.; Sun, J.; Zhang, W.; Xiu, Y.; Feng, H.; Lin, Y. A machine learning nowcasting method based on real-time reanalysis data. *J. Geophys. Res. Atmos.* **2017**, *122*, 4038–4051. [CrossRef]
40. Tran, Q.K.; Song, S.K. Computer Vision in Precipitation Nowcasting: Applying Image Quality Assessment Metrics for Training Deep Neural Networks. *Atmosphere* **2019**, *10*, 244. [CrossRef]
41. Scikit-Learn. Machine Learning in Python. 2021. Available online: http://scikit-learn.org/stable/ (accessed on 1 May 2021).
42. Mel, R.A.; Viero, D.P.; Carniello, L.; D'Alpaos, L. Optimal floodgate operation for river flood management: The case study of Padova (Italy). *J. Hydrol. Reg. Stud.* **2020**, *30*, 100702. [CrossRef]

Article

Statistical Machine Learning in Model Predictive Control of Nonlinear Processes

Zhe Wu [1], David Rincon [1], Quanquan Gu [2] and Panagiotis D. Christofides [1,3,*]

[1] Department of Chemical and Biomolecular Engineering, University of California, Los Angeles, CA 90095-1592, USA; wuzhe@g.ucla.edu (Z.W.); fdrinconc@gmail.com (D.R.)
[2] Department of Computer Science, University of California, Los Angeles, CA 90095-1592, USA; qgu@cs.ucla.edu
[3] Department of Electrical and Computer Engineering, University of California, Los Angeles, CA 90095-1592, USA
* Correspondence: pdc@seas.ucla.edu

Abstract: Recurrent neural networks (RNNs) have been widely used to model nonlinear dynamic systems using time-series data. While the training error of neural networks can be rendered sufficiently small in many cases, there is a lack of a general framework to guide construction and determine the generalization accuracy of RNN models to be used in model predictive control systems. In this work, we employ statistical machine learning theory to develop a methodological framework of generalization error bounds for RNNs. The RNN models are then utilized to predict state evolution in model predictive controllers (MPC), under which closed-loop stability is established in a probabilistic manner. A nonlinear chemical process example is used to investigate the impact of training sample size, RNN depth, width, and input time length on the generalization error, along with the analyses of probabilistic closed-loop stability through the closed-loop simulations under Lyapunov-based MPC.

Keywords: generalization error; recurrent neural networks; machine learning; model predictive control; nonlinear systems

1. Introduction

Modeling large-scale, complex nonlinear processes has been a long-standing research problem in process systems engineering. The traditional approaches to modeling nonlinear processes include data-driven modeling approach with parameters identified from industrial/simulation data [1,2], and first-principles modeling approach based on a fundamental understanding of the underlying physico-chemical phenomena. While traditional first-principles modeling approach has been used extensively in monitoring, control and optimization of chemical processes, it can be time-demanding and inaccurate to model complex nonlinear processes using first-principle modeling tools. Machine learning methods have been increasingly adopted to model complex nonlinear systems due to their ability to model a rich set of nonlinear functions and handle efficiently with big datasets from processes [3–10]. Among many machine learning modeling techniques, recurrent neural network (RNN) is widely used to model nonlinear dynamic systems using time-series data [11–13]. While the history of machine learning methods in chemical process control can be traced back to 1990s [14–18], machine learning has become popular again this decade due to a number of reasons such as cheaper computation (mature and efficient libraries/hardware), availability of large datasets, and advanced learning algorithms. Designing MPC systems that utilize machine learning models with well-characterized accuracy is a new frontier in control systems that will impact the next generation of industrial control systems.

Despite the success of machine learning methods in modeling nonlinear chemical processes in the context of MPC, there remain fundamental challenges that limit the implementation of machine-learning-based MPC to real chemical processes. One important challenge is to characterize the generalization ability on unseen data for machine learning models trained using finite training samples. Furthermore, a theoretical analysis of closed-loop stability for MPC using machine learning models needs to be developed via machine learning and control theory. Typically, theoretical developments on machine-learning-based MPC derived closed-loop stability properties based on the assumption of bounded modeling errors. For example, in [9], a Lyapunov-based MPC scheme using RNN models as the prediction model has been developed with guaranteed closed-loop stability by assuming that the RNN models are able to obtain a sufficiently small and bounded testing error. Similarly, a neural Lyapunov MPC that trains a stabilizing nonlinear MPC based on surrogate model and neural-network-based terminal cost was proposed in [19] with stability properties derived by assuming the boundedness of modeling error. Additionally, in [20], a nonparametric machine learning model is implemented together with MPC in which input-to-state stability is evaluated. In [21], a learning-based MPC targeting deterministic linear models is proposed in which safety, stability, and robustness are proved. However, the fundamental question regarding the generalization accuracy of machine learning models in MPC has not been addressed.

Probably approximately correct (PAC) learning theory is a framework that mathematically analyze the generalization ability of machine learning models [22]. Specifically, in PAC learning, given a set of training data, the learner is supposed to choose the optimal hypothesis (i.e., machine learning model) that yields a low generalization error with high probability from a certain class of hypotheses. Therefore, PAC learning theory provides a useful tool that demonstrates under what conditions a learning algorithm will probably output an approximately correct hypothesis. For example, in [23], PAC learning theory was used to study the learnability of compression learning algorithm for the optimization problem of stochastic MPC using a finite number of realizations of the uncertainty. In [24], PAC learning was used to analyze the generalization performance of a convex piecewise linear classifier that classifies the thermal comfort in a HVAC system. However, to the best of our knowledge, the use of statistical machine learning theory in analyzing stability properties of machine learning models in MPC, and guiding machine learning model structure and training data collection have not been fully explored.

Many recent works have been developed characterizing learnability of neural networks in terms of sample complexity and generalization error [25–32]. Generalization error bound is a common methodology in statistical machine learning for evaluating the predictive performance of machine learning algorithms [33]. This bound depends on a number of factors such as the number of data samples, the number of layers and neurons, bounds of weight matrices, initialization method, among others. For example, in [29], a generalization error bound was developed for a family of RNN models including vanilla RNNs, long short term memory and minimal gated unit. The generalization error bound was established for multiclass classification problems, and was dependent on the total number of network parameters and the spectral norms of the weight matrices. In [27], a sample complexity bound that was fully independent of network depth and width under some assumptions was developed for feedforward neural networks. In [34], an expected risk bound was developed for RNNs that model single-output nonlinear dynamic systems. However, at this stage, generalization error bounds for RNNs that model multiple-input and multiple-output (MIMO) nonlinear dynamic systems using time-series data have not been studied.

Motivated by the above, in this work, we develop the methodological framework of generalization error bounds from machine learning theory for the development and verification of RNN models with specific theoretical accuracy guarantees and integrate these models into model predictive control system design for nonlinear chemical processes. Specifically, in Section 2, the class of nonlinear systems, the formulation of RNNs, along with some general assumptions on system stabilizability and RNN development are presented. In Section 3, preliminaries including some important definitions and lemmas are first presented, followed by the development of a probabilistic generalization error bound for RNN models accounting for the impact of training data size and the number of neurons and layers on accuracy and guiding network structure selection and training. In Section 4, the RNN models are incorporated in the MPC formulation, under which probabilistic closed-loop stability is derived based on the RNN generalization error bound. Finally, in Section 5, a chemical reactor example is used to demonstrate the impact of training sample size, RNN depth and width, input time length on its generalization error. Additionally, closed-loop simulations are carried out to analyze the probabilistic closed-loop stability and performance.

2. Preliminaries

2.1. Notation

The Frobenius norm of A is denoted by $\|A\|_F$. The Euclidean norm of a vector is denoted by the operator $|\cdot|$ and the weighted Euclidean norm of a vector is denoted by the operator $|\cdot|_Q$ where Q is a positive definite matrix. $\mathbf{R}_+$ denotes nonnegative real numbers. $\mathbf{x}^T$ denotes the transpose of $\mathbf{x}$. The notation $L_f V(\mathbf{x})$ denotes the standard Lie derivative $L_f V(\mathbf{x}) := \frac{\partial V(\mathbf{x})}{\partial x} f(\mathbf{x})$. Set subtraction is denoted by "\", i.e., $A \backslash B := \{x \in \mathbf{R}^n \mid x \in A, x \notin B\}$. A function $f(\cdot)$ is of class $\mathcal{C}^1$ if it is continuously differentiable. A continuous function $\alpha : [0, a) \to [0, \infty)$ belongs to class $\mathcal{K}$ if it is strictly increasing and is zero only when evaluated at zero. A function $f : \mathbf{R}^n \to \mathbf{R}^m$ is said to be L-Lipschitz, $L \geq 0$, if $|f(a) - f(b)| \leq L|a - b|$ for all $a, b \in \mathbf{R}^n$. $\mathbb{P}(A)$ denotes the probability that event A will occur. $\mathbb{E}[X]$ denotes the expected value of a random variable X.

2.2. Class of Systems

The class of continuous-time nonlinear systems considered is described by the following state-space form:

$$\dot{x} = F(x, u) := f(x) + g(x)u, \; x(t_0) = x_0 \tag{1}$$

where $x \in \mathbf{R}^n$ and $u \in \mathbf{R}^k$ are the sate vector, and the manipulated input vector. The control action is constrained by $u \in U := \{u_{\min} \leq u \leq u_{\max}\} \subset \mathbf{R}^k$, where $u_{\min}$ and $u_{\max}$ represent the minimum and the maximum value vectors of inputs allowed, respectively. $f(\cdot)$ and $g(\cdot)$ are sufficiently smooth vector and matrix functions of dimensions $n \times 1$, and $n \times k$, respectively. Without loss of generality, the initial time t_0 is taken to be zero ($t_0 = 0$), and it is assumed that $f(0) = 0$, and thus, the origin is a steady-state of the system of Equation (1).

We assume the system of Equation (1) is stabilizable in the sense that there exists a stabilizing controller $u = \Phi(x) \in U$ that renders the origin exponentially stable. The stabilizability assumption implies that there exists a $\mathcal{C}^1$ control Lyapunov function $V(x)$ such that for all x in an open neighborhood D around the origin, the following inequalities hold:

$$c_1 |x|^2 \leq V(x) \leq c_2 |x|^2, \tag{2}$$

$$\frac{\partial V(x)}{\partial x} F(x, \Phi(x)) \leq -c_3 |x|^2, \tag{3}$$

$$\left| \frac{\partial V(x)}{\partial x} \right| \leq c_4 |x| \tag{4}$$

where c_1, c_2, c_3 and c_4 are positive constants. Additionally, the Lipschitz property of $F(x, u)$ and the boundedness of u implies there exist positive constants M_F, L_x, L'_x such that the following inequalities hold for all $x, x' \in D$ and $u \in U$:

$$|F(x,u)| \leq M_F \tag{5}$$

$$|F(x,u) - F(x',u)| \leq L_x|x - x'| \tag{6}$$

$$\left|\frac{\partial V(x)}{\partial x}F(x,u) - \frac{\partial V(x')}{\partial x}F(x',u)\right| \leq L'_x|x - x'| \tag{7}$$

Following the data generation method in [9], open-loop simulations of the nonlinear system of Equation (1) are first conducted to generate a large dataset that captures the system dynamics for $x \in \Omega_\rho$ and $u \in U$, where $\Omega_\rho := \{x \in \mathbf{R}^n \mid V(x) \leq \rho\}, \rho > 0$, is a compact set within which the system stability is guaranteed using the controller $u = \Phi(x) \in U$. Specifically, we sweep over all the values that (x, u) can take by running extensive open-loop simulations of the system of Equation (1) under various $x_0 \in \Omega_\rho$ and inputs u to generate a large number of dynamic trajectories. The open-loop simulation of the continuous system of Equation (1) under a sequence of inputs $u \in U$ is carried out in a sample-and-hold fashion (i.e., the inputs are fed into the system of Equation (1) as a piecewise constant function, $u(t) = u(t_k), \forall t \in [t_k, t_{k+1})$, where $t_{k+1} := t_k + \Delta$, and Δ is the sampling period). The nonlinear system of Equation (1) is integrated via explicit Euler method with a sufficiently small integration time step $h_c < \Delta$. Using the open-loop simulation data, recurrent neural network (RNN) models are developed to predict future states for (at least) one sampling period based on the current state measurements, and the manipulated inputs that will be applied for the next sampling period. In other words, the RNN model is developed to predict $x(t), \forall t \in [t_k, t_{k+1})$ based on the measurements $x(t_k)$ and the inputs $u \in [t_k, t_{k+1})$. Finally, the time-series dataset is partitioned into three subsets for the purposes of training, validation and testing.

2.3. Recurrent Neural Network Model

Consider an RNN model that approximates the nonlinear dynamics of the system of Equation (1) with m sequences of T-time-length data points $(\mathbf{x}_{i,t}, \mathbf{y}_{i,t})$ where $\mathbf{x}_{i,t} \in \mathbf{R}^{d_x}$ is the RNN input, and $\mathbf{y}_{i,t} \in \mathbf{R}^{d_y}$ is the RNN output, $i = 1, ..., m$ and $t = 1, ..., T$ (Figure 1). It should be noted that the RNN inputs and outputs do not necessarily represent the nonlinear system inputs and states/outputs in Equation (1). Therefore, to differentiate the notations for RNN inputs/outputs from those for the nonlinear system of Equation (1), all the vectors for RNN models are written in boldface. Additionally, to simplify the discussion, the RNN model of Equations (8) and (9) is developed to predict states over one sampling period with total time steps $T = \frac{\Delta}{h_c}$ (i.e., the RNN model is to predict future states for all the integration time step h_c within one sampling period Δ). As a result, the RNN input $\mathbf{x}_{i,t}$ consists of the current state measurements and manipulated inputs that will be applied over $t = 1, ..., T$, and the RNN output $\mathbf{y}_{i,t}$ consists of the predicted states over $t = 1, ..., T$. Note that $\mathbf{x}_{i,t}$ remains unchanged over $t = 1, ..., T$ due to the sample-and-hold implementation of manipulated inputs.

The dataset is developed consisting of m data sequences drawn independently from some underlying distribution over $\mathbf{R}^{d_x \times T} \times \mathbf{R}^{d_y \times T}$. In this work, we consider a one-hidden-layer RNN with hidden states $\mathbf{h}_i \in \mathbf{R}^{d_h}$ computed as follows:

$$\mathbf{h}_{i,t} = \sigma_h(U\mathbf{h}_{i,t-1} + W\mathbf{x}_{i,t}) \tag{8}$$

where σ_h is the element-wise nonlinear activation function (e.g., ReLU). $U \in \mathbf{R}^{d_h \times d_h}$ and $W \in \mathbf{R}^{d_h \times d_x}$ are weight matrices connected to the hidden states and input vector, respectively. The output layer $\mathbf{y}_{i,t}$ is computed as follows:

$$\mathbf{y}_{i,t} = \sigma_y(V\mathbf{h}_{i,t}) \tag{9}$$

where $V \in \mathbf{R}^{d_y \times d_h}$ is the weight matrix, and σ_y is the element-wise activation function in the output layer (typically linear unit for regression problems).

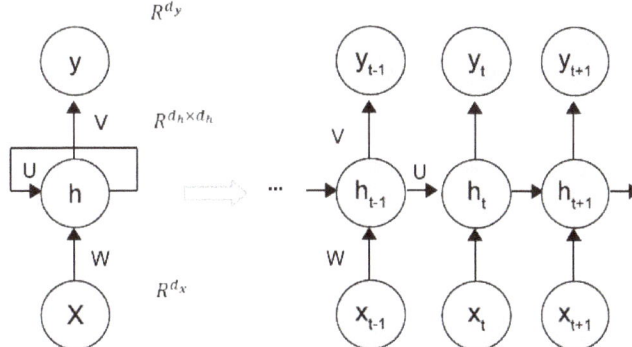

Figure 1. Recurrent neural network structure.

We consider the loss function $L(\mathbf{y}, \bar{\mathbf{y}})$ which calculates the squared difference between the true value $\bar{\mathbf{y}}$ and the predicted value $\mathbf{y}$ (i.e., L_2 loss). Without loss of generality, we have the following assumptions on the RNN model and dataset.

Assumption 1. *The RNN inputs are bounded, i.e., $|\mathbf{x}_{i,t}| \leq B_X$, for all $i = 1, ..., m$ and $t = 1, ..., T$.*

Assumption 2. *The Frobenius norms of all the weight matrices are bounded as follows:*

$$\|U\|_F \leq B_{U,F}, \|V\|_F \leq B_{V,F}, \|W\|_F \leq B_{W,F} \tag{10}$$

Assumption 3. *Training, validation, and testing datasets are drawn from the same distribution.*

Assumption 4. *The nonlinear activation function σ_h is 1-Lipschitz continuous, and is positive-homogeneous, i.e., $\sigma_h(\alpha z) = \alpha \sigma_h(z)$ for all $\alpha \geq 0$ and $z \in \mathbf{R}$.*

Remark 1. *All the assumptions made are standard in machine learning theory, and can be presented in system-theoretic language as follows. Assumption 1 assumes that the RNN inputs are bounded, which is consistent with the fact that the process states x and inputs u are bounded by $x \in \Omega_\rho$ and $u \in U$. Assumption 2 requires the RNN weight matrices to be bounded, which implies that only a finite class of neural network hypotheses are considered for modeling the nonlinear system of Equation (1). Assumption 3 is a natural and necessary assumption for generalization performance analysis. It implies that the machine learning models built from industrial operation data will be applied to the same process with the same data distribution. An example of activation function that satisfies Assumption 4 is Rectified Linear Unit (ReLu), which is a nonlinear activation function that has gained popularity in the machine learning domain.*

3. RNN Generalization Error

Since any learning algorithms are evaluated on finite training samples only, and do not provide any information on their predictive performance for unseen data, generalization error provides an important measure of how accurately a neural network model is able to predict output values for input data that has not been used in training. To implement machine learning models into real chemical processes, it is necessary to demonstrate that models are developed with a desired generalization error such that they can be applied for any reasonable operating conditions beyond those in the training dataset while maintaining a sufficiently small modeling error. In this section, we develop an upper bound for the generalization error of RNN models, and demonstrate that this error can be bounded with high probability provided that the training data samples and neural network structure meet a few requirements.

3.1. Preliminaries

We first present some important definitions and lemmas that will be used in the derivation of RNN generalization error. The random variables satisfying sub-Gaussian distribution, which is a probability distribution with strong tail decay, are defined as follows:

Definition 1. *A centered random variable $x \in \mathbf{R}$ is said to be sub-Gaussian with variance proxy σ^2, if $\mathbb{E}[x] = 0$, and the moment generating function satisfies*

$$\mathbb{E}[\exp(aX)] \leq \exp\left(\frac{a^2\sigma^2}{2}\right), \forall a \in \mathbf{R} \tag{11}$$

Lemma 1 (McDiarmid's inequality [35]). *Consider independent random variables $X_1, ..., X_n \in X$ and a function $f : X^n \to \mathbf{R}$ with bounded difference property, i.e., there exist positive numbers c_i such that the following inequality holds for all $x_1, ..., x_n, x'_i \in X$, $x_i \neq x'_i$ and $i \in \{1, ..., n\}$:*

$$|f(x_1, ..., x_i, ..., x_n) - f(x_1, ..., x'_i, ..., x_n)| \leq c_i \tag{12}$$

then the following probability holds for any $a > 0$:

$$\mathbb{P}(f(X_1, ..., X_n) - \mathbb{E}[f(X_1, ..., X_n)] \geq a) \leq \exp\left(-\frac{2a^2}{\sum_{i=1}^{n} c_i^2}\right) \tag{13}$$

Let $L(\mathbf{y}_t, \bar{\mathbf{y}}_t)$ be the loss function, where $\mathbf{y}_t = h(\mathbf{x}_t)$ is the predicted RNN output, and $h(\cdot)$ represents the RNN functions in the hypothesis class $\mathcal{H}$ mapping input $\mathbf{x} \in \mathbf{R}^{d_x}$ to output $\mathbf{y} \in \mathbf{R}^{d_y}$. The following error definitions are commonly used in machine learning theory.

Definition 2. *Given a function h that predicts output values y for each input x, and an underlying distribution D, the **expected loss/error** or **generalization error** is*

$$L_D(h) \triangleq \mathbb{E}[L(h(x), y)] = \int_{X \times Y} L(h(x), y) \rho(x, y) dxdy \tag{14}$$

where $\rho(x, y)$ is joint probability distribution for x and y, and X, Y are the vector space of all possible inputs, and outputs, respectively.

Since in general the joint probability distribution ρ is unknown, we use the data samples drawn from this unknown probability distribution to compute empirical error, which is a proxy measure for the expected loss.

Definition 3. *Given a dataset with m data samples $S = (s_1, ..., s_m)$, where $s_i = (x_i, y_i)$, the **empirical error** or **risk** is*

$$\hat{\mathbb{E}}_S[L(h(x), y)] = \frac{1}{m} \sum_{i=1}^{m} L(h(x_i), y_i) \tag{15}$$

The RNN model is developed by minimizing the empirical risk of Equation (15) using a set of m data sequences. To ensure that the RNN model achieves a desired generalization performance in the sense that it well captures the nonlinear dynamics of the system of Equation (1) for various operation conditions, the objective of this work is to show that the generalization error $\mathbb{E}[L(h(\mathbf{x}), \mathbf{y})]$ can be bounded provided that the empirical risk is sufficiently small and bounded.

We consider the mean squared error (MSE) as loss function in this work. It is readily shown that the MSE loss function $L(\mathbf{y}, \bar{\mathbf{y}})$ is not Lipschitz continuous for all $\mathbf{y}, \bar{\mathbf{y}} \in \mathbf{R}^{d_y}$. However, since we consider a finite hypothesis class that satisfies Assumptions 1–4, we can show that the RNN output is bounded. This is consistent with the fact that the nonlinear system of Equation (1) is operated in the stability region Ω_ρ, and therefore, the RNN outputs are bounded within a compact set.

Let $r_t > 0$ denote the upper bound of $\mathbf{y}_t$, i.e., $|\mathbf{y}_t| \leq r_t$, $t = 1, \ldots, T$. Without loss of generality, we assume that the true outputs are also bounded by r_t. Therefore, the MSE loss function is a locally Lipschitz continuous function satisfying the following inequality for all $|\mathbf{y}_t|, |\bar{\mathbf{y}}_t| \leq r_t$.

$$|L(\mathbf{y}_1, \bar{\mathbf{y}}) - L(\mathbf{y}_2, \bar{\mathbf{y}})| \leq L_r |\mathbf{y}_1 - \mathbf{y}_2| \qquad (16)$$

where L_r is the local Lipschitz constant.

The generalization error of a neural network function h_S chosen from a hypothesis class $\mathcal{H}$ based on a certain learning algorithm and a training dataset S drawn from distribution D can be decomposed into the approximation error and the estimation error as follows:

$$L_D(h_S) - L_D(h^*) = (\min_{h \in \mathcal{H}} L_D(h) - L_D(h^*)) + (L_D(h_S) - \min_{h \in \mathcal{H}} L_D(h)) \qquad (17)$$

where the first and second terms in parentheses represent **approximation error** and **estimation error**, respectively. Specifically, $L_D(h_S)$ represents the error evaluated using the hypothesis h_S over the underlying data distribution D. h^* represents the optimal hypothesis (maybe outside of the finite hypothesis class $\mathcal{H}$) for the data distribution D. $\min_{h \in \mathcal{H}} L_D(h)$ is the optimal hypothesis within $\mathcal{H}$ that minimizes the loss functions over the distribution D. It can be seen that the approximation error depends on how close the hypothesis class $\mathcal{H}$ is to the optimal hypothesis h^*. In other words, a larger hypothesis class $\mathcal{H}$ generally leads to a lower approximation error since it is more likely that the optimal hypothesis h^* is included in $\mathcal{H}$. The estimation error depends on both the hypothesis class size and training data, and characterizes how good the selected hypothesis h_S associated with the training dataset S is with respect to the best hypothesis $\min_{h \in \mathcal{H}} L_D(h)$ within hypothesis class $\mathcal{H}$. As a result, a larger hypothesis class $\mathcal{H}$ may in turn lead to a higher estimation error since it is more difficult to find the optimal hypothesis within $\mathcal{H}$ over the distribution D. From the error decomposition of Equation (17), we demonstrate the dependencies of generalization error on the training dataset size and the complexity of hypothesis class. In the next section, we will take advantage of Rademacher complexity technique to derive a generalization error bound accounting for its dependencies on the above factors in a quantitative aspect. The results will also provide a guide for the design of neural network structures and the collection of training data in order to achieve a desired generalization performance for a specific modeling task.

3.2. Rademacher Complexity Bound

Rademacher complexity quantifies the richness of a class of functions, and is often used in machine learning theory to bound the generalization error. The definition of empirical Rademacher complexity is given below.

Definition 4 (Empirical Rademacher Complexity). *Given a hypothesis class $\mathcal{F}$ of real-valued functions, and a set of data samples $S = \{s_1, \ldots, s_m\}$, the empirical Rademacher complexity of $\mathcal{F}$ is defined as*

$$\mathcal{R}_S(\mathcal{F}) = \mathbb{E}_\epsilon \left[\sup_{f \in \mathcal{F}} \frac{1}{m} \sum_{i=1}^m \epsilon_i f(s_i) \right] \qquad (18)$$

where $\epsilon = (\epsilon_1, \ldots, \epsilon_m)^T$ with ϵ_i being independent and identically distributed (i.i.d.) Rademacher random variables satisfying $\mathbb{P}(\epsilon_i = 1) = \mathbb{P}(\epsilon_i = -1) = 0.5$.

We also have the following contraction inequality for the hypothesis class $\mathcal{H}$ of vector-valued functions $h \in \mathbf{R}^{d_y}$.

Lemma 2 (c.f. Corollary 4 in [36]). *Consider a hypothesis class $\mathcal{H}$ of vector-valued functions $h \in \mathbf{R}^{d_y}$, and a set of data samples $S = \{s_1, ..., s_m\}$. Let $L(\cdot)$ be a L_r-Lipschitz function mapping $h \in \mathbf{R}^{d_y}$ to $\mathbf{R}$, then we have*

$$\mathbb{E}_\epsilon \left[\sup_{h \in \mathcal{H}} \sum_{i=1}^m \epsilon_i L(h(x_i), y_i) \right] \leq \sqrt{2} L_r \mathbb{E}_\epsilon \left[\sup_{h \in \mathcal{H}} \sum_{i=1}^m \sum_{k=1}^{d_y} \epsilon_{ik} h_k(x_i) \right] \qquad (19)$$

where $h_k(\cdot)$ is the k-th component in the vector-valued function $h(\cdot)$, and ϵ_{ik} is an $m \times d_y$ matrix of independent Rademacher variables. In the following text, we will omit the subscript ϵ of expectation for simplicity.

Since the RHS of Equation (19) is generally difficult to compute, we can reduce it to scalar classes, and derive the following bound [36]:

$$\mathbb{E} \left[\sup_{h \in \mathcal{H}} \sum_{i=1}^m \sum_{k=1}^{d_y} \epsilon_{ik} h_k(x_i) \right] \leq \sum_{k=1}^{d_y} \mathbb{E} \left[\sup_{h \in \mathcal{H}_k} \sum_{i=1}^m \epsilon_i h(x_i) \right] \qquad (20)$$

where $\mathcal{H}_k, k = 1, ..., d_y$, are classes of scalar-valued functions that correspond to the components of vector-valued functions in $\mathcal{H}$. Equation (20) will later be used in the derivation of the generalization error bound for RNN models approximating the nonlinear system of Equation (1).

Let $\mathcal{G}_t$ be the family of loss functions associated to $\mathcal{H}$ mapping the first t-time-step inputs $\{x_1, x_2, ..., x_t\} \in \mathbf{R}^{d_x \times t}$ to the t-th output $y_t \in \mathbf{R}^{d_y}$.

$$\mathcal{G}_t = \{g_t : (\mathbf{x}, \bar{\mathbf{y}}) \to L(h(\mathbf{x}), \bar{\mathbf{y}}), h \in \mathcal{H}\} \qquad (21)$$

where $\mathbf{x}$ is the RNN input vector, and $\bar{\mathbf{y}}$ is the true output vector. The following lemma characterizes the upper bound for the generalization error using Rademacher complexity $\mathcal{R}_S(\mathcal{G}_t)$.

Lemma 3 (c.f. Theorem 3.3 in [37]). *Given a set of m i.i.d. data samples, with probability at least $1 - \delta$ over samples $S = (x_{i,t}, y_{i,t})_{t=1}^T$, $i = 1, ..., m$, the following inequality holds for all $g_t \in \mathcal{G}_t$:*

$$\mathbb{E}[g_t(x, y)] \leq \frac{1}{m} \sum_{i=1}^m g_t(x_i, y_i) + 2\mathcal{R}_S(\mathcal{G}_t) + 3\sqrt{\frac{\log(\frac{2}{\delta})}{2m}} \qquad (22)$$

Proof. While the full proof can be found in many machine learning books, e.g., [37], a proof sketch is presented below to help readers understand the derivation of Equation (22). To simplify the notations, let $\mathbb{E}[g_t]$ and $\hat{\mathbb{E}}_S[g_t]$ denote the expected loss $\mathbb{E}[g_t(\mathbf{x}, \mathbf{y})]$ and the empirical loss $\frac{1}{m}\sum_{i=1}^m g_t(x_i, y_i)$ based on a dataset S with m data samples, respectively. Additionally, we assume that $g_t(\mathbf{x}, \mathbf{y})$ is bounded in $[0, 1]$ (if not, we can scale the RNN output layer or loss function) without loss of generality. We define $\beta(S)$ to be a function of data samples $S = (s_1, s_2, ..., s_m)$ as follows, where s_i represents each data sample $(x_{i,t}, y_{i,t})$, $i = 1, ..., m$.

$$\beta(S) = \sup_{g_t \in \mathcal{G}_t} \left(\mathbb{E}[g_t] - \hat{\mathbb{E}}_S[g_t] \right) \qquad (23)$$

Given two datasets $S = (s_1, ..., s_i, ..., s_m)$ and $S' = (s_1, ..., s'_i, ..., s_m)$ with only one different data point, i.e., $s_i \neq s'_i$, the following inequality holds for any $g_t(\mathbf{x}_i, \mathbf{y}_i) \in [0, 1]$:

$$\begin{aligned}
|\beta(S) - \beta(S')| &= \left| \sup_{g_t \in \mathcal{G}_t} (\mathbb{E}[g_t] - \hat{\mathbb{E}}_S[g_t]) - \sup_{g_t \in \mathcal{G}_t} (\mathbb{E}[g_t] - \hat{\mathbb{E}}_{S'}[g_t]) \right| \\
&\leq \left| \sup_{g_t \in \mathcal{G}_t} (\hat{\mathbb{E}}_{S'}[g_t] - \hat{\mathbb{E}}_S[g_t]) \right| \\
&= \left| \sup_{g_t \in \mathcal{G}_t} \frac{g_t(s'_i) - g_t(s_i)}{m} \right| \\
&\leq \frac{1}{m}
\end{aligned} \quad (24)$$

Then, using the McDiarmid's inequality in Lemma 1 and letting $a \geq \sqrt{\frac{\log(\frac{2}{\delta})}{2m}}$, we have

$$\mathbb{P}[\beta(S) - \mathbb{E}_S[\beta(S)] \geq a] \leq \exp\left(\frac{-2a^2}{\sum_{i=1}^m \frac{1}{m^2}}\right) = \exp(-2a^2 m) \leq \frac{\delta}{2} \quad (25)$$

where $\mathbb{E}_S[\beta(S)]$ denotes the expectation of $\beta(S)$ with respect to the dataset S of m data samples. Equivalently, the following inequality holds with probability at least $1 - \frac{\delta}{2}$, for any $\delta > 0$, :

$$\beta(S) \leq \mathbb{E}_S[\beta(S)] + \sqrt{\frac{\log(\frac{2}{\delta})}{2m}} \quad (26)$$

Next, we derive the upper bound for $\mathbb{E}_S[\beta(S)]$ as follows:

$$\begin{aligned}
\mathbb{E}_S[\beta(S)] &= \mathbb{E}_S\left[\sup_{g_t \in \mathcal{G}_t} (\mathbb{E}[g_t] - \hat{\mathbb{E}}_S[g_t])\right] \\
&\leq \mathbb{E}_{S,S'}\left[\sup_{g_t \in \mathcal{G}_t} (\hat{\mathbb{E}}_{S'}[g_t] - \hat{\mathbb{E}}_S[g_t])\right] \\
&= \mathbb{E}_{\epsilon,S,S'}\left[\sup_{g_t \in \mathcal{G}_t} \left(\frac{1}{m} \sum_{i=1}^m \epsilon_i(g_t(s'_i) - g_t(s_i))\right)\right] \\
&\leq \mathbb{E}_{\epsilon,S'}\left[\sup_{g_t \in \mathcal{G}_t} \left(\frac{1}{m} \sum_{i=1}^m \epsilon_i g_t(s'_i)\right)\right] + \mathbb{E}_{\epsilon,S}\left[\sup_{g_t \in \mathcal{G}_t} \left(\frac{1}{m} \sum_{i=1}^m -\epsilon_i g_t(s_i)\right)\right] \\
&= 2\mathbb{E}_{\epsilon,S}\left[\sup_{g_t \in \mathcal{G}_t} \frac{1}{m} \sum_{i=1}^m \epsilon_i g_t(s_i)\right] = 2\mathbb{E}_{\epsilon,S}[\mathcal{R}_S(\mathcal{G}_t)]
\end{aligned} \quad (27)$$

where the first line is by substituting the definition of Equation (23) into $\mathbb{E}_S[\beta(S)]$. The second line is derived using the fact that $\mathbb{E}[g_t] = \mathbb{E}_{S'}[\hat{\mathbb{E}}_{S'}(g_t)]$ and the property of supremum function: $\sup_{g_t \in \mathcal{G}_t} \mathbb{E}_{S'}(f(S', g_t)) \leq \mathbb{E}_{S'}[\sup_{g_t \in \mathcal{G}_t} f(S', g_t)]$ for any function f. The third line is derived by introducing Rademacher variables ϵ_i, which do not affect its outcome since ϵ_i are i.i.d. randome variables taking values in $\{-1, +1\}$. The fourth line is obtained by separating the supremum function as $\sup(f + g) \leq \sup(f) + \sup(g)$, and the last line is derived using the fact that Rademacher variables ϵ_i have a symmetric distribution. Note that $\mathbb{E}_{\epsilon,S}[\mathcal{R}_S(\mathcal{G}_t)]$ in the last line of Equation (27) represents the expectation of the empirical Rademacher complexity, $\mathcal{R}_S(\mathcal{G}_t)$, over all samples of size m drawn from the same distribution. In order to bound this term, we apply McDiarmid's inequality again using confidence $\frac{\delta}{2}$, which yields a similar result as in Equation (25). Finally, using union bound which states that $\mathbb{P}(\cup_i A_i) \leq \sum_i \mathbb{P}(A_i)$ holds for any finite or countable set of events A_i, $i = 1, 2, ...$, the following inequality holds with probability at least $1 - \delta$:

$$\beta(S) \leq 2\left(\mathcal{R}_S(\mathcal{G}_t) + \sqrt{\frac{\log \frac{2}{\delta}}{2m}}\right) + \sqrt{\frac{\log \frac{2}{\delta}}{2m}}$$
$$= 2\mathcal{R}_S(\mathcal{G}_t) + 3\sqrt{\frac{\log \frac{2}{\delta}}{2m}} \tag{28}$$

By substituting the definition of $\beta(S)$ of Equation (23) into the above equation, we obtain the result in Equation (22). This completes the proof of Lemma 3. □

It can be seen from Equation (22) that the generalization error bound depends on the empirical error (the first term), Rademacher complexity (the second term), and an error function associated with the confidence δ and the number of samples m (the last term). Since the first and last terms are known given a set of m training data, in order to characterize the upper bound for the generalization error $\mathbb{E}[g_t(\mathbf{x}, \mathbf{y})]$, we need to determine the upper bound for the Rademacher complexity $\mathcal{R}_S(\mathcal{G}_t)$. Since most of the established results of Rademacher complexity are with respect to feedforward neural networks modeling real-valued functions only, we will start with a lemma for the hypothesis class of real-valued functions.

Lemma 4. *Given a hypothesis class $\mathcal{H}_k$ of real-valued functions corresponding to the k-th component of vector-valued function class $\mathcal{H}$, and a set of m i.i.d. data samples $S = (\mathbf{x}_{i,t}, \mathbf{y}_{i,t})_{t=1}^T$, $i = 1, ..., m$, the following inequality holds for the scaled empirical Rademacher complexity $m\mathcal{R}_S(\mathcal{H}_k) = \mathbb{E}[\sup_{h \in \mathcal{H}_k} \sum_{i=1}^m \epsilon_i h(\mathbf{x}_i)]$.*

$$m\mathcal{R}_S(\mathcal{H}_k) = \frac{1}{\lambda} \log \exp\left(\lambda \mathbb{E}\left[\sup_{h \in \mathcal{H}_k} \sum_{i=1}^m \epsilon_i h(\mathbf{x}_i)\right]\right)$$
$$\leq \frac{1}{\lambda} \log\left(\mathbb{E}\left[\sup_{h \in \mathcal{H}_k} \exp(\lambda \sum_{i=1}^m \epsilon_i h(\mathbf{x}_i))\right]\right) \tag{29}$$

where $\lambda > 0$ is an arbitrary parameter.

Proof. Equation (29) can be readily proved by using Jensen's inequality which states that given a random variable X and a convex function $\beta(\cdot)$, it holds that $\beta(\mathbb{E}[X]) \leq \mathbb{E}[\beta(X)]$. Equation (29) will be used in the derivation of the upper bound for the Rademacher complexity $R_S(\mathcal{H})$ in Lemma 7. □

We can see from the definition of Rademacher complexity of Equation (18) that the value of $\mathcal{R}_S(\mathcal{G}_t)$ depends on the complexity of hypothesis class $\mathcal{G}_t$. However, since the RNN model of Equations (8) and (9) is a complex nonlinear function which is difficult to measure its learning capacity, we need to peel off the nonlinear activation functions and weight matrices through layers. The following lemma shows the "peeling" step used in the derivation of Rademacher complexity for the output layer of RNNs.

Lemma 5 (c.f. Lemma 1 in [27]). *Given a hypothesis class $\mathcal{H}$ of vector-valued functions that map the RNN inputs $\mathbf{x} \in \mathbf{R}^{d_x}$ to the hidden states $\mathbf{h} \in \mathbf{R}^{d_h}$, and any convex and monotonically increasing function $p : \mathbf{R} \to \mathbf{R}_+$, the following inequality holds for the RNN model of Equations (8) and (9) with a 1-Lipschitz, positive-homogeneous activation function $\sigma_y(\cdot)$:*

$$\mathbb{E}\left[\sup_{h \in \mathcal{H}, \|V\|_F \leq B_{V,F}} p\left(\left|\sum_{i=1}^m \epsilon_i \sigma_y(V h_i)\right|\right)\right] \leq 2 \cdot \mathbb{E}\left[\sup_{h \in \mathcal{H}} p\left(B_{V,F} \cdot \left|\sum_{i=1}^m \epsilon_i h_i\right|\right)\right] \tag{30}$$

Proof. The proof is omitted here as it is similar to the proof for the next lemma, which will be presented in detail. Interested readers can refer to [27] for the proof of Lemma 5. □

Lemma 5 peels off the weight matrix V between the RNN hidden layer and output layer. To further peel off the weight matrices in the RNN hidden layers, we provide the following lemma.

Lemma 6. *Given a hypothesis class $\mathcal{H}$ of vector-valued functions that map the RNN inputs $x \in \mathbf{R}^{d_x}$ to the hidden states $h \in \mathbf{R}^{d_h}$, and any convex and monotonically increasing function $p : \mathbf{R} \to \mathbf{R}_+$, the following equation holds for the RNN model of Equations (8) and (9) with a 1-Lipschitz, positive-homogeneous activation function $\sigma_h(\cdot)$:*

$$\mathbb{E}\left[\sup_{h\in\mathcal{H}, \|U\|_F \leq B_{U,F}, \|W\|_F \leq B_{W,F}} p\left(\left|\sum_{i=1}^m \epsilon_i \mathbf{h}_{i,t}\right|\right)\right]$$
$$= \mathbb{E}\left[\sup_{h\in\mathcal{H}, \|U\|_F \leq B_{U,F}, \|W\|_F \leq B_{W,F}} p\left(\left|\sum_{i=1}^m \epsilon_i \sigma_h(U\mathbf{h}_{i,t-1} + W\mathbf{x}_{i,t})\right|\right)\right] \quad (31)$$
$$\leq 2\mathbb{E}\left[\sup_{h\in\mathcal{H}} p\left(B_{U,F}\left|\sum_{i=1}^m \epsilon_i \mathbf{h}_{i,t-1}\right| + B_{W,F}\left|\sum_{i=1}^m \epsilon_i \mathbf{x}_{i,t}\right|\right)\right]$$

Proof. We first define an augmented weight matrix $Z = [U|W] \in \mathbf{R}^{d_h \times (d_h + d_x)}$, and an augmented vector $\bar{\mathbf{h}}_{i,t} = [\mathbf{h}_{i,t-1}|\mathbf{x}_{i,t}] \in \mathbf{R}^{d_h + d_x}$. To simplify the discussion, we assume that the Frobenius norm of the matrix Z is bounded by $\|Z\|_F \leq B_{Z,F}$, given that both U and W are bounded by $\|U\|_F \leq B_{U,F}$ and $\|W\|_F \leq B_{W,F}$. Then, the hidden layer vector at t-th time step, $\mathbf{h}_{i,t}$, can be written as follows:

$$\mathbf{h}_{i,t} = \sigma_h(U\mathbf{h}_{i,t-1} + W\mathbf{x}_{i,t}) = \sigma_h(Z\bar{\mathbf{h}}_{i,t}) \quad (32)$$

Letting $\mathbf{z}_1, \mathbf{z}_2, ..., \mathbf{z}_h$ denote the rows of the matrix Z, we have

$$\left|\sum_{i=1}^m \epsilon_i \mathbf{h}_{i,t}\right|^2 = \sum_{j=1}^{d_h} |\mathbf{z}_j|^2 \left(\sum_{i=1}^m \epsilon_i \sigma_h\left(\frac{\mathbf{z}_j^T}{|\mathbf{z}_j|}\bar{\mathbf{h}}_{i,t}\right)\right)^2 \quad (33)$$

The supremum of Equation (33) over all the weight matrix $Z = [\mathbf{z}_1\, \mathbf{z}_2\, ...\, \mathbf{z}_h]$ that satisfies $\|Z\|_F \leq B_{Z,F}$ is obtained when $|\mathbf{z}_j| = B_{Z,F}$ for some j, and $|\mathbf{z}_i| = 0$ for all $i \neq j$. Therefore, we have

$$\mathbb{E}\left[\sup_{h\in\mathcal{H}, \|U\|_F \leq B_{U,F}, \|W\|_F \leq B_{W,F}} p\left(\left|\sum_{i=1}^m \epsilon_i \mathbf{h}_{i,t}\right|\right)\right]$$
$$= \mathbb{E}\left[\sup_{h\in\mathcal{H}, |\mathbf{z}|=B_{Z,F}} p\left(\left|\sum_{i=1}^m \epsilon_i \sigma_h(\mathbf{z}^T \bar{\mathbf{h}}_{i,t})\right|\right)\right] \quad (34)$$

Since $p(\cdot)$ is a convex and monotonically increasing function, $p(|a|) \leq p(a) + p(-a)$ holds, and the above equation can be further bounded as follows:

$$\mathbb{E}\left[\sup_{h\in\mathcal{H}, |\mathbf{z}|=B_{Z,F}} p\left(\left|\sum_{i=1}^m \epsilon_i \sigma_h(\mathbf{z}^T \bar{\mathbf{h}}_{i,t})\right|\right)\right] \leq \mathbb{E}\left[\sup_{h\in\mathcal{H}, |\mathbf{z}|=B_{Z,F}} p\left(\sum_{i=1}^m \epsilon_i \sigma_h(\mathbf{z}^T \bar{\mathbf{h}}_{i,t})\right)\right]$$
$$+ \mathbb{E}\left[\sup_{h\in\mathcal{H}, |\mathbf{z}|=B_{Z,F}} p\left(-\sum_{i=1}^m \epsilon_i \sigma_h(\mathbf{z}^T \bar{\mathbf{h}}_{i,t})\right)\right] \quad (35)$$
$$= 2\mathbb{E}\left[\sup_{h\in\mathcal{H}, |\mathbf{z}|=B_{Z,F}} p\left(\sum_{i=1}^m \epsilon_i \sigma_h(\mathbf{z}^T \bar{\mathbf{h}}_{i,t})\right)\right]$$

where the last equality is derived from the fact that the random variables ϵ_i have a symmetric distribution, i.e., $\mathbb{P}(\epsilon_i = 1) = \mathbb{P}(\epsilon_i = -1) = 0.5$. Following the proof in [27] and Theorem 4.12 in [38], the RHS of Equation (35) can be further bounded by

$$2\mathbb{E}\left[\sup_{h\in\mathcal{H}, |\mathbf{z}|=B_{Z,F}} p\left(\sum_{i=1}^{m}\epsilon_i\sigma_h(\mathbf{z}^T\bar{\mathbf{h}}_{i,t})\right)\right] \leq 2\mathbb{E}\left[\sup_{h\in\mathcal{H}, |\mathbf{z}|=B_{Z,F}} p\left(\sum_{i=1}^{m}\epsilon_i\mathbf{z}^T\bar{\mathbf{h}}_{i,t}\right)\right]$$

$$\leq 2\mathbb{E}\left[\sup_{h\in\mathcal{H}, |\mathbf{u}|=B_{U,F}, |\mathbf{w}|=B_{W,F}} p\left(|\mathbf{u}|\left|\sum_{i=1}^{m}\epsilon_i\mathbf{h}_{i,t-1}\right| + |\mathbf{w}|\left|\sum_{i=1}^{m}\epsilon_i\mathbf{x}_{i,t}\right|\right)\right] \quad (36)$$

$$= 2\mathbb{E}\left[\sup_{h\in\mathcal{H}} p\left(B_{U,F}\left|\sum_{i=1}^{m}\epsilon_i\mathbf{h}_{i,t-1}\right| + B_{W,F}\left|\sum_{i=1}^{m}\epsilon_i\mathbf{x}_{i,t}\right|\right)\right]$$

□

Based on Lemmas 5 and 6, the following lemma provides an upper bound for the Rademacher complexity of the RNN hypothesis class.

Lemma 7. *Let* $\mathcal{H}_{k,t}$, $k = 1, ..., d_y$, *be the class of real-valued functions that corresponds to the k-th component of the RNN output at t-th time step, with weight matrices and activation functions satisfying Assumptions 1–4. Given a set of m i.i.d. data samples* $S = (\mathbf{x}_{i,t}, \mathbf{y}_{i,t})_{t=1}^{T}$, $i = 1, ..., m$, *the following equation holds for the Rademacher complexity:*

$$\mathcal{R}_S(\mathcal{H}_{k,t}) \leq \frac{M(\sqrt{2\log(2)t}+1)B_X}{\sqrt{m}} \quad (37)$$

where $M = B_{V,F} B_{W,F} \frac{(B_{U,F})^t - 1}{B_{U,F} - 1}$.

Proof. Let $\mathbf{v}_k$ be the k-th row in the weight matrix V. Using Equations (29) and (30), the scaled Rademacher complexity $m\mathcal{R}_S(\mathcal{H}_k)$ can be bounded as follows:

$$m\mathcal{R}_S(\mathcal{H}_{k,t}) = \mathbb{E}\left[\sup_{h\in\mathcal{H}_{k,t}, \|V\|_F \leq B_{V,F}} \sum_{i=1}^{m}\epsilon_i\sigma_y(\mathbf{v}_k\mathbf{h}_{i,t})\right]$$

$$\leq \frac{1}{\lambda}\log\mathbb{E}\left[\sup_{h\in\mathcal{H}_{k,t}, \|V\|_F \leq B_{V,F}} \exp\left(\lambda\sum_{i=1}^{m}\epsilon_i\sigma_y(\mathbf{v}_k\mathbf{h}_{i,t})\right)\right] \quad (38)$$

$$\leq \frac{1}{\lambda}\log\mathbb{E}\left[\sup_{h\in\mathcal{H}_{k,t}} \exp\left(B_{V,F}\lambda\left|\sum_{i=1}^{m}\epsilon_i\mathbf{h}_{i,t}\right|\right)\right]$$

where $\exp(\cdot)$ corresponds to the monotonically increasing function $p(\cdot)$ in Lemmas 5 and 6. Then, we use Equation (31) and further derive the bound for the RHS of the above equation as follows:

$$\frac{1}{\lambda}\log\mathbb{E}\left[\sup_{h\in\mathcal{H}_{k,t}} \exp\left(B_{V,F}\lambda\left|\sum_{i=1}^{m}\epsilon_i\mathbf{h}_{i,t}\right|\right)\right]$$

$$\leq \frac{1}{\lambda}\log\left(2\cdot\mathbb{E}\left[\sup_{h\subset\mathcal{H}_{k,t-1}} \exp\left(B_{V,F}\lambda\cdot\left(B_{U,F}\left|\sum_{i=1}^{m}\epsilon_i\mathbf{h}_{i,t-1}\right| + B_{W,F}\left|\sum_{i=1}^{m}\epsilon_i\mathbf{x}_{i,t}\right|\right)\right)\right]\right) \quad (39)$$

Assuming that the initial hidden states $\mathbf{h}_{i,0} = 0$, by recursively applying Lemma 6 to the term $|\sum_{i=1}^{m}\epsilon_i\mathbf{h}_{i,t-1}|$ in Equation (39), we obtain that

$$mR_S(\mathcal{H}_{k,t}) \leq \frac{1}{\lambda}\log\left(2^t \cdot \mathbb{E}\left[\exp\left(B_{V,F}\lambda \cdot \left(B_{W,F} \cdot \left|\sum_{i=1}^{m}\epsilon_i\mathbf{x}_{i,t}\right| \cdot \sum_{j=0}^{t-1}(B_{U,F})^j\right)\right)\right]\right)$$
$$= \frac{1}{\lambda}\log\left(2^t \cdot \mathbb{E}\left[\exp\left(B_{V,F}\lambda \cdot \left(B_{W,F} \cdot \left|\sum_{i=1}^{m}\epsilon_i\mathbf{x}_{i,t}\right| \cdot \frac{(B_{U,F})^t - 1}{B_{U,F} - 1}\right)\right)\right]\right) \tag{40}$$

It is noted that the RNN model in this work is developed to predict one sampling time, for which the RNN inputs $\mathbf{x}_{i,t}$ remain the same. If the RNN inputs are varying over time, Equation (40) can be modified by taking the maximum value of $|\sum_{i=1}^{m}\epsilon_i\mathbf{x}_{i,t}|$ within the prediction period. Subsequently, we define the following random variable q

$$q = M\left|\sum_{i=1}^{m}\epsilon_i\mathbf{x}_{i,t}\right| \tag{41}$$

where the randomness comes from the Rademacher variables ϵ_i, and M denotes the product of all weight matrices, i.e., $M = B_{V,F}B_{W,F}\frac{(B_{U,F})^t - 1}{B_{U,F} - 1}$. Then, Equation (40) can be written as

$$mR_S(\mathcal{H}_{k,t}) \leq \frac{1}{\lambda}\log(2^t \cdot \mathbb{E}[\exp(\lambda q)])$$
$$= \frac{t\log(2)}{\lambda} + \frac{1}{\lambda}\log(\mathbb{E}[\exp(\lambda(q - \mathbb{E}[q]))]) + \mathbb{E}[q] \tag{42}$$

Using Jensen's inequality, we can bound $\mathbb{E}[q]$ as follows:

$$\mathbb{E}[q] = \mathbb{E}\left[M\left|\sum_{i=1}^{m}\epsilon_i\mathbf{x}_{i,t}\right|\right] \leq M\sqrt{\mathbb{E}\left[\left|\sum_{i=1}^{m}\epsilon_i\mathbf{x}_{i,t}\right|^2\right]} = M\sqrt{\sum_{i=1}^{m}|\mathbf{x}_{i,t}|^2} \leq \sqrt{m}MB_X \tag{43}$$

where the second equality comes from the fact that ϵ_i are i.i.d. Rademacher random variables, and the last inequality is due to the assumption that $|\mathbf{x}_{i,t}| \leq B_X$. Subsequently, following the results in [38], we can show q is sub-Gaussian with the following variance factor v since q satisfies a bounded-difference condition with respect to its random variables ϵ_i, i.e., $q(\epsilon_1, ..., \epsilon_i, ..., \epsilon_m) - q(\epsilon_1, ..., -\epsilon_i, ..., \epsilon_m) \leq 2M|\mathbf{x}_{i,t}|$.

$$v = \frac{1}{4}\sum_{i=1}^{m}(2M|\mathbf{x}_{i,t}|)^2 = M^2\sum_{i=1}^{m}|\mathbf{x}_{i,t}| \tag{44}$$

According to the property of sub-Gaussian random variables in Definition 1, the following inequality holds for q:

$$\frac{1}{\lambda}\log(\mathbb{E}[\exp(\lambda(q - \mathbb{E}[q]))]) \leq \frac{\lambda M^2 \sum_{i=1}^{m}|\mathbf{x}_{i,t}|}{2} \tag{45}$$

Let $\lambda = \frac{\sqrt{2\log(2)t}}{M\sqrt{\sum_{i=1}^{m}|\mathbf{x}_{i,t}|^2}} > 0$. The Rademacher complexity $mR_S(\mathcal{H}_{k,t})$ in Equation (42) can be bounded as follows:

$$mR_S(\mathcal{H}_{k,t}) \leq \frac{t\log(2)}{\lambda} + \frac{1}{\lambda}\log(\mathbb{E}[\exp(\lambda(q - \mathbb{E}[q]))]) + \mathbb{E}[q]$$
$$\leq M(\sqrt{2\log(2)t} + 1)\sqrt{\sum_{i=1}^{m}|\mathbf{x}_{i,t}|^2} \tag{46}$$
$$\leq M(\sqrt{2\log(2)t} + 1)\sqrt{m}B_X$$

□

Lemma 7 develops the Rademacher complexity upper bound for the hypothesis class $\mathcal{H}_k$ of real-valued functions that map RNN inputs to the k-th output. Subsequently, we derive the generalization bound for the loss function associated with the vector-valued functions that map the RNN inputs to the output vector by taking advantage of the contraction inequality of Equations (19) and (20).

Theorem 1. *Let $\mathcal{G}_t$ be the family of loss function associated to the hypothesis class $\mathcal{H}_t$ of vector-valued functions that map the RNN inputs to the RNN output at t-th time step, with weight matrices and activation functions satisfying Assumptions 1–4. Given a set of m i.i.d. data samples $S = (x_{i,t}, y_{i,t})_{t=1}^T$, $i = 1, ..., m$, with probability at least $1 - \delta$ over S, we have r*

$$\mathbb{E}[g_t(x,y)] \leq \frac{1}{m}\sum_{i=1}^{m} g_t(x_i, y_i) + 3\sqrt{\frac{\log(\frac{2}{\delta})}{2m}} + \mathcal{O}\left(L_r d_y \frac{M(\sqrt{2\log(2)t} + 1)B_X}{\sqrt{m}}\right) \quad (47)$$

where $M = B_{V,F} B_{W,F} \frac{(B_{U,F})^t - 1}{B_{U,F} - 1}$.

Proof. Using the results in Lemma 7 and Equations (19) and (20), we can derive the following upper bound for the loss function $L(h(x_i), y_i)$ with $h(x_i)$ being vector-valued functions:

$$\mathcal{R}_S(\mathcal{G}_t) = \mathbb{E}\left[\sup_{h \in \mathcal{H}} \frac{1}{m}\sum_{i=1}^{m} \epsilon_i L(h(x_i), y_i)\right] \leq \sqrt{2}L_r \mathbb{E}\left[\sup_{h \in \mathcal{H}} \frac{1}{m}\sum_{i=1}^{m}\sum_{k=1}^{d_y} \epsilon_{ik} h_k(x_i)\right]$$
$$\leq \sqrt{2}L_r d_y \frac{M(\sqrt{2\log(2)t} + 1)B_X}{\sqrt{m}} \quad (48)$$

Then, substituting Equation (48) into Equation (22), we derive the generalization error bound in Equation (47). □

Remark 2. *As stated in [27], the assumption of positive-homogeneity for the nonlinear activation function can be loosened in some cases, under which a similar result of generalization error bound can be derived. Interested readers are referred to Lemma 2 and Theorem 2 in [27].*

Remark 3. *The generalization error bound of Equation (47) implies that the following attempts can be taken to reduce the generalization error: (1) minimize the empirical loss $\frac{1}{m}\sum_{i=1}^{m} g_t(x_i, y_i)$ over the training data samples S through a careful design of neural network, and (2) increase the number of training samples m. Additionally, as discussed in the error decomposition of Equation (17), increasing the complexity hypothesis class in terms of larger weight matrices bounds M could decrease the approximation error, but may also increase the estimation error, which corresponds to the last term $\mathcal{O}(\cdot)$ in Equation (47). Therefore, in practice, we generally start with a simple neural network and gradually increase it complexity in terms of more neurons, layers and larger weight matrices bounds to improve the training and testing performance. The whole process stops when the testing error starts increasing, which indicates the occurrence of overfitting.*

Remark 4. *While the actual generalization error is difficult to obtain due to unknown data distribution and complexity of hypothesis class, Equation (47) characterizes the upper bound for the gap between the generalization error and empirical error by moving the term $\frac{1}{m}\sum_{i=1}^{m} g_t(x_i, y_i)$ to the LHS of Equation (47). Since the neural network training process itself is to minimize the training error only, this generalization gap is more useful in practice by showing how good the neural network will be for unseen data under the same data distribution. In terms of modeling the nonlinear system of Equation (1), this generalization gap provides an upper bound for the modeling error for all the states in the operating region, and can be used in the design of model-based controllers that probabilistically ensure closed-loop stability accounting for bounded modeling errors.*

Remark 5. *It is noticed that the generalization error bound also depends on the time length t of RNN inputs, which is different from the results derived for the feedforward neural networks in [27]. Additionally, unlike other deep neural networks which utilize different parameters for each hidden layer, RNNs share the same weight matrix U at each time step, and therefore, the bound for the product of weight matrices is derived in the form of $M = B_{V,F} B_{W,F} \frac{(B_{U,F})^t - 1}{B_{U,F} - 1}$. From Equation (47), it can be seen that as the data sequence length t increases, the network hypothesis becomes more complex, which leads to a larger generalization error bound. Therefore, a shorter time sequence prediction is preferred from the perspective of prediction accuracy. However, it does not necessarily mean a short prediction period is always desirable from the control perspective, especially in model predictive control (MPC) schemes. In Section 5, we will demonstrate that the RNN models predicting a short period of time achieved desired prediction performance in open-loop tests, but perform poorly in closed-loop simulation due to the error accumulated during successive execution of RNN predictions within MPC prediction horizon.*

4. RNN-Based MPC with Probabilistic Stability Analysis

In this section, we present the formulation of Lyapunov-based MPC (LMPC) that uses RNN models to predict evolution of future states, along with the closed-loop stability analysis showing the boundedness of closed-loop state of Equation (1) in the stability region for all times in probability.

4.1. Lyapunov-Based Control Using RNN Models

To simplify the discussion of RNN stability properties for the continuous-time nonlinear system of Equation (1), we represent the RNN model in the following continuous-time form [9]:

$$\dot{\hat{x}} = F_{nn}(\hat{x}, u) := A\hat{x} + \Theta^T z \tag{49}$$

where $\hat{x} \in \mathbf{R}^n$ and $u \in \mathbf{R}^k$ are the RNN state vector and the manipulated input vector, respectively. $z = [z_1, ..., z_n, z_{n+1}, ..., z_{k+n}] = [\sigma(\hat{x}_1), ..., \sigma(\hat{x}_n), u_1, ..., u_k] \in \mathbf{R}^{n+k}$ is a vector of both the input u and the network state $\hat{x}$, where $\sigma(\cdot)$ represents the nonlinear activation function. A is a diagonal coefficient matrix with all diagonal elements being negative, and $\Theta = [\theta_1, ..., \theta_n] \in \mathbf{R}^{(k+n) \times n}$ with $\theta_i = b_i[w_{i1}, ..., w_{i(k+n)}]$, $i = 1, ..., n$, where w_{ij} denotes the weight connecting the jth input to the ith neuron, $i = 1, ..., n$ and $j = 1, ..., (k+n)$. The weight matrices and activation functions satisfy Assumptions 1–4. To simplify the notation, we use Equation (49) to represent one-hidden-layer RNN model, and bias terms are not explicitly included in Equation (49); however, it is noted that the results that we will derive in this section are not restricted to one-hidden-layer RNN models, and can be extended to deep RNNs with multiple hidden layers.

We assume that there exists a stabilizing feedback controller $u = \Phi_{nn}(x) \in U$ that can render the origin of the RNN model of Equation (49) exponentially stable in an open neighborhood $\hat{D}$ around the origin. The stabilizability assumption implies the existence of a $\mathcal{C}^1$ control Lyapunov function $\hat{V}(x)$ such that the following inequalities hold for all x in $\hat{D}$:

$$\hat{c}_1 |x|^2 \leq \hat{V}(x) \leq \hat{c}_2 |x|^2, \tag{50}$$

$$\frac{\partial \hat{V}(x)}{\partial x} F_{nn}(x, \Phi_{nn}(x)) \leq -\hat{c}_3 |x|^2, \tag{51}$$

$$\left| \frac{\partial \hat{V}(x)}{\partial x} \right| \leq \hat{c}_4 |x| \tag{52}$$

where $\hat{c}_1$, $\hat{c}_2$, $\hat{c}_3$, $\hat{c}_4$ are positive constants. The closed-loop stability region for the RNN model of Equation (49) is characterized as a level set of Lyapunov function embedded in $\hat{D}$ as follows: $\Omega_{\hat{\rho}} := \{ x \in \hat{D} \mid \hat{V}(x) \leq \hat{\rho} \}$, where $\hat{\rho} > 0$. Additionally, there exist positive constants M_{nn} and L_{nn} such that the following inequalities hold for all $x, x' \in \Omega_{\hat{\rho}}$ and $u \in U$:

$$|F_{nn}(x, u)| \leq M_{nn} \tag{53}$$

$$\left|\frac{\partial \hat{V}(x)}{\partial x} F_{nn}(x, u) - \frac{\partial \hat{V}(x')}{\partial x} F_{nn}(x', u)\right| \leq L_{nn}|x - x'| \tag{54}$$

Due to the model mismatch between the nonlinear system of Equation (1) and the RNN model of Equation (49), the following proposition is developed to demonstrate that the feedback controller $u = \Phi_{nn}(x) \in U$ is able to stabilize the system of Equation (1) with high probability if the modeling error is sufficiently small.

Proposition 1. *Consider the RNN model trained using a set of m i.i.d. data samples $S = (x_{i,t}, y_{i,t})_{t=1}^T$, $i = 1, ..., m$, and satisfying Assumptions 1–4. Under the assumption that the feedback controller $u = \Phi_{nn}(x) \in U$ renders the the origin of the RNN system of Equation (49) exponentially stable for all $x \in \Omega_{\hat{\rho}}$, if for all $x \in \Omega_{\hat{\rho}}$ and $u \in U$, the modeling error can be constrained by $|F(x, u) - F_{nn}(x, u)| \leq \gamma|x|$, where γ is a positive real number satisfying $\gamma < \hat{c}_3/\hat{c}_4$, then the controller $u = \Phi_{nn}(x) \in U$ also renders the origin of the nonlinear system of Equation (1) exponentially stable with probability at least $1 - \delta$ for all $x \in \Omega_{\hat{\rho}}$.*

Proof. To demonstrate that the origin of the nominal system of Equation (1) can be rendered exponentially stable $\forall x \in \Omega_{\hat{\rho}}$ with probability at least $1 - \delta$ under the controller $u = \Phi_{nn}(x) \in U$ designed for the RNN model of Equation (49), we prove that the time-derivative of $\hat{V}$ associated with the state x of Equation (1) can be rendered negative in probability under $u = \Phi_{nn}(x) \in U$. Based on Equations (51) and (52), $\dot{V}$ is derived as follows:

$$\begin{aligned}\dot{\hat{V}} &= \frac{\partial \hat{V}(x)}{\partial x} F(x, \Phi_{nn}(x)) \\ &= \frac{\partial \hat{V}(x)}{\partial x} (F_{nn}(x, \Phi_{nn}(x)) + F(x, \Phi_{nn}(x)) - F_{nn}(x, \Phi_{nn}(x))) \\ &\leq -\hat{c}_3|x|^2 + \hat{c}_4|x| \cdot |(F(x, \Phi_{nn}(x)) - F_{nn}(x, \Phi_{nn}(x)))|\end{aligned} \tag{55}$$

where the last term $|F(x, \Phi_{nn}(x)) - F_{nn}(x, \Phi_{nn}(x))|$ represents the error between the RNN model and the process model of Equation (1). Since the RNN model is trained using sampled data with a sufficiently small time interval (i.e., integration time step h_c), the modeling error term for the same initial state $x(t) = \hat{x}(t)$ can be approximated as follows:

$$\begin{aligned}|F(x, \Phi_{nn}(x)) &- F_{nn}(x, \Phi_{nn}(x))| \\ &\leq \left|\frac{x(t + h_c) - x(t)}{h_c} - \frac{\hat{x}(t + h_c) - \hat{x}(t)}{h_c}\right| + \mathcal{O}(h_c) \\ &\leq \left|\frac{x(t + h_c) - \hat{x}(t + h_c)}{h_c}\right| + \mathcal{O}(h_c)\end{aligned} \tag{56}$$

where $\hat{x}$ is the predicted state by RNN model, and x is the state of actual nonlinear system of Equation (1). $\mathcal{O}(h_c)$ is the truncation error from finite difference method. Since $|x(t + h_c) - \hat{x}(t + h_c)|$ represents the Euclidean norm of the prediction error, while the generalization error bound is derived using MSE as loss function in Theorem 1, the modeling error can be bounded as follows:

$$|F(x, \Phi_{nn}(x)) - F_{nn}(x, \Phi_{nn}(x))| \leq E_M \tag{57}$$

where

$$E_M = \frac{1}{h_c}\sqrt{\frac{1}{m}\sum_{i=1}^{m}g(x_i,y_i) + 3\sqrt{\frac{\log(\frac{2}{\delta})}{2m}} + \mathcal{O}\left(L_r d_y \frac{M(\sqrt{2\log(2)h_c}+1)B_X}{\sqrt{m}}\right) + \mathcal{O}(h_c)} \tag{58}$$

By choosing the number of samples $m \geq m_N(\delta, h_c, |x|)$, where $m_N(\delta, h_c, |x|)$ is the minimum data sample size satisfying $E_M \leq \gamma |x|$, $\gamma < \hat{c}_3/\hat{c}_4$, we have the following equation showing that $\dot{\hat{V}}$ can be rendered negative for all $x \in \Omega_\rho$ and $x \neq 0$ with probability at least $1 - \delta$, i.e., $\mathbb{P}[\dot{\hat{V}} \leq 0] \geq 1 - \delta$,

$$\begin{aligned}\dot{\hat{V}} &\leq -\hat{c}_3|x|^2 + \hat{c}_4|x|\cdot|F(x,\Phi_{nn}(x)) - F_{nn}(x,\Phi_{nn}(x))| \\ &\leq -\hat{c}_3|x|^2 + \hat{c}_4|x|\frac{\hat{c}_3|x|}{\hat{c}_4} \\ &= -\tilde{c}_3|x|^2 \\ &\leq 0\end{aligned} \tag{59}$$

where $\tilde{c}_3 = -\hat{c}_3 + \hat{c}_4\gamma < 0$ for any $\gamma < \hat{c}_3/\hat{c}_4$. Therefore, with probability at least $1 - \delta$, the closed-loop state of the system of Equation (1) converges to the origin under $u = \Phi_{nn}(x) \in U$ for all $x_0 \in \Omega_{\hat{\rho}}$. □

Remark 6. *The modeling error constraint $E_M \leq \gamma|x|$, $\forall x \in \Omega_{\hat{\rho}}$ implies that more data is needed for states closer to the origin. This is because when x approaches the origin, the upper bound $\gamma|x|$ is close to zero, and therefore, the prediction of $\hat{x}$ should be more accurate in order to yield a desired approximation of system dynamics $\dot{x} = F(x,u)$ using numerical methods. As a result, it seems that an infinite number of data samples may be needed when state converges to the origin (i.e., x is infinitely close to zero). However, we will show in the next subsection that the requirement of such a large dataset for the states around a small neighborhood around the origin is not necessary for operation under MPC. This is because under sample-and-hold implementation of control actions, the states are forced to be bounded in a small ball around the origin, instead of converging to the exact steady-state. Therefore, the modeling error constraint $E_M \leq \gamma|x|$, $\forall x \in \Omega_{\hat{\rho}}$ can be loosened for states in this small ball, which could improve computational efficiency of training process.*

4.2. Stabilization of Nonlinear System under Lyapunov-Based Controller

Subsequently, the following propositions are developed to demonstrate the impact of sample-and-hold implementation of control actions on system stability. Specifically, Proposition 2 demonstrates that in the presence of mismatch between the plant model of Equation (1) and the RNN models of Equation (49), the error between the predicted state and the actual state is bounded in a finite period of time. Then, we consider the Lyapunov-based controller $u = \Phi_{nn}(x)$ applied to the nonlinear system of Equation (1) in sample-and-hold fashion, and demonstrate in Proposition 3 that with high probability, the nonlinear system of Equation (1) can be stabilized using the controller $u = \Phi_{nn}(x)$ designed for the RNN model of Equation (49).

Proposition 2 (c.f. Proposition 3 in [9]). *Consider the nonlinear system $\dot{x} = F(x,u)$ of Equation (1) and the RNN model $\dot{\hat{x}} = F_{nn}(\hat{x},u)$ of Equation (49) with the same initial condition $x_0 = \hat{x}_0 \in \Omega_{\hat{\rho}}$. There exists a class $\mathcal{K}$ function $f_w(\cdot)$ and a positive constant κ such that the following inequalities hold $\forall x, \hat{x} \in \Omega_{\hat{\rho}}$:*

$$|x(t) - \hat{x}(t)| \leq f_w(t) := \frac{E_M}{L_x}(e^{L_x t} - 1) \tag{60}$$

$$\hat{V}(x) \leq \hat{V}(\hat{x}) + \frac{\hat{c}_4\sqrt{\hat{\rho}}}{\sqrt{\hat{c}_1}}|x - \hat{x}| + \kappa|x - \hat{x}|^2 \tag{61}$$

Proof. The proof can be found in [9], and is omitted here. Note that the proof in [9] considers the nonlinear system subject to bounded disturbances, while in this work, we consider the nominal system without disturbances only. However, the stability results derived in this section can be readily generalized to the disturbed systems provided that the disturbances are sufficiently small and bounded. Additionally, the modeling error term in [9] is replaced by E_M (see the definition of E_M in Equation (58)) in Equations (60) and (61) which accounts for the RNN generalization error derived in a probabilistic manner. □

The following proposition is developed to show probabilistic closed-loop stability of the nonlinear system of Equation (1) under sample-and-hold implementation of the controller $u = \Phi_{nn}(x) \in U$.

Proposition 3 (c.f. Proposition 4 in [9]). *Consider the nonlinear system of Equation (1) with the controller $u = \Phi_{nn}(\hat{x}) \in U$ that meets the conditions of Equations (50)–(52), and the RNN model of Equation (49) that meets all the conditions in Theorem 1. Under the sample-and-hold implementation of control actions, i.e., $u(t) = \Phi_{nn}(\hat{x}(t_k))$, $\forall t \in [t_k, t_{k+1})$, where $t_{k+1} := t_k + \Delta$. there exist $\epsilon_w > 0$, $\Delta > 0$ and $\hat{\rho} > \rho_{min} > \rho_{nn} > \rho_s$ that satisfy*

$$-\frac{\tilde{c}_3}{\tilde{c}_2}\rho_s + L'_x M_F \Delta \leq -\epsilon_w \tag{62}$$

and

$$\rho_{nn} := \max\{\hat{V}(\hat{x}(t+\Delta)) \mid \hat{x}(t) \in \Omega_{\rho_s}, u \in U\} \tag{63}$$

$$\rho_{min} \geq \rho_{nn} + \frac{\hat{c}_4\sqrt{\hat{\rho}}}{\sqrt{\hat{c}_1}} f_w(\Delta) + \kappa(f_w(\Delta))^2 \tag{64}$$

such that for any $x(t_k) \in \Omega_{\hat{\rho}} \setminus \Omega_{\rho_s}$, with probability at least $1 - \delta$, the following inequality holds:

$$\hat{V}(x(t)) \leq \hat{V}(x(t_k)), \forall t \in [t_k, t_{k+1}) \tag{65}$$

and the state $x(t)$ of the nonlinear system of Equation (1) is bounded in $\Omega_{\hat{\rho}}$ for all times and ultimately bounded in $\Omega_{\rho_{min}}$.

Proof. The key steps for the proof of Proposition 3 are presented below, and the full proof is omitted here as it is similar to the proof of Proposition 4 in [9]. The only difference is that Equation (65) now holds in probability due to the probabilistic nature of the modeling error bound.

To show that the state will move towards Ω_{ρ_s}, which is a sufficiently small level set of $\hat{V}$ around the origin, we show that the time derivative of $\hat{V}$ can be rendered negative for any $x(t_k) \in \Omega_{\hat{\rho}} \setminus \Omega_{\rho_s}$ under $u = \Phi_{nn}(x) \in U$.

$$\begin{aligned}\dot{\hat{V}}(x(t)) &= \frac{\partial \hat{V}(x(t))}{\partial x} F(x(t), \Phi_{nn}(x(t_k))) \\ &= \frac{\partial \hat{V}(x(t_k))}{\partial x} F(x(t_k), \Phi_{nn}(x(t_k))) + \frac{\partial \hat{V}(x(t))}{\partial x} F(x(t), \Phi_{nn}(x(t_k))) \\ &\quad - \frac{\partial \hat{V}(x(t_k))}{\partial x} F(x(t_k), \Phi_{nn}(x(t_k)))\end{aligned} \tag{66}$$

As shown in Proposition 1, by choosing the number of samples $m \geq m_N(\delta, h_c, |x|)$ such that $E_M \leq \gamma|x|$, where $\gamma < \hat{c}_3/\hat{c}_4$, it holds that $\mathbb{P}[\dot{\hat{V}} \leq 0] \geq 1 - \delta$ under $u = \Phi_{nn}(x) \in U$.

Then, using the Lipschitz condition in Equations (5)–(7) and the condition for Lyapounov function in Equations (50)–(52), Equation (66) can be further bounded as follows:

$$\dot{\hat{V}}(x(t)) \leq -\frac{\tilde{c}_3}{\hat{c}_2}\rho_s + \frac{\partial \hat{V}(x(t))}{\partial x}F(x(t), \Phi_{nn}(x(t_k))) - \frac{\partial \hat{V}(x(t_k))}{\partial x}F(x(t_k), \Phi_{nn}(x(t_k)))$$
$$\leq -\frac{\tilde{c}_3}{\hat{c}_2}\rho_s + L'_x|x(t) - x(t_k)| \quad (67)$$
$$\leq -\frac{\tilde{c}_3}{\hat{c}_2}\rho_s + L'_x M_F \Delta$$

Therefore, if Equation (62) is satisfied, we can find a negative real number $-\epsilon_w$ that bounds the time derivative of $\hat{V}$. This implies that for any state $x(t_k) \in \Omega_{\hat{\rho}} \setminus \Omega_{\rho_s}$, with probability at least $1 - \delta$, the Lyapunov function value will decrease in one sampling time, and therefore, the state can ultimately reach the set Ω_{ρ_s} under $u = \Phi_{nn}(x) \in U$ with a certain probability. Additionally, since $\dot{\hat{V}}$ may not be rendered negative within Ω_{ρ_s} under sample-and-hold implementation of Lyapunov-based control law $u = \Phi_{nn}(x) \in U$, the predicted state of the RNN model of Equation (49) is only required to be bounded in $\Omega_{\rho_{nn}}$, which is a slightly larger level set that includes Ω_{ρ_s} (see definition of $\Omega_{\rho_{nn}}$ in Equation 63). In this case, we can show that the state of the actual nonlinear system of Equation (49) is bounded in $\Omega_{\rho_{min}}$, which is a superset of $\Omega_{\rho_{nn}}$ that accounts for the modeling error within one sampling period (see definition of $\Omega_{\rho_{min}}$ in Equation (64)). As a result, we do not impose any constraints on $\dot{\hat{V}}$ for $x \in \Omega_{\rho_s}$. This explains why the modeling error constraint $E_M \leq \gamma |x|$ is not necessary for $x \in \Omega_{\rho_s}$ as stated in Remark 6. □

4.3. Lyapunov-Based MPC Using RNN Models for Nonlinear Systems

The Lyapunov-based model predictive control design is given by the following optimization problem [9,10]:

$$\mathcal{J} = \min_{u \in S(\Delta)} \int_{t_k}^{t_{k+N}} L_{MPC}(\tilde{x}(t), u(t)) dt \quad (68)$$

$$\text{s.t.} \quad \dot{\tilde{x}}(t) = F_{nn}(\tilde{x}(t), u(t)) \quad (69)$$

$$u(t) \in U, \; \forall t \in [t_k, t_{k+N}) \quad (70)$$

$$\tilde{x}(t_k) = x(t_k) \quad (71)$$

$$\dot{\hat{V}}(x(t_k), u) \leq \dot{\hat{V}}(x(t_k), \Phi_{nn}(x(t_k))), \text{ if } x(t_k) \in \Omega_{\hat{\rho}} \setminus \Omega_{\rho_{nn}} \quad (72)$$

$$\hat{V}(\tilde{x}(t)) \leq \rho_{nn}, \; \forall t \in [t_k, t_{k+N}), \text{ if } x(t_k) \in \Omega_{\rho_{nn}} \quad (73)$$

where $\tilde{x}$, N and $S(\Delta)$ are the predicted states, the prediction horizon length, and the set of piecewise constant functions with period Δ, respectively. We use $\dot{\hat{V}}(x, u)$ to represent the time derivative of Lyapunov function $\hat{V}$, i.e., $\dot{\hat{V}}(x, u) = \frac{\partial \hat{V}(x)}{\partial x}(F_{nn}(x, u))$. After solving the optimization problem of Equations (68)–(73) at $t = t_k$, we apply the first control action $u(t)$, $t \in [t_k, t_{k+1})$ from the optimal input trajectory $u^*(t)$, $t \in [t_k, t_{k+N})$ to the system of Equation (1). Then the horizon is rolled one sampling period forward, and the LMPC is resolved at the next sampling time with new state measurements available at $t = t_{k+1}$.

The optimization problem of Equations (68)–(73) minimizes the objective function of Equation (68), which is the integral of $L_{MPC}(\tilde{x}(t), u(t))$ over the prediction horizon, subject to the constraints of Equations (69)–(73). The RNN model of Equation (49) is used to predict state evolution over $t \in [t_k, t_{k+N})$ given the state measurements at $t = t_k$ in Equation (71). In the constraint of Equation (69), the RNN model of Equation (49) is used to predict the states of the closed-loop system. The constraint of Equation (70) ensures that the input are bounded over the entire prediction horizon. Finally, the constraints of Equations (72)–(73) drives the predicted state towards the origin and ultimately maintain it inside $\Omega_{\rho_{nn}}$. It should be noted that despite the probabilistic nature of the RNN generalization error bound, the neural network prediction of Equation (69) is deterministic after training is completed. In other words, given the same initial state $x(t_k)$, and the manipulated inputs $u(t)$, $\forall t \in [t_k, t_{k+N})$, the RNN model of Equation (69) produces deterministic results that statistically approximate the evolution of states over $t \in [t_k, t_{k+N})$. This is different from stochastic MPC which uses a stochastic process model in the MPC formulation, and therefore, requires calculation of uncertainty prorogation and accounts for probabilistic constraint satisfaction. The LMPC formulation of Equations (68)–(73) is solved with a deterministic RNN model, based on which recursive feasibility is guaranteed, and probabilistic stability results can be developed.

The following theorem is established to demonstrate that LMPC ensures closed-loop stability for the nonlinear system of Equation (1) with high probability provided that the RNN model is well constructed that satisfies the modeling error constraint in Proposition 1.

Theorem 2. *Consider the closed-loop system of Equation (1) under the LMPC of Equations (68)–(73) based on the controller $\Phi_{nn}(x)$ that satisfies Equations (50)–(52). Let $\Delta > 0$, $\epsilon_w > 0$ and $\hat{\rho} > \rho_{min} > \rho_{nn} > \rho_s$ satisfy Equations (62)–(64). Then, given any initial state $x_0 \in \Omega_{\hat{\rho}}$, if the RNN model is developed satisfying the conditions in Proposition 2 and Proposition 3, there always exists a feasible solution for the optimization problem of Equations (68)–(73) . Additionally, by choosing the number of samples $m \geq m_N(\delta, h_c, |x|)$ such that $E_M \leq \gamma |x|$ holds, then for each time step, with probability at least $1 - \delta$, closed-loop stability is guaranteed for the system of Equation (1) under the LMPC of Equations (68)–(73) in the sense that $x(t) \in \Omega_{\hat{\rho}}$, $\forall t \geq 0$, and $x(t)$ ultimately converges to $\Omega_{\rho_{min}}$.*

Proof. The proof consists of two parts. In the first part, we prove recursive feasibility of the LMPC optimization problem of Equations (68)–(73) . The proof of this part follows closely the proof of Theorem 2 in [9], which shows that the stabilizing controller $u(t) = \Phi_{nn}(x(t)) \in U$, $t = [t_k, t_{k+N})$ is a feasible solution to the LMPC optimization problem. Specifically, when $x(t_k) \in \Omega_{\hat{\rho}} \setminus \Omega_{\rho_{nn}}$ at $t = t_k$, it is readily shown that the control action $u(t) = \Phi_{nn}(x(t_k))$ is a feasible solution that satisfies the constraint of Equation (72) by taking the equal sign. When $x(t_k) \in \Omega_{\rho_{nn}}$, as shown in [9], $u(t) = \Phi_{nn}(x(t)) \in U$, $t = [t_k, t_{k+N})$ again are feasible solutions that maintain predicted states within $\Omega_{\rho_{nn}}$ within the prediction horizon.

In the second part, we prove that closed-loop stability is guaranteed in probability for the nonlinear system of Equation (1) under LMPC. Specifically, when $x(t_k) \in \Omega_{\hat{\rho}} \setminus \Omega_{\rho_{nn}}$ at $t = t_k$, we have shown in Proposition 3 that for each sampling time, $\hat{V}(x(t)) \leq \hat{V}(x(t_k))$ holds under $u(t) = \Phi_{nn}(x(t)) \in U$ for $t \in [t_k, t_{k+1})$ with probability at least $1 - \delta$. This implies that the state of the actual nonlinear system of Equation (1) can be driven towards the origin under the LMPC using RNN models for prediction provided that the modeling error is sufficiently small and satisfies $E_M \leq \gamma |x|$, $\forall x \in \Omega_{\hat{\rho}}$. When $x(t_k) \in \Omega_{\rho_{nn}}$, the input sequences are optimized to minimize the objective function of Equation (68) while meeting the constraint of Equation (73). However, due to the existence of modeling error, the true states may leave $\Omega_{\rho_{nn}}$ while the predicted states remain inside $\Omega_{\rho_{nn}}$. In Proposition 3, we have shown that with probability at least $1 - \delta$, the true state of the system of Equation (1) can be bounded within $\Omega_{\rho_{min}}$, which is a superset of $\Omega_{\rho_{nn}}$ designed accounting for the modeling error within one sampling period. Additionally, it is noted that depending on the prediction horizon of RNN models, we may need to perform RNN predictions successively to obtain the full prediction of the state trajectory over the entire prediction horizon, $t \in [t_k, t_{k+N})$. For example, in this work, the RNN model of Equation (49) is developed to predict one sampling period forward, and thus, in order to predict state trajectory over $t \in [t_k, t_{k+N})$, we need to carry out RNN predictions N times. After the initial prediction at $t = t_k$, each prediction uses the previous predicted state as the initial state, along with the manipulated input u to predict the state at the next sampling time. This inevitably accumulates the modeling error over calculation, which may lead to a probability lower than $1 - \delta$ for the final state prediction error to be bounded by $E_M \leq \gamma |x|$.

As a result, the true states may further deviate from predicted states, and ultimately leave $\Omega_{\rho_{min}}$ within finite time. Despite the degradation of prediction performance over time, closed-loop stability is not affected since LMPC is implemented in a rolling horizon manner with feedback state measurements available every sampling time. The input sequences are re-optimized using new state measurements at every sampling time to meet desired closed-loop performance. Additionally, since the modeling error condition $E_M \leq \gamma|x|$ holds for the first sampling period, the state of the actual nonlinear system of Equation (1) is guaranteed to not leave $\Omega_{\rho_{min}}$ within one sampling period with probability at least $1-\delta$ as shown in Proposition 3. At the next sampling period, the constraints of Equation (72) and of Equation (73) will be activated depending on the measurement of $x(t_{k+1})$. Regardless of where $x(t_{k+1})$ is, the LMPC of Equations (68)–(73) will drive the predicted state into $\Omega_{\rho_{nn}}$, and correspondingly, maintain the true state within $\Omega_{\rho_{min}}$ in probability. Therefore, for any state $x(t_k) \in \Omega_{\hat{\rho}}$, with probability at least $1-\delta$, the closed-loop state of the system of Equation (1) is bounded in $\Omega_{\hat{\rho}}$ for each sampling time, and is ultimately bounded within $\Omega_{\rho_{min}}$. This completes the proof of Theorem 2. □

Remark 7. *It is noted that in Theorem 2, the probability of closed-loop stability (i.e., at least $1-\delta$) is derived for each sampling time since the probability of the modeling error bounded by $\gamma|x|$ is at least $1-\delta$ for one sampling period only. It is difficult to compute the overall probability of closed-loop stability for the entire state trajectory because given an initial state $x_0 \in \Omega_{\hat{\rho}}$, we do not know how many times steps it will take to drive the state into $\Omega_{\rho_{min}}$ beforehand. Additionally, the actual probability of closed-loop stability for each time step could be higher than the lower bound $1-\delta$ due to many reasons. For example, 1) the RNN model is well trained that yields a modeling error far below its upper bound, and 2) closed-loop stability may be unaffected if the next state does not leave $\Omega_{\hat{\rho}}$ even if the modeling error exceeds its upper bound during one sampling period. Therefore, the probability $1-\delta$ is conservative in many cases, and only provides a lower bound for the probability of closed-loop stability.*

5. Application to a Chemical Process Example

We use the same chemical process example as in [10] to illustrate the application of LMPC using RNN models. However, in this work, we will primarily demonstrate the use of generalization error bound framework to provide estimates of their accuracy in the development of RNN models for nonlinear dynamic processes. Specifically, we carry out five case studies to evaluate the relation between RNN generalization error and a number of factors such as data sample size, RNN depth/width, and data time length that impact its performance. Additionally, after the RNN model is incorporated in the LMPC formulation, we will demonstrate the closed-loop performances under the RNN models developed with different data sample size and structures, and evaluate their probabilistic closed-loop stability properties. We consider a well-mixed, non-isothermal continuous stirred tank reactor (CSTR) with an irreversible second-order exothermic reaction in this example. The reaction transforms a reactant A to a product B ($A \rightarrow B$), where C_{A0}, T_0 and F denote the inlet concentration of A, the inlet temperature and feed volumetric flow rate of the reactor, respectively. A heating jacket is used to supply/remove heat to/from the CSTR at a rate Q. The CSTR dynamic model is represented by the following material and energy balance equations:

$$\begin{aligned}\frac{dC_A}{dt} &= \frac{F}{V}(C_{A0} - C_A) - k_0 e^{\frac{-E}{RT}} C_A^2 \\ \frac{dT}{dt} &= \frac{F}{V}(T_0 - T) + \frac{-\Delta H}{\rho_L C_p} k_0 e^{\frac{-E}{RT}} C_A^2 + \frac{Q}{\rho_L C_p V}\end{aligned} \quad (74)$$

where C_A and T are the concentration of reactant A and temperature in the reactor, respectively. Q denotes the heat input rate, and V is the volume of the reacting liquid in the reactor. F, T_0, and C_{A0} are the volumetric flow rate, the feed temperature and the feed concentration of reactant A, respectively. We assume that the reacting liquid has a constant density of ρ_L and a heat capacity of C_p. ΔH, k_0, E, and R represent the enthalpy of reaction, pre-exponential constant, activation energy, and ideal gas constant, respectively. The list of process parameter values can be found in [10].

The objective of LMPC is to stabilize the CSTR at its unstable equilibrium point $(C_{As}, T_s) = (1.95\ kmol/m^3, 402\ K)$ corresponding to $(C_{A0_s}, Q_s) = (4\ kmol/m^3, 0\ kJ/hr)$ by manipulating the inlet concentration of species A and the heat input rate. All the process states (C_A, T) and manipulated inputs (C_{A0}, Q) are represented in the deviation variables form, i.e., $\Delta C_{A0} = C_{A0} - C_{A0_s}$, $\Delta Q = Q - Q_s$, $\Delta C_A = C_A - C_{As}$, and $\Delta T = T - T_s$. To simplify the notation, we use $x^T = [\Delta C_A\ \Delta T]$ and $u^T = [\Delta C_{A0}\ \Delta Q]$ to represent CSTR states and inputs, respectively. By using deviation variables, the equilibrium point of the CSTR of Equation (74) is at the origin of the state-space. The following positive definite P matrix is used to characterize the closed-loop stability region $\Omega_{\hat{\rho}}$ (i.e., a level set of Lyapunov function $V(x) = x^T P x$) with $\hat{\rho} = 368$:

$$P = \begin{bmatrix} 1060 & 22 \\ 22 & 0.52 \end{bmatrix} \quad (75)$$

Additionally, the manipulated inputs are required to be bounded as follows: $|\Delta C_{A0}| \leq 3.5\ kmol/m^3$ and $|\Delta Q| \leq 5 \times 10^5$ kJ/hr to meet physical constraints. The integration of RNN models in MPC follows the method in [10,39]. Specifically, the RNN models are developed offline using Keras (version 2.4) [40], and then used to predict future states based on the state measurement at each sampling time in the real-time implementation of MPC. Then, the nonlinear optimization problem of the LMPC of Equations (68)–(73) is solved under the sampling period $\Delta = 10^{-2}$ hr using PyIpopt, which is the python module of the IPOPT software package (version 3.9.1) [41]. The dynamic model of Equation (74) is integrated using numerical method, i.e., explicit Euler method, with a sufficiently small integration time step of $h_c = 10^{-4}$ hr.

5.1. RNN Generalization Performance

In this section, we carry out a number of RNN trainings with different RNN structures and data samples to show the relation between RNN generalization performance and a number of factors such as RNN input length, width, depth, weight bounds and data sample size.

5.1.1. Case Study 1: Data Sample Size

In the first case study, we trained RNN models using different data sample sizes. Specifically, we follow data generation method in [10] to initially generate a large dataset from open-loop simulation of Equation (74) under various control actions $u \in U$ and initial conditions within the stability region, i.e., $x_0 \in \Omega_{\hat{\rho}}$. The dataset consists of 200,000 time-series data samples, and is separated into 140,000 training, 30,000 validation, and 30,000 testing samples. The RNN models are developed by gradually increasing the training sample size, and is tested using unseen data from the testing dataset. It should be noted that only the data sample size is changed in this case study, while all the other parameters such as the RNN structure (i.e., number of layers, neurons, and other hyper-parameters) and training algorithm remain the same for all RNN models. The RNN models are developed with one hidden layer of 50 neurons, and using mean squared errors (MSE) as loss function.

Figure 2 shows the variation of RNN training and testing performances with respect to the training sample size. In the top figure of Figure 2, it is observed that both the testing and training MSEs increase as training data becomes less; in the bottom figure, we show the generalization gap $\mathbb{E}[g_t(\mathbf{x}, \mathbf{y})] - \frac{1}{m}\sum_{i=1}^{m} g_t(\mathbf{x}_i, \mathbf{y}_i)$ in Equation (47), where the expected error $\mathbb{E}[g_t(\mathbf{x}, \mathbf{y})]$ is approximated using the testing dataset. The trend in Figure 2 is consistent with the result in Theorem 1, which demonstrates that more training data is needed in order to obtain a lower generalization gap between expected loss and training loss. Additionally, it is noticed when the training sample size is greater than 3000, both training and testing MSEs approach zero, and no significant improvement is observed for the models using more training data. The trend in Figure 2 also follows the relation between generalization error and data sample size in Equation (47), i.e., the generalization gap $\mathbb{E}[g_t(\mathbf{x},\mathbf{y})] - \frac{1}{m}\sum_{i=1}^{m} g_t(\mathbf{x}_i,\mathbf{y}_i)$ is roughly proportional to $\frac{1}{\sqrt{m}}$, which shows that the generalization gap initially decreases fast when the sample size m starts increasing from zero, and changes slowly when m becomes large.

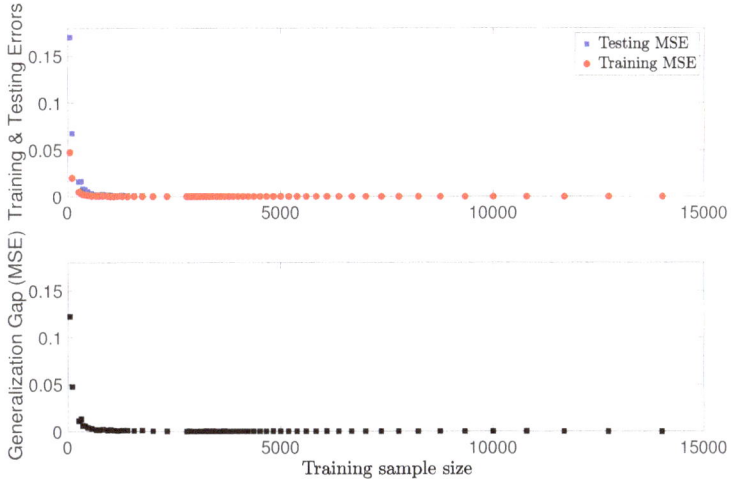

Figure 2. RNN generalization performance vs. training sample size.

5.1.2. Case Study 2 : RNN Depth and Width

In the second case study, we train RNN models with various depths and widths. 1400 training data, 300 validation data, and 300 testing data are used for all models. We first develop RNN models by fixing the network depth as one hidden layer, and increasing the number of neurons. As shown in Figure 3, both training and testing errors decrease as the network width increases up to 250 neurons. However, as more neurons are added (i.e., 270 and 280 neurons in Figure 3), the testing MSE increases while the training MSE remains close to zero all the time, which implies that overfitting has occurred during training. As a result, the generalization gap in Figure 3 shows a similar pattern, which decreases initially and increases again when a large number of neurons are used. While theoretically the expected error of Equation (47) does not explicitly depend on the network width, the results in Figure 3 are consistent with the fact that increasing the capacity of a model by adding more layers and/or more nodes to layers can improve the network learnability, but may also lead to overfitting.

Subsequently, we train RNN models by increasing the number of layers, and fixing five neurons each layer. Figure 4 shows that the testing MSE starts at around 0.02 for one hidden layer, gradually decreases with more layers, and finally increases again as the neural network becomes deeper. Meanwhile, the training MSE remains close to zero at the beginning, yet also slightly increases as the number of hidden layers increase. From Figure 4, it is concluded that one hidden layer is not sufficient to learn the process dynamics well, and with two, three, and four layers, the RNN models achieve the best training and generalization performance among all the models. Similar to Figure 3, the increase of generalization gap in Figure 4 implies that deeper RNN models are overfitting the training data. Additionally, it is also interesting to notice that the training error slightly increases in deeper networks. While in general, the generalization performance deteriorates and the training error remains unaffected when increasing the capacity of a model, the worse training performance in Figure 4 are actually common in neural network development due to the difficulty of training deep networks. Specifically, the optimization problem of neural network training is highly non-convex, and may get stuck at some local minima as the network becomes deeper. This is noticed during the training of RNN models in Figure 4, where both the training and validation losses exhibit a sharp increase at a certain epoch and then get stuck around that point until the end of epochs. Additionally, with more hidden layers, the number of parameters to be trained grows exponentially, which could lead to a poor training performance without a careful tuning of other hyperparameters.

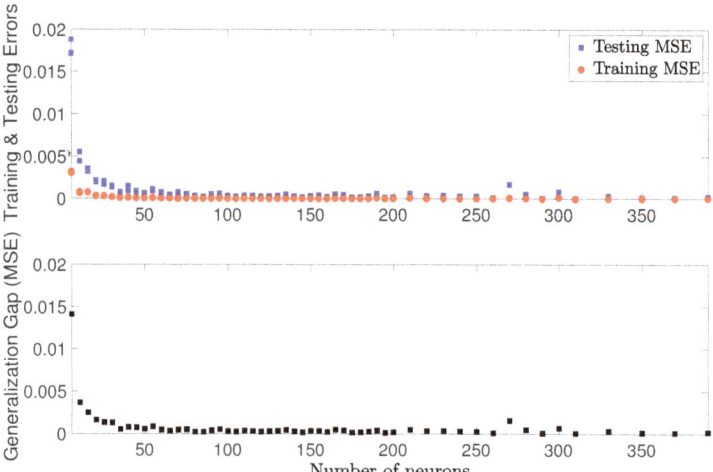

Figure 3. RNN generalization performance vs. RNN width (One hidden layer with increasing number of neurons).

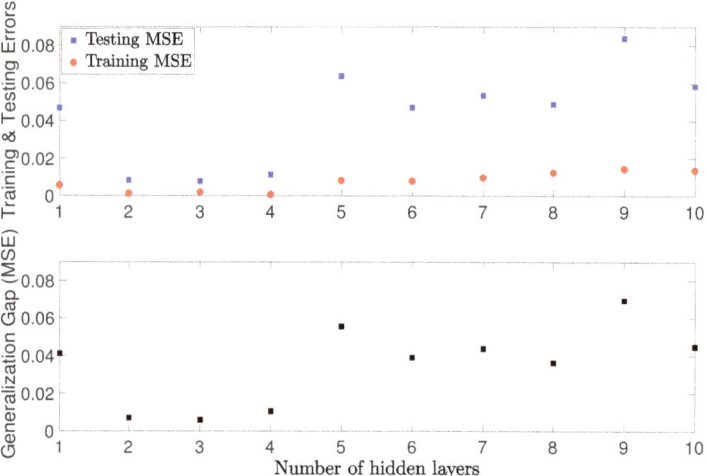

Figure 4. RNN generalization performance vs. RNN depth (Increasing the number of hidden layers and fixing 5 neurons for each layer).

Remark 8. *At first glance, the generalization error trend in Figures 3 and 4 seems in contrast to the results in Equation (47), which shows the generalization error bound is proportional to the complexity of RNN hypothesis class. However, it should be noted that Equation (47) only gives the upper bound for the generalization error of RNN models from the hypothesis class. It does not mean all the RNN models from the hypothesis class have a generalization error as large as* its upper bound. From the error decomposition of Equation (17) showing the interplay between approximation and estimation errors, we have learned that as we enlarge the hypothesis class, the approximation error decreases, but the estimation error may increase. In this case study, by increasing the complexity of RNN hypothesis class in terms of more layers and neurons, overall the generalization performance improves; however, as the RNN models become deeper, overfitting also occurs due to a large estimation error. Therefore, in practice, we can do a grid search such as Figures 3 and 4 to determine the optimal number of layers and neurons.

5.1.3. Case Study 3: Different Regions in $\Omega_{\hat{\rho}}$

As discussed in Remark 6, to meet the modeling error constraint $E_M \leq \gamma|x|$, $\forall x \in \Omega_{\hat{\rho}}$, more data is needed as the state approaches the origin, i.e., $x \to 0$. It is equivalent to show that under the same data density for different regions within the stability region $\Omega_{\hat{\rho}}$, a larger constant γ is needed to bound the modeling error $E_M \leq \gamma|x|$ for the states close to the origin. Therefore, in this case study, we develop multiple RNN models for different regions inside $\Omega_{\hat{\rho}}$ with the same data density, and demonstrate the variation of generalization performances. Specifically, we choose 9 level sets of Lyapunov function $\Omega_{\rho_i} := \{x \in \mathbf{R}^n \mid \hat{V}(x) \leq \rho_i\}$, $i = 0, ..., 8$, within $\Omega_{\hat{\rho}}$, with $\hat{\rho} = 368$ and $\rho_i = [40, 88, 115, 138, 159, 177, 195, 213, 244]$. For example, the first RNN model (model 0 with ρ_0) is developed and tested using the data within Ω_{ρ_0}, the second RNN model (model 1 with ρ_1) uses the data between Ω_{ρ_0} and Ω_{ρ_1}, and so on. Figure 5 shows a schematic of the training regions considered for the CSTR of Equation (74), where x_s is the steady-state, and $\Omega_{\hat{\rho}}$ is the stability region. The training datasets are generated for each region (i.e., elliptical annuli in Figure 5) with the same data density, where the data density is defined as the ratio of sample size to the area of each elliptical annulus. Similarly, in this case study, we use data from different regions within $\Omega_{\hat{\rho}}$ to build RNN models, while all the other parameters remain the same. The RNN models are developed with one hidden layer of 20 neurons, and using MSE as loss function.

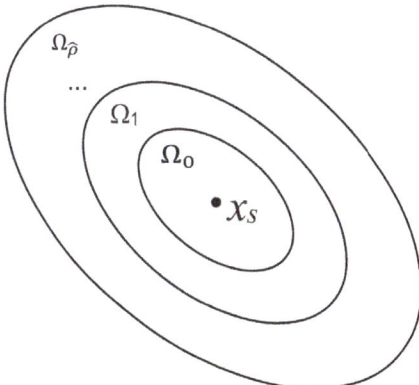

Figure 5. Schematic of different regions inside $\Omega_{\hat{\rho}}$.

To compute the modeling error $|F(x, u) - F_{nn}(x, u)| = |\frac{dx}{dt} - \frac{d\hat{x}}{dt}|$ where x and $\hat{x}$ denote the true state and predicted state, respectively, we carry out prediction for one integration time step, and use finite difference method to approximate the derivatives following Equation (56). Specifically, we first calculate the training and testing mean absolute errors (MAE) and divide them by the integration time step h_c, i.e., $|\frac{x(t+h_c) - \hat{x}(t+h_c)}{h_c}|$. Subsequently, to obtain an approximated value of γ for each model, i.e., $\frac{E_M}{|x|} \leq \gamma$, we divide those MAEs by the maximum value of $|x|$ in each elliptical annulus in Figure 5. Figure 6 shows the training and testing errors for the RNN models trained for different regions inside $\Omega_{\hat{\rho}}$.

It is observed that under the same data density, the models trained for the regions close to the origin (i.e., Models 0, 1, and 2 for Ω_{ρ_0}, Ω_{ρ_1} and Ω_{ρ_2}) produce larger generalization gaps. This implies a larger γ, or equivalently, more data is needed to meet the constraint $E_M \leq \gamma |x|$ for x in these regions. Additionally, it is observed that the generalization gap settles at around 2×10^{-5} for model 4 and after because those RNN models have achieved the best they can do under the current neural network training settings and data density.

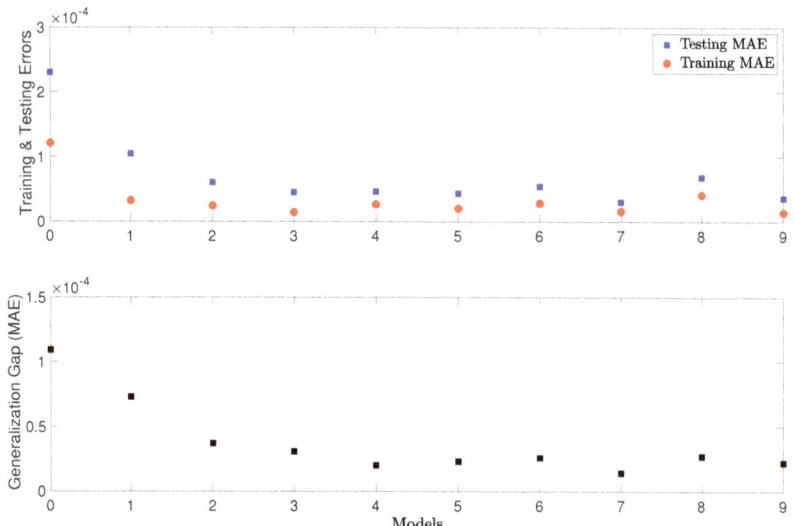

Figure 6. RNN generalization performance vs. different regions in Ω_ρ.

5.1.4. Case Study 4: Weight Matrix Bound

From Equation (48), it is seen that the generalization gap also depends on the weight matrix bound. To evaluate the relation between generalization performance and weight matrix bound, in this case study, we train RNN models with different weight matrix bounds. Specifically, we impose an upper bound constraint for each element in the RNN weight matrices with the following values $[0.8, 1.3, 1.8, 2.5, 3.0, 3.4, 3.9, 4.3]$.

The Frobenius norms of all the weight matrices are therefore also bounded. The training and testing errors are calculated following the approach in Case study 1, and are shown in Figure (7). It is observed that as the weight matrix bound becomes larger, the generalization gap gradually increases and settles at around 8×10^{-4}. This behavior implies that the RNN model is over-fitting when training with a large weight bound. The reason for the trend in Figure 7 is similar to that for Case study 2, which demonstrates that as the size of neural network hypothesis class becomes larger with increasing weight bounds, it is easier to find a hypothesis that fits training data well, but could also lead to large testing error (i.e., over-fitting).

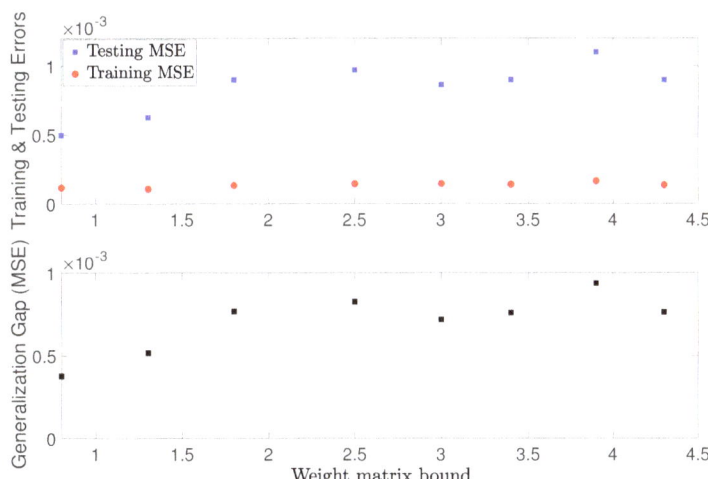

Figure 7. RNN generalization performance vs. weight matrix bound.

5.1.5. Case Study 5: RNN Input Length

Lastly, we study the dependency of RNN generalization error on the input time length t according to Equation (47). If we unfold a vanilla RNN over time to form a multi-layer feedfoward neural network, then this relation can also be interpreted in the way that a deep feedforward neural network has a large generalization error. In this example, we train RNN models with different input time length as follows: $t = 10^{-3} \times [1, 2, 3, 4, 5, 6, 7, 8, 9, 10]$ hr.

Figure 8 shows the training and testing errors for different time lengths. Specifically, as RNN input time length increases, it is seen that the training error remains at a very low level for all models, but the testing error gradually increases and finally settles at around 6×10^{-3}. It is concluded from Figure 8 that a shorter input sequence yields better generalization performance, which is consistent with the theoretical result shown in Equation (47). However, it should be noted that a shorter input sequence does not necessarily yield better prediction in the formulation of MPC because as discussed in Theorem 2, in order to predict future states for a long prediction horizon, the RNN prediction needs to be executed successively, which inevitably accumulates the error during calculations. Therefore, when used in MPC, the RNN input length should be carefully chosen to account for MPC prediction horizon and maintain a desired generalization performance simultaneously.

Remark 9. *A small training dataset was chosen in Case studies 2–5 for demonstration purposes. Specifically, it was demonstrated in Case study 1 that with more than 3000 data samples, both training and testing errors are rendered sufficiently small. Therefore, to better demonstrate the relation between RNN generalization error bound and RNN depth/width, and data time length in other case studies, we chose a small training dataset such that significant differences can be observed by varying RNN depths, widths, time sequence length. However, it is noted that in practice, the sample size and all the other factors studied in this manuscript should be carefully chosen in order to improve the RNN generalization performance.*

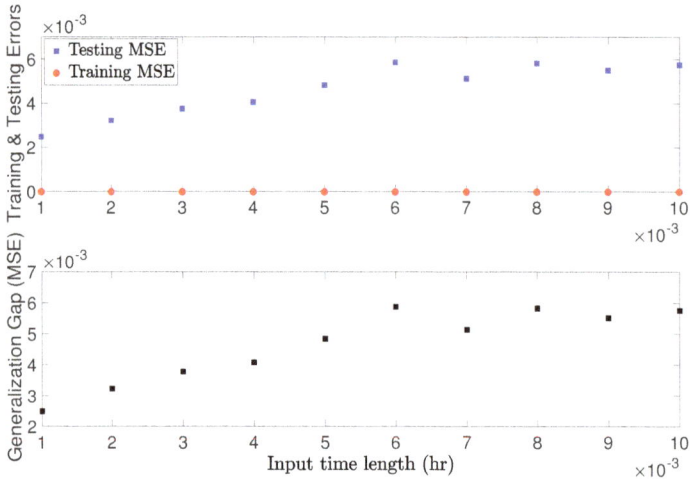

Figure 8. RNN generalization performance vs. input time length.

5.2. Closed-Loop Performance Analysis

In this section, we carry out closed-loop simulations of CSTR under the LMPC of Equations (68)–(73) using the different RNN models derived from the previous case studies. Additionally, we demonstrate the probabilistic closed-loop stability properties of RNN-based LMPC through extensive closed-loop simulations for the CSTR of Equation (74) with different initial conditions.

Figures 9–12 show the simulation results using 48 different initial conditions within $\Omega_{\hat{\rho}}$ for a few RNN models trained in Case study 1. Specifically, we first discretize the stability region $\Omega_{\hat{\rho}}$ and choose 48 initial conditions $x_0 \in \Omega_{\hat{\rho}}$ that are evenly spread within the stability region. Then, we run closed-loop simulations for all initial conditions using the following settings: (1) the whole simulation period t_p is twenty sampling periods (i.e., $20 \times 0.01 = 0.2$ hr), (2) the stability region $\Omega_{\hat{\rho}}$ and the terminal region $\Omega_{\rho_{min}}$ are characterized as $\hat{\rho} = 368$ and $\rho_{min} = 2$, respectively, and (3) the simulations are carried out using UCLA Hoffman 2 cluster and the optimization problem is solved using the python module of the IPOPT software package (i.e., PyIpopt). After obtaining the closed-loop profiles for each initial condition, the following policies are utilized to determine whether the closed-loop system is stable or not. Specifically, the closed-loop system is considered unstable if (1) the closed-loop state leaves the stability region $\Omega_{\hat{\rho}}$ at any point during the simulation, or (2) the closed-loop state remains inside $\Omega_{\hat{\rho}}$, but stays outside of $\Omega_{\rho_{min}}$ until the end of simulation or leaves $\Omega_{\rho_{min}}$ after entering for the first time.

Figure 9 shows the probability of closed-loop stability calculated following the above policies. It is seen that with more training data, the probability of the CSTR of Equation (74) being stabilized at its steady-state becomes higher, and the probability settles at around 0.78 for a sufficiently large dataset. The probability results in Figure 9 for RNN models in Case study 1 are consistent with its generalization performance plot in Figure 2, which shows that the generalization error decreases with more data used for training. In addition to the calculation of the probability for closed-loop stability, we also use the MPC cost function of Equation (68) as an indicator for comparing control performance in terms of the convergence speed and energy consumption. Specifically, the MPC cost function of Equation (68) in this example is designed in the following form:

$$L_{MPC}(x,u) = x^T P x + u^T Q u \qquad (76)$$

where $P = [1000\ 0;\ 0\ 1]$ and $Q = [1\ 0;\ 0\ 3 \times 10^{-10}]$ are chosen such that the two states and the two inputs are in the same order of magnitude, respectively. Also, in this example, we put more penalty on the states x to allow the states to be driven to the steady-state more quickly. For each RNN model, we calculate the total costs $\int_{t=0}^{t_p} L_{MPC}(x,u)dt$ over the entire simulation period $t_p = 0.2$ hr, and sum up the cost values for all the trajectories initiated from 48 different initial conditions. Figure 10 shows the MPC total costs for the RNN models trained with different data sample sizes. It is demonstrated that with less training data, the MPC achieves a higher total cost, representing a slower convergence to the steady-state and/or a higher energy consumption. With a large number of training data (i.e., ≥ 6000), the MPC total costs remain at around 1420, and no significant improvement is noticed with more data added in training. Additionally, Figures 11 and 12 show the closed-loop state trajectory and state profiles for one of the initial condition out of 48 initial conditions. As shown in Figure 11, the state trajectory using the RNN model trained with 50 training data (dashed line) leaves the stability region due to poor predictions in solving the MPC optimization problem. On the contrary, the state trajectory using the RNN model with 14,000 training data (solid line) moves towards the steady-state smoothly and is ultimately bounded in the terminal set $\Omega_{\rho_{min}}$. This can also be seen in the closed-loop state profiles of Figure 12, where the temperature under 50 training data shows a sharp increase at 0.03 hr.

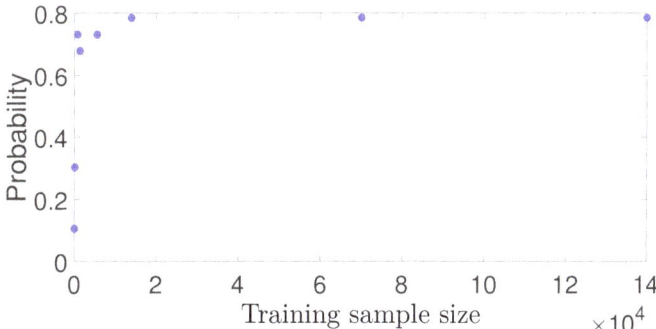

Figure 9. Probability of closed-loop stability vs. training sample sizes.

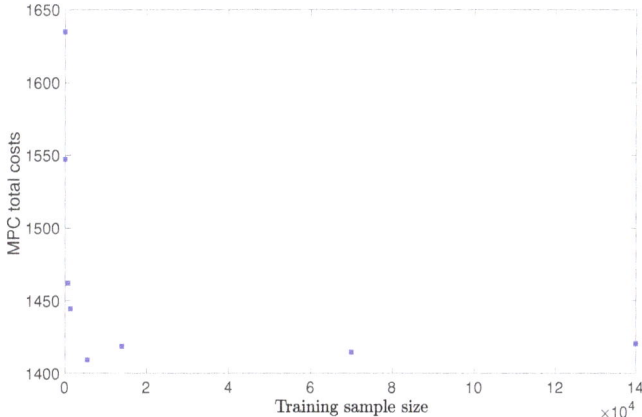

Figure 10. MPC total costs vs. training sample size.

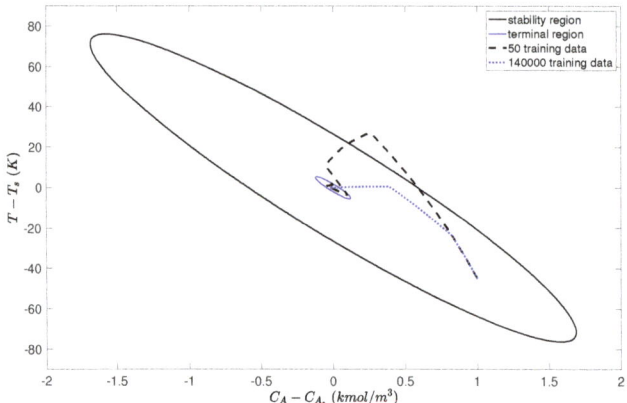

Figure 11. Closed-loop state trajectory under LMPC using two RNN models trained with different data sample sizes.

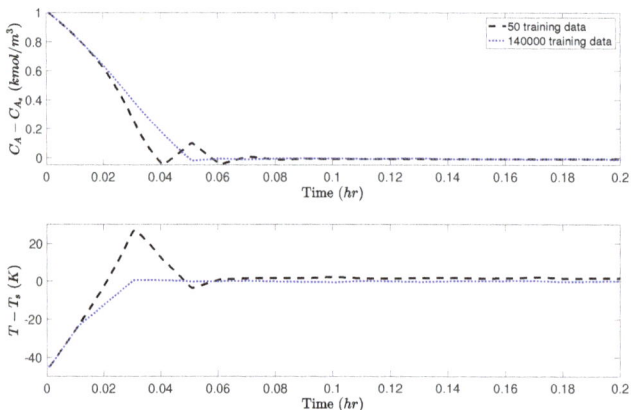

Figure 12. Closed-loop state profiles under LMPC using two RNN models trained with different data sample sizes.

Similar to the analysis for Case study 1, Figures 13–16 show the probability of closed-loop stability, MPC total costs, as well as the state-space trajectory and state profiles for one of the initial condition for the RNN models in Case study 2. In Figure 13, it is shown that the probability starts from 0.5, and settles at around 0.7 for wider RNN models (i.e., more neurons). Figure 14 shows the MPC total costs for different models, from which it is demonstrated that the first model with only 5 neurons has a extremely high value, and all the other models achieve a total cost around 1500. Figures 13 and 14 demonstrate that all the RNN models except the first one achieve desired closed-loop performance in terms of high probability of closed-loop stability and low total costs. This is due to the low generalization error (around 0.005) for nearly all the models in Figure 3. Figure 15 shows the comparison of the closed-loop state trajectories under the two RNN models using 5 and 350 neurons, respectively, from which it is demonstrated that the model with 5 neurons (dashed line) drives the state out of the stability region, while the one with 350 neurons successfully stabilizes the system in the terminal set. The corresponding state profiles (i.e., $C_A - C_{As}$ and $T - T_s$) can be found in Figure 16.

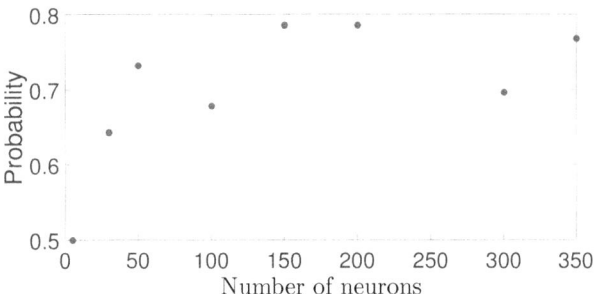

Figure 13. Probability of closed-loop stability vs. RNN width.

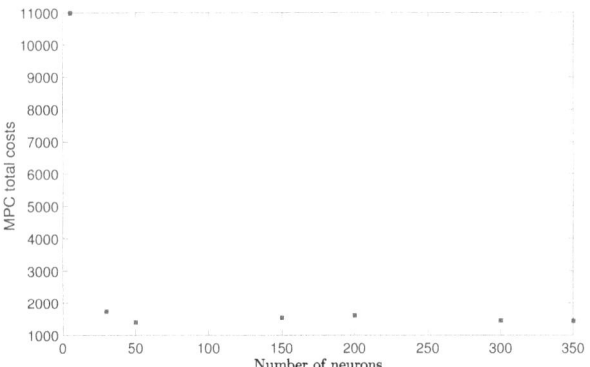

Figure 14. MPC total costs vs. RNN width.

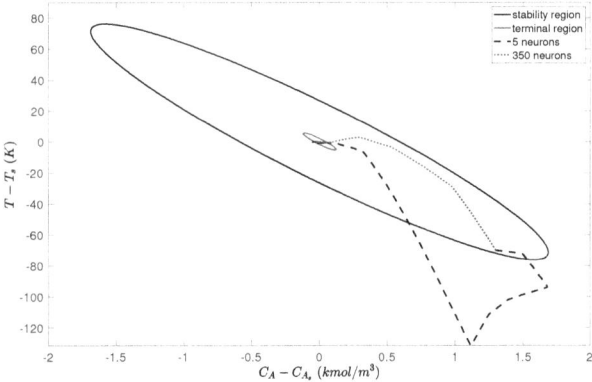

Figure 15. Closed-loop state trajectory under LMPC using two RNN models trained with different widths.

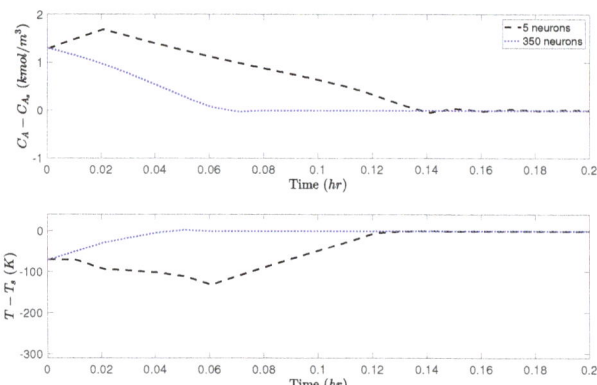

Figure 16. Closed-loop state profiles under LMPC using two RNN models trained with different widths.

To simplify the discussion for the remaining case studies, we will show the probability plot and MPC total cost plot only. Figure 17 shows the probability of closed-loop stability with respect to different RNN depths. It is demonstrated that the probability starts close to zero for one layer, increases up to 0.7 for four layers, and then decreases to almost zero for six layer and after. This trend follows exactly the generalization error plot in Figure 4, which shows the model with two, three and four layers achieve the lowest generalization error, and the models with more than five layers show worse generalization performance due to overfitting. Comparing to the closed-loop results for the RNNs with various widths in Figures 13 and 14, it is not surprising to see that the overall probability of closed-loop stability in this case study is worse because the open-loop generalization performance for the RNNs developed with different depths (Figure 4) is worse than that for the RNNs developed with different widths (Figure 3). Additionally, in Figure 18, we observe a similar pattern showing that the MPC total costs have the lowest values for two, three and four layers, and rise up for more layers.

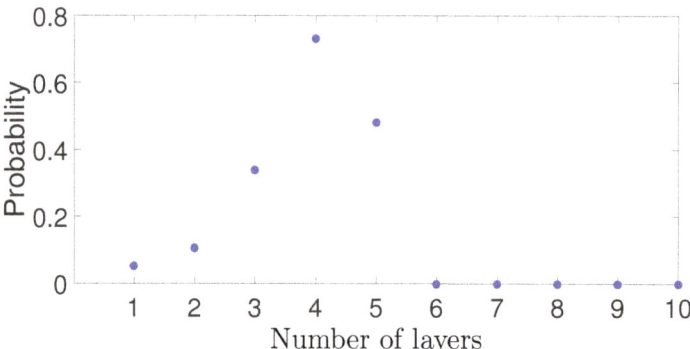

Figure 17. Probability of closed-loop stability vs. RNN depth.

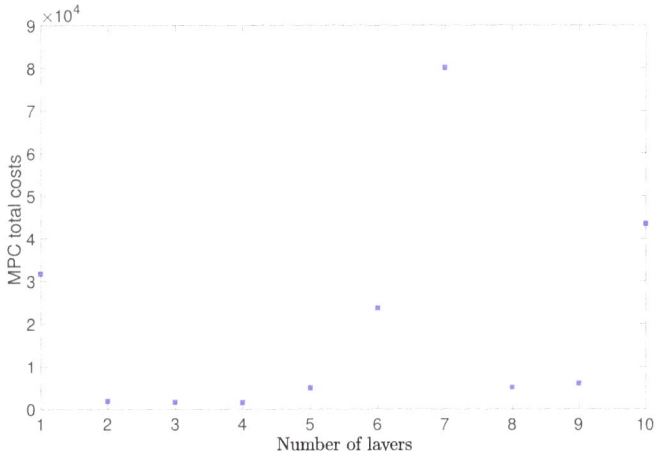

Figure 18. MPC total costs vs. RNN depth.

Closed-loop simulations for Case study 3 of different regions in $\Omega_{\hat{\rho}}$ are not carried out in this work, since the MPC formulation of Equations (68)–(73) only uses a single RNN model for prediction. Additionally, it is demonstrated from previous case studies that a single RNN model is sufficient to capture the process dynamics in the stability region, and therefore, there is no need to use different RNN models for different regions in $\Omega_{\hat{\rho}}$ from the control perspective.

Figure 19 shows the probability of closed-loop stability for the RNN models with different weight matrix bounds in Case study 4. It is shown that all the RNN models achieve a probability up to 0.7. The high probability of closed-loop stability is expected since in the open-loop generalization error plot in Figure 7, it is shown that all the models with different weight matrix bounds have a sufficiently small generalization error around 8×10^{-4}. As a result, the MPC total costs in Figure 20 are stable around 1000 for all models.

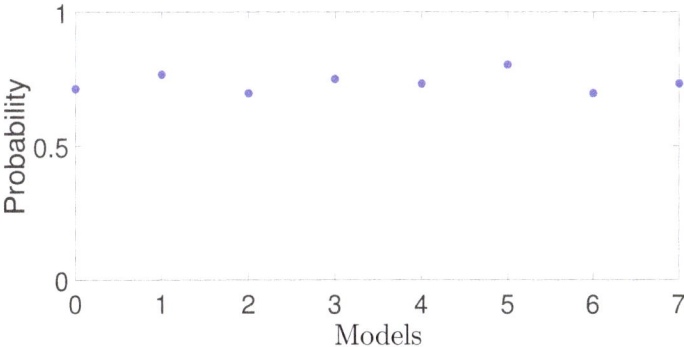

Figure 19. Probability of closed-loop stability vs. weight matrix bound.

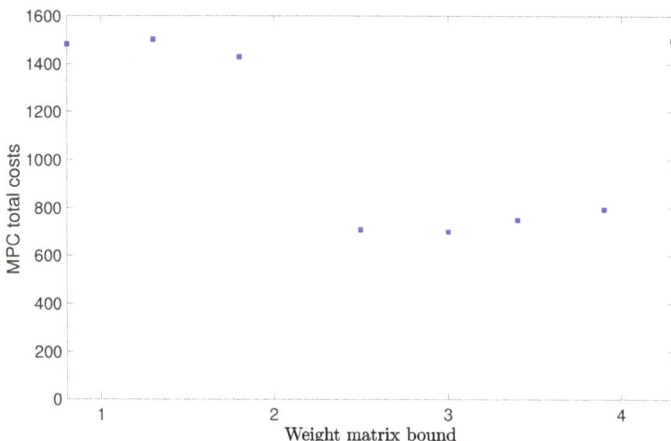

Figure 20. MPC total costs vs. weight matrix bound.

Lastly, Figures 21 and 22 show the closed-loop simulation results for Case study 5. As shown in Figure 21, the probability of closed-loop stability increases as the RNN input time length increases, and settles at around 0.9 for input time length greater than 6×10^{-3} hr. This seems inconsistent with the generalization performance of Figure 8 which shows the generalization error increases for longer input sequences at first glance. However, as we have discussed earlier, a low open-loop generalization error for short input sequences does not guarantee a desired closed-loop performance under MPC. Specifically, with shorter input sequences, the RNN prediction needs to be executed successively in each MPC iteration to predict all the future states within the prediction horizon. For example, in order to predict one sampling time $\Delta = 10^{-2}$ hr, the first RNN model with 1×10^{-3} input length in Figure 21 needs to run 10 times, and each time uses the previous predicted state as the initial state. The error accumulates during the calculation, which ultimately leads to poorer closed-loop performance. Therefore, for RNN models used in MPC, the input time length should be chosen carefully accounting for the system sampling time and MPC prediction horizon. Additionally, Figure 22 shows the MPC total costs with respect to different RNN input time lengths. It is seen that the first RNN model achieves the worst cost value, and all the other models have similar cost values around 2000. Through the closed-loop simulation of all the case studies investigated in the previous section, we demonstrate that the closed-loop performance is consistent with the open-loop generalization performance in the way that lower generalization errors typically leads to higher probability of closed-loop stability and lower MPC total costs. Therefore, the generalization error bound proposed in this work provides an efficient method for choosing neural network structure and data sample size to meet the closed-loop stability requirements.

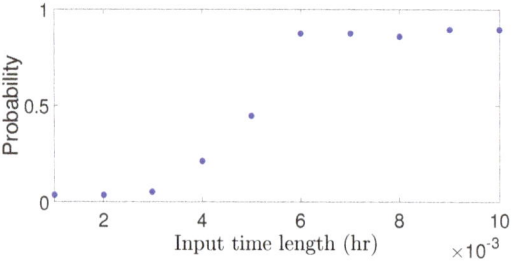

Figure 21. Probability of closed-loop stability vs. input time length.

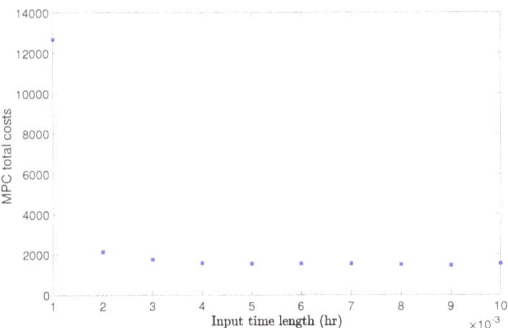

Figure 22. MPC total costs vs. input time length.

Remark 10. *The RNN models are trained offline, and the RNN-based MPC is solved in real time with new state measurements available at each sampling time. The averaged computation time for solving RNN-based MPC per sampling step is around 10 s, which is less than one sampling period $\Delta = 0.01\ hr = 36\ s$ in this example. Therefore, the RNN-based MPC scheme can be implemented in real time without any computational issues.*

6. Conclusions

In this work, we developed a generalization probabilistic error bound for RNN models by taking advantage of the Rademacher complexity method for vector-valued functions. The RNN models were incorporated in the design of MPC, and probabilistic closed-loop stability properties were derived based on the RNN generalization error bounds. A number of case studies were simulated using a nonlinear chemical reactor example to demonstrate the impact of training sample size, the number of neurons and layers, regions where the data was generated, and input time length on the RNN generalization performance. Closed-loop simulation were carried out to further demonstrate the probabilistic closed-loop stability properties derived by the RNN-based LMPC.

Author Contributions: Z.W. developed the main results, performed the simulation studies and prepared the initial draft of the paper. D.R. contributed to the simulation studies in this manuscript. Q.G. and P.D.C. developed the idea of RNN generalization error, oversaw all aspects of the research and revised this manuscript. All authors have read and agreed to the published version of the manuscript.

Funding: This research received no external funding.

Institutional Review Board Statement: Not applicable.

Informed Consent Statement: Not applicable.

Data Availability Statement: Not applicable.

Conflicts of Interest: The authors declare that they have no conflict of interest regarding the publication of the research article.

References

1. Cozad, A.; Sahinidis, N.V.; Miller, D.C. A combined first-principles and data-driven approach to model building. *Comput. Chem. Eng.* **2015**, *73*, 116–127. [CrossRef]
2. Wilson, Z.T.; Sahinidis, N.V. The ALAMO approach to machine learning. *Comput. Chem. Eng.* **2017**, *106*, 785–795. [CrossRef]
3. Ali, J.M.; Hussain, M.A.; Tade, M.O.; Zhang, J. Artificial Intelligence techniques applied as estimator in chemical process systems–A literature survey. *Expert Syst. Appl.* **2015**, *42*, 5915–5931.
4. Han, H.; Wu, X.; Qiao, J. Real-time model predictive control using a self-organizing neural network. *IEEE Trans. Neural Netw. Learn. Syst.* **2013**, *24*, 1425–1436. [PubMed]
5. Wang, T.; Gao, H.; Qiu, J. A combined adaptive neural network and nonlinear model predictive control for multirate networked industrial process control. *IEEE Trans. Neural Netw. Learn. Syst.* **2016**, *27*, 416–425. [CrossRef] [PubMed]

6. Wang, Y. A new concept using LSTM Neural Networks for dynamic system identification. In Proceedings of the American Control Conference 2017, Seattle, WA, USA, 24–26 May 2017; pp. 5324–5329.
7. Wong, W.; Chee, E.; Li, J.; Wang, X. Recurrent Neural Network-Based Model Predictive Control for Continuous Pharmaceutical Manufacturing. *Mathematics* **2018**, *6*, 242. [CrossRef]
8. Shahnazari, H.; Mhaskar, P.; House, J.M.; Salsbury, T.I. Modeling and fault diagnosis design for HVAC systems using recurrent neural networks. *Comput. Chem. Eng.* **2019**, *126*, 189–203. [CrossRef]
9. Wu, Z.; Tran, A.; Rincon, D.; Christofides, P.D. Machine Learning-Based Predictive Control of Nonlinear Processes. Part I: Theory. *AIChE J.* **2019**, *65*, e16729. [CrossRef]
10. Wu, Z.; Tran, A.; Rincon, D.; Christofides, P.D. Machine Learning-Based Predictive Control of Nonlinear Processes. Part II: Computational Implementation. *AIChE J.* **2019**, *65*, e16734. [CrossRef]
11. Pan, Y.; Wang, J. Nonlinear model predictive control using a recurrent neural network. In Proceedings of the IEEE International Joint Conference on Neural Networks 2008, Hong Kong, China, 1–8 June 2008; pp. 2296–2301.
12. Pan, Y.; Wang, J. Model predictive control of unknown nonlinear dynamical systems based on recurrent neural networks. *IEEE Trans. Ind. Electron.* **2011**, *59*, 3089–3101. [CrossRef]
13. Xu, J.; Li, C.; He, X.; Huang, T. Recurrent neural network for solving model predictive control problem in application of four-tank benchmark. *Neurocomputing* **2016**, *190*, 172–178. [CrossRef]
14. Hoskins, J.; Himmelblau, D. Process control via artificial neural networks and reinforcement learning. *Comput. Chem. Eng.* **1992**, *16*, 241–251. [CrossRef]
15. Vepa, R. A review of techniques for machine learning of real-time control strategies. *Intell. Syst. Eng.* **1993**, *2*, 77–90. [CrossRef]
16. Hussain, M. Review of the applications of neural networks in chemical process control—Simulation and online implementation. *Artif. Intell. Eng.* **1999**, *13*, 55–68. [CrossRef]
17. Hewing, L.; Wabersich, K.; Menner, M.; Zeilinger, M. Learning-based model predictive control: Toward safe learning in control. *Annu. Rev. Control Robot. Auton. Syst.* **2020**, *3*, 269–296. [CrossRef]
18. Venkatasubramanian, V. The promise of artificial intelligence in chemical engineering: Is it here, finally? *AIChE J.* **2019**, *65*, 466–478. [CrossRef]
19. Mittal, M.; Gallieri, M.; Quaglino, A.; Salehian, S.; Koutník, J. Neural lyapunov model predictive control. *arXiv* **2020**, arXiv:2002.10451.
20. Limon, D.; Calliess, J.; Maciejowski, J. Learning-based nonlinear model predictive control. *IFAC-PapersOnLine* **2017**, *50*, 7769–7776. [CrossRef]
21. Aswani, A.; Gonzalez, H.; Sastry, S.; Tomlin, C. Provably safe and robust learning-based model predictive control. *Automatica* **2013**, *49*, 1216–1226. [CrossRef]
22. Valiant, L.G. A theory of the learnable. *Commun. ACM* **1984**, *27*, 1134–1142. [CrossRef]
23. Lygeros, J.; Margellos, K.; Prandini, M. Compression learning for chance constrained stochastic MPC. *IFAC-PapersOnLine* **2015**, *48*, 286–293. [CrossRef]
24. Zhou, Y.; Li, D.; Spanos, C. Learning optimization friendly comfort model for HVAC model predictive control. In Proceedings of the 2015 IEEE International Conference on Data Mining Workshop (ICDMW), Atlantic City, NJ, USA, 14–17 November 2015; pp. 430–439.
25. Bartlett, P.; Foster, D.J.; Telgarsky, M. Spectrally-normalized margin bounds for neural networks. *arXiv* **2017**, arXiv:1706.08498.
26. Zhang, Y.; Lee, J.; Wainwright, M.; Jordan, M.I. On the learnability of fully-connected neural networks. In Proceedings of the Artificial Intelligence and Statistics PMLR 2017, Fort Lauderdale, FL, USA, 20–22 April 2017; pp. 83–91.
27. Golowich, N.; Rakhlin, A.; Shamir, O. Size-independent sample complexity of neural networks. In Proceedings of the Conference On Learning Theory PMLR 2018, Stockholm, Sweden, 5–9 July 2018; pp. 297–299.
28. Cao, Y.; Gu, Q. Tight sample complexity of learning one-hidden-layer convolutional neural networks. *arXiv* **2019**, arXiv:1911.05059.
29. Chen, M.; Li, X.; Zhao, T. On generalization bounds of a family of recurrent neural networks. *arXiv* **2019**, arXiv:1910.12947.
30. Cao, Y.; Gu, Q. Generalization bounds of stochastic gradient descent for wide and deep neural networks. *arXiv* **2019**, arXiv:1905.13210.
31. Zou, D.; Gu, Q. An improved analysis of training over-parameterized deep neural networks. *arXiv* **2019**, arXiv:1906.04688.
32. Akpinar, N.; Kratzwald, B.; Feuerriegel, S. Sample complexity bounds for recurrent neural networks with application to combinatorial graph problems. *arXiv* **2019**, arXiv:1901.10289.
33. Reid, M. Generalization bounds. In *Encyclopedia of Machine Learning*; Springer: Boston, MA, USA, 2010; pp. 447–454._328. [CrossRef]
34. Hanson, J.; Raginsky, M.; Sontag, E. Learning Recurrent Neural Net Models of Nonlinear Systems. *arXiv* **2020**, arXiv:2011.09573.
35. Sammut, C.; Webb, G.I. *Encyclopedia of Machine Learning*; Springer Science & Business Media: Berlin/Heidelberg, Germany, 2011.
36. Maurer, A. A vector-contraction inequality for rademacher complexities. In *Lecture Notes in Computer Science, Proceedings of the International Conference on Algorithmic Learning Theory, Bari, Italy, 19–21 October 2016*; Springer: Berlin/Heidelberg, Germany, 2016; pp. 3–17.
37. Mohri, M.; Rostamizadeh, A.; Talwalkar, A. *Foundations of Machine Learning*; MIT Press: Cambridge, MA, USA, 2018.

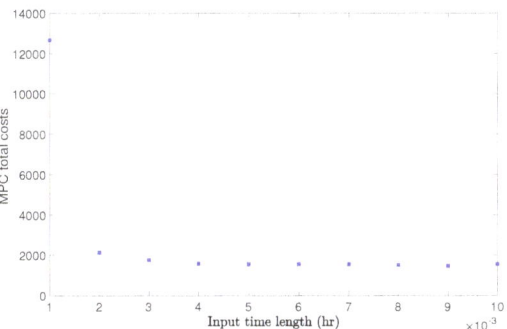

38. Ledoux, M.; Talagrand, M. *Probability in Banach Spaces: Isoperimetry and Processes*; Springer Science & Business Media: Berlin/Heidelberg, Germany, 2013.
39. Wu, Z.; Rincon, D.; Christofides, P.D. Process structure-based recurrent neural network modeling for model predictive control of nonlinear processes. *J. Process Control* **2020**, *89*, 74–84. [CrossRef]
40. Keras. Available online: https://keras.io (accessed on 1 August 2015)
41. Wächter, A.; Biegler, L.T. On the implementation of an interior-point filter line-search algorithm for large-scale nonlinear programming. *Math. Program.* **2006**, *106*, 25–57. [CrossRef]

Article

NICE: Noise Injection and Clamping Estimation for Neural Network Quantization

Chaim Baskin [1,*,†], Evgenii Zheltonozhkii [1,†], Tal Rozen [2,†], Natan Liss [2], Yoav Chai [3], Eli Schwartz [3], Raja Giryes [3], Alexander M. Bronstein [1] and Avi Mendelson [1]

1. Department of Computer Science, Technion, Haifa 3200003, Israel; evgeniizh@campus.technion.ac.il (E.Z.); bron@cs.technion.ac.il (A.M.B.); avi.mendelson@cs.technion.ac.il (A.M.)
2. Department of Electrical Engineering, Technion, Haifa 3200003, Israel; tal.rozen@campus.technion.ac.il (T.R.); lissnatan@campus.technion.ac.il (N.L.)
3. School of Electrical Engineering, Tel-Aviv University, Tel-Aviv 6997801, Israel; yoavchai1@mail.tau.ac.il (Y.C.); eliyahus@mail.tau.ac.il (E.S.); raja@tauex.tau.ac.il (R.G.)
* Correspondence: chaimbaskin@cs.technion.ac.il
† These authors contributed equally to this work.

Abstract: Convolutional Neural Networks (CNNs) are very popular in many fields including computer vision, speech recognition, natural language processing, etc. Though deep learning leads to groundbreaking performance in those domains, the networks used are very computationally demanding and are far from being able to perform in real-time applications even on a GPU, which is not power efficient and therefore does not suit low power systems such as mobile devices. To overcome this challenge, some solutions have been proposed for quantizing the weights and activations of these networks, which accelerate the runtime significantly. Yet, this acceleration comes at the cost of a larger error unless spatial adjustments are carried out. The method proposed in this work trains quantized neural networks by noise injection and a learned clamping, which improve accuracy. This leads to state-of-the-art results on various regression and classification tasks, e.g., ImageNet classification with architectures such as ResNet-18/34/50 with as low as 3 bit weights and activations. We implement the proposed solution on an FPGA to demonstrate its applicability for low-power real-time applications. The quantization code will become publicly available upon acceptance.

Keywords: neural networks; low power; quantization; CNN architecture

1. Introduction

Deep neural networks are important tools in the machine learning arsenal. They have shown spectacular success in a variety of tasks in a broad range of fields such as computer vision, computational and medical imaging, signal, image, speech, and language processing [1–3].

However, while deep learning models' performance is impressive, the computational and storage requirements of both training and inference are harsh. For example, ResNet-50 [4], a popular choice for image detection, has 98 MB parameters and requires 4 GFLOPs of computations for a single inference. Common devices do not have such resources, which makes deep learning infeasible especially when it comes to low-power devices such as smartphones and the Internet of Things (IoT).

In an attempt to solve these problems, many researchers have recently proposed less demanding models, often at the expense of more complicated training procedures. Since the training is usually performed on servers with significantly larger resources, this is usually an acceptable trade-off. Some methods include pruning weights and feature maps, which reduce the model's memory print and compute resources [5,6], low-rank decomposition that removes the redundancy of parameters and feature maps [7,8], and efficient architecture design that requires less communication and has more feasible deployment [9,10].

One prominent approach is to quantize the networks. This approach reduces the size of memory needed to keep a large number of parameters while also reducing the computation resources. The default choice for the data type of the neural networks' weights and feature maps (activations) is 32 bit (single-precision) floating point. Gupta et al. [11] have shown that quantizing the pre-trained weights to a 16 bit fixed point has almost no effect on the accuracy of the networks. Moreover, minor modifications allow performing an integer-only 8 bit inference with reasonable performance degradation [12], which is utilized in DL frameworks, such as TensorFlow. One of the current challenges in network quantization is reducing the precision even further, up to 1–5 bits per value. In this case, straightforward techniques may result in unacceptable quality degradation.

Contribution. This paper introduces a novel simple approach denoted NICE (noise injection and clamping estimation) for neural network quantization that relies on the following two easy-to-implement components: (i) noise injection during training that emulates the quantization noise introduced at inference time and (ii) statistics-based initialization of parameter and activation clamping for faster model convergence. In addition, activation clamp is learned during train time. We also propose an integer-only scheme for an FPGA on a regression task [13].

Our proposed strategy for network training leads to an improvement over the state-of-the-art quantization techniques in the performance vs. complexity trade-off. Our approach can be applied directly to existing architectures without the need to modify them at training (as opposed, for example, to the teacher–student approaches [14] that require to train a bigger network, or the XNOR networks [15] that typically increase the number of parameters by a significant factor in order to meet accuracy goals).

Moreover, our new technique allows quantizing all the parameters in the network to fixed point (integer) values. This includes the batch-norm component that is usually not quantized in other works. Thus, our proposed solution allows the integration of neural networks in dedicated hardware devices such as FPGA and ASIC easier. As a proof of concept, we present also a case study of such an implementation on hardware.

2. Related Work

Expressiveness-based methods. The quantization of neural networks to extremely low-precision representations (up to 2 or 3 possible values) has been actively studied in recent years [15–18]. To overcome the accuracy reduction, some works proposed to use a wider network [14,19,20], which compensates the expressiveness reduction of the quantized networks. For example, 32 bit feature maps were regarded as 32 binary ones. Another way to improve expressiveness, adopted by Zhu et al. [19] and Zhou et al. [21] is to add a linear scaling layer after each of the quantized layers.

Keeping a full-precision copy of quantized weights. Lately, the most common approach to training a quantized neural network [15,16,22–24] is to keep two sets of weights—forward pass is performed with quantized weights, and updates are performed on full precision ones, i.e., approximating gradients with the straight-through estimator (STE) [25]. For quantizing the parameters, either a stochastic or deterministic function can be used.

Distillation. One of the leading approaches used today for quantization relies on the idea of distillation [26]. In distillation a teacher–student setup is used, where the teacher is either the same or a larger full precision neural network, and the student is the quantized one. The student network is trained to imitate the output of the teacher network. This strategy is successfully used to boost the performance of existing quantization methods [14,27,28].

Model parametrization. Zhang et al. [18] proposed to represent the parameters with learned basis vectors that allow acquiring an optimized non-uniform representation. In this case, MAC operations can be computed with bitwise operations. Choi et al. [29] proposed to learn the clamping value of the activations to find the balance between clamping and quantization errors. In this work, we also learn this value but with the difference that we are learning the clamps value directly using STE backpropagation method without any regulations on the loss. Jung et al. [28] created a more complex parameterization of both weights and activations and approximated them with symmetric piecewise linear function, learning both the domains and the parameters directly from the loss function of the network.

Optimization techniques. Zhou et al. [21] and Dong et al. [30] used the idea of not quantizing all the weights simultaneously but rather gradually increasing the number of quantized weights to improve the convergence. McKinstry et al. [31] demonstrated that 4 bit fully integer neural networks can achieve full-precision performance by applying simple techniques to combat variance of gradients: larger batches and proper learning rate annealing with longer training time. However, 8 bit and 32 bit integer representations were used for the multiplicative (i.e., batch normalization) and additive constants (biases), respectively.

Generalization bounds. Interestingly, the quantization of neural networks has been used recently as a theoretical tool to understand better the generalization of neural networks. It has been shown that while the generalization error does not scale with the number of parameters in over-parameterized networks, it does so when these networks are being quantized [32].

Hardware implementation complexity. While the quantization of CNN parameters leads to a reduction of power and area, it can also generate unexpected changes in the balance between communication and computation. Karbachevsky et al. [33] studied the impact of CNN quantization on hardware implementation of computational resources. It combines the research conducted in Baskin et al. [34] to propose a computation and communication analysis for quantized CNN.

3. Method

In this work, we propose a training scheme for quantized neural networks designed for fast inference on hardware with integer-only arithmetic. To achieve maximum performance, we applied a combination of several well-known and novel techniques. Firstly, in order to emulate the effect of quantization, we injected additive random noise into the network weights. Uniform noise distribution is known to approximate well the quantization error for fine quantizers; however, our experiments show that it is also suitable for relatively coarse quantization. As seen in Figure 1 the distribution of noise is almost uniform for 4 and 5 bits and only starts to deviate from the uniform model in 3 bits, which corresponds to only 8 bins.

Figure 1. Weight quantization error histogram for a range of bitwidths.

Furthermore, some amount of random weight perturbation seems to have a regularization effect beneficial for the overall convergence of the training algorithm. Secondly, we used a gradual training scheme to minimize the perturbation of network parameters performed simultaneously. In order to give the quantized layers as many gradient updates as possible, we used the STE approach to pass the gradients to the quantized layers. After the gradual phase, the whole network was quantized and trained for a number of fine-tuning epochs. Thirdly, we propose to clamp both the activations and weights in order to reduce the quantization bin size (and, thus, the quantization error) at the expense of some sacrifice of the dynamic range. The clamping values were initialized using the statistics of each layer. In order to truly optimize the trade-off between the reduction of the quantization error vs. that of the dynamic range, we learned optimal clamping values by defining a loss on the quantization error.

Lastly, following the common approach proposed by Zhou et al. [23], we did not quantize the first and last layers of the networks, which have significantly higher impacts on network performance.

Algorithm 1 summarizes the proposed training method for network quantization. The remainder of the section details these main ingredients of our method.

Algorithm 1 Training a neural network with NICE. N denotes the number of layers; S is the number of epochs in which each layer's weights are noised; T is the total number of training epochs; c is the current noised layer; i denotes the ith layer; W is the weights of the layer; f denotes the layer's function, i.e, convolution or fully connected; and α and β are hyper-parameters.

1: **procedure** NICE (Network) ▷ Training using NICE method
2:   **for** $i = 0$ to N **do**
3:     $mean_{w_i} \leftarrow Running\ mean(W_i)$ ▷ Weights of each layer
4:     $std_{w_i} \leftarrow Running\ std(W_i)$ ▷ Weights of each layer
5:     $mean_{a_i} \leftarrow Running\ mean(Act_i)$ ▷ Activations of each layer on training set
6:     $std_{a_i} \leftarrow Running\ std(Act_i)$ ▷ Activations of each layer on training set
7:     $c_{w_i} \leftarrow mean_{w_i} + \alpha \times std_{w_i}$ ▷ Weight clamp
8:     $c_{a_i} \leftarrow mean_{a_i} + \beta \times std_{a_i}$ ▷ Activations clamp
9:   **end for**
10:   $e \leftarrow 0$
11:   $c \leftarrow 0$
12:   **while** $e \neq T$ **do**
13:     **for** $i = 0$ to N **do**
14:       **for** $s = 0$ to S **do**
15:         $W_{i=c} \leftarrow Noise(W_{i=c})$ ▷ Adding uniform noise to weights
16:         $W_i \leftarrow Clamp(W_i, -c_{w_i}, c_{w_i})$
17:         $W_{i<c} \leftarrow Quantize(W_{i<c})$
18:         $Act_i \leftarrow f(W_i, Act_{i-1})$
19:         $Clamped_i \leftarrow Clamp(Act_i, 0, c_{a_i})$
20:         $Act_i \leftarrow Quantize(Clamped_i)$
21:         $Learn(c_{a_i})$ ▷ Backpropagation
22:       **end for**
23:       $c \leftarrow c + 1$
24:     **end for**
25:     $e \leftarrow e + 1$
26:   **end while**
27: **end procedure**

3.1. Uniform Noise Injection

We propose to inject uniform additive noise to weights and biases during model training to emulate the effect of quantization incurred at inference. Prior works have investigated the behavior of quantization error [35,36] and concluded that in sufficiently fine-grain quantizers, it can be approximated as a uniform random variable. We observed the same phenomena and empirically verified it for weight quantization as coarse as 5 bits.

The advantage of the proposed method is that the updates performed during the backward pass immediately influence the forward pass, in contrast to strategies that directly quantize the weights, where small updates often leave them in the same bin, thus, effectively unchanged.

In order to achieve a dropout-like effect in the noise injection, we use a Bernoulli distributed mask M, quantizing part of the weights and adding noise to the others. From empirical evidence, we chose $M \sim Ber(0.05)$ as it gave the best results for the range of bitwidths in our experiments. Instead of using the quantized value $\hat{w} = \mathcal{Q}_\Delta(w)$ of a weight

w in the forward pass, $\hat{w} = (1 - M)\mathcal{Q}_\Delta(w) + M(w - e)$ is used with $e \sim \text{Uni}(-\Delta/2, \Delta/2)$, where Δ denotes the size of the quantization bin.

3.2. Gradual Quantization

In order to improve the scalability of the method for deeper networks, it is desirable to avoid the significant change of the network behavior due to quantization. Thus, we start by gradually adding a subset of weights to the set of quantized parameters, allowing the rest of the network to adapt to the changes.

The gradual quantization is performed in the following way: the network is split into N equally-sized blocks of layers $\{B_1, \ldots, B_N\}$. At the i-th stage, we inject the noise into the weights of the layers from block B_i. The previous blocks $\{B_1, \ldots, B_{i-1}\}$ are quantized, while the following blocks $\{B_{i+1}, \ldots, B_N\}$ remain at full precision. We apply the gradual process only once, i.e., when the N-th stage finishes, in the remaining training epochs we quantize and train all the layers using the STE approach.

This gradual process of increasing the number of quantized layers is similar to the one proposed by Xu et al. [37]. This gradual process reduces, via the number of parameters, the amount of simultaneously injected noise and improves convergence. Since we start from the earlier blocks, the later ones have an opportunity to adapt to the quantization error affecting their inputs, and thus, the network does not change drastically during any phase of quantization. After finishing the training with the noise injection into the block of layers B_N, we continue the training of the fully quantized network for several epochs until convergence. In the case of a pre-trained network destined for quantization, we have found that the optimal block size is a single layer with the corresponding activation, while using more than one epoch of training with the noise injection per block does not improve performance.

3.3. Clamping and Quantization

In order to quantize the network weights, we clamp their values in the range $[-c_w, c_w]$:

$$w_c = \text{Clamp}(w, -c_w, c_w) = \max\left(-c_w, \min\left(x, c_w\right)\right). \tag{1}$$

The parameter c_w is defined per layer and is initialized with $c_w = \text{mean}(w) + \beta \times \text{std}(w)$, where w values are the weights of the layer, and β is a hyper-parameter. Given c_w, we uniformly quantize the clamped weight into B_w bits according to

$$\hat{w} = \left[w_c \frac{2^{B_w - 1} - 1}{c_w} \right] \frac{c_w}{2^{B_w - 1} - 1},$$

where $[\cdot]$ denotes the rounding operation.

The quantization of the network activations is performed in a similar manner. The conventional ReLU activation function in CNNs is replaced by the clamped ReLU,

$$a_c = \text{Clamp}(a, 0, c_a), \tag{2}$$

where a denotes the output of the linear part of the layer, a_c is the nonnegative value of the clamped activation prior to quantization, and c_a is the clamping range. The constant c_a is set as a local parameter of each layer and is learned with the other parameters of the network via backpropagation. We used the initialization $c_a = \text{mean}(a) + \alpha \times \text{std}(a)$ with the statistics computed on the training dataset and α set as a hyper-parameter.

A quantized version of the truncated activation is obtained by quantizing a_c uniformly to B_a bits,

$$\hat{a} = \left[a_c \frac{2^{B_a} - 1}{c_a} \right] \cdot \frac{c_a}{2^{B_a} - 1}. \tag{3}$$

Since the Round function is non-differentiable, we used the STE approach to propagate the gradients through it to the next layer. For the update of c_a, we calculated the derivative of $\hat{a}$ with respect to c_a as

$$\frac{\partial \hat{a}}{\partial a_c} = \begin{cases} 1, & a_c \in [0, c_a] \\ 0, & \text{otherwise.} \end{cases} \quad (4)$$

Figure 2 depicts the evolution of the activation clamp values throughout the epochs. In this experiment, α was set to 5. It can be seen that activation clamp values converge to values smaller than the initialization. This shows that the layer prefers to shrink the dynamic range of the activations, which can be interpreted as a form of regularization similar in its purpose to weight decay on weights.

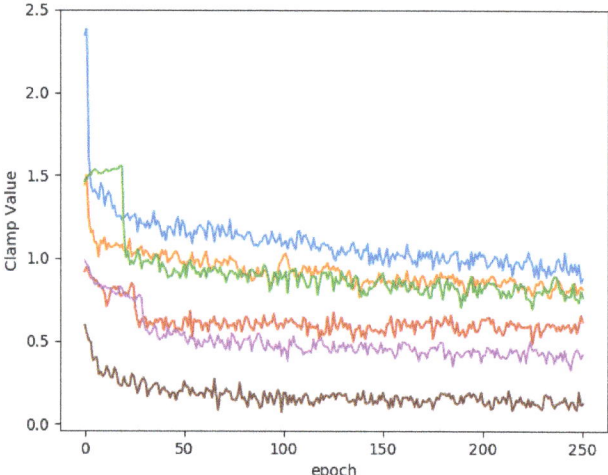

Figure 2. Activation clamp values during ResNet-18 training on CIFAR10 dataset.

The quantization of the layer biases is more complex, since their scale depends on the scales of both the activations and the weights. For each layer, we initialize the bias clamping value as

$$c_b = \left(\underbrace{\frac{c_a}{2^{B_a}-1}}_{\text{Activation scale}} \cdot \underbrace{\frac{c_w}{2^{B_w-1}-1}}_{\text{Weight scale}} \right) \cdot \left(\underbrace{2^{B_b-1}-1}_{\text{Maximal bias value}} \right), \quad (5)$$

where B_b denotes the bias bitwidth. The biases are clamped and quantized in the same manner as the weights.

4. Results

To demonstrate the effectiveness of our method, we implemented it in PyTorch and evaluated it using image classification datasets (ImageNet and CIFAR-10) and a regression scenario (the MSR joint denoising and demosaicing dataset [38]). In all the experiments, we used a pre-trained FP32 model, which was then quantized using NICE.

4.1. CIFAR-10

We tested NICE with ResNet-18 on CIFAR-10 for various quantization levels of the weights and activations. Table 1 reports the results. Notice that for the case of 3 bit weights activities, we obtain the same accuracy and for the 2 bit case, only a small degradation. Moreover, observe that when we quantize only the weights or activations, we get a nice regularization effect that improves the achieved accuracy.

Table 1. NICE accuracy (% top-1) on CIFAR-10 for range of bitwidths.

		Activation Bits			
		1	2	3	32
Weight bits	2	89.5	92.53	92.69	92.71
	3	91.32	92.74	93.01	93.26
	32	91.87	93.04	93.15	93.02

4.2. ImageNet

For quantizing the ResNet-18/34/50 networks for ImageNet, we fine-tuned a given pre-trained network using NICE. We trained a network for a total of 120 epochs, following the gradual process described in Section 3.2 with the number of stages N set to the number of trainable layers. We used an SGD optimizer with a learning rate of 10^{-4}, momentum of 0.9, and weight decay of 4×10^{-5}.

Table 2 compares NICE with other leading approaches to low-precision quantization [18,28,29,31]. Various quantization levels of the weights and activations are presented. As a baseline, we used a pre-trained full-precision model.

Table 2. ImageNet comparison. We report top-1, top-5 accuracy on ImageNet compared with state-of-the-art prior methods. For each DNN architecture, rows are sorted in number of bits. Baseline results were taken from PyTorch model zoo. Compared methods: JOINT [28], PACT [29], LQ-Nets [18], FAQ [31].

Network	Method	Precision (w,a)	Accuracy (% Top-1)	Accuracy (% Top-5)
ResNet-18	baseline	32,32	69.76	89.08
ResNet-18	FAQ	8,8	70.02	89.32
ResNet-18	NICE (Ours)	5,5	**70.35**	**89.8**
ResNet-18	PACT	5,5	69.8	89.3
ResNet-18	NICE (Ours)	4,4	69.79	**89.21**
ResNet-18	JOINT	4,4	69.3	-
ResNet-18	PACT	4,4	69.2	89.0
ResNet-18	FAQ	4,4	**69.81**	89.10
ResNet-18	LQ-Nets	4,4	69.3	88.8
ResNet-18	JOINT	3,3	**68.2**	-
ResNet-18	NICE (Ours)	3,3	67.68	**88.2**
ResNet-18	LQ-Nets	3,3	**68.2**	87.9
ResNet-18	PACT	3,3	68.1	**88.2**
ResNet-34	baseline	32,32	73.30	91.42
ResNet-34	FAQ	8,8	73.71	**91.63**
ResNet-34	NICE (Ours)	5,5	**73.72**	91.60
ResNet-34	NICE (Ours)	4,4	**73.45**	**91.41**
ResNet-34	FAQ	4,4	73.31	91.32
ResNet-34	LQ-Nets	3,3	71.9	88.15
ResNet-34	NICE (Ours)	3,3	71.74	**90.8**
ResNet-50	baseline	32,32	76.15	92.87
ResNet-50	FAQ	8,8	76.52	93.09
ResNet-50	PACT	5,5	76.7	93.3
ResNet-50	NICE (Ours)	5,5	**76.73**	**93.31**
ResNet-50	NICE (Ours)	4,4	**76.5**	93.3
ResNet-50	LQ-Nets	4,4	75.1	92.4
ResNet-50	PACT	4,4	**76.5**	93.2
ResNet-50	FAQ	4,4	76.27	92.89
ResNet-50	NICE (Ours)	3,3	75.08	92.35
ResNet-50	PACT	3,3	**75.3**	**92.6**
ResNet-50	LQ-Nets	3,3	74.2	91.6

Our approach achieves state-of-the-art results for 4 and 5 bits quantization and comparable results for 3 bits quantization, on the different network architectures. Moreover, notice that our results for the 5,5 setup, on all the tested architectures, have slightly outperformed the FAQ 8,8 results.

4.3. Regression—Joint Denoising and Demosaicing

In addition to the classification tasks, we apply NICE on a regression task—namely, joint image denoising and demosaicing. The network we used is the one proposed in [13]. We slightly modified it by adding to it Dropout with $p = 0.05$, removing the tanh activations, and adding skip connections between the input and the output images. These skip connections improve the quantization results as, in this case, the network only needs to learn the necessary modifications to the input image. Figure 3 shows the whole network, where the modifications are marked in red. The three channels of the input image are quantized to 16 bits, while the output of each convolution, when followed by an activation, are quantized to 8 bits (marked in Figure 3). The first and last layers are also quantized.

We applied NICE on a full-precision pre-trained network for 500 epochs with Adam optimizer with learning rate of 3×10^{-5}. The data were augmented with random horizontal and vertical flipping. Since we are not aware of any other work of quantization for this task, we implemented WRPN [17] as a baseline for comparison. Table 3 reports the test set PSNR for the MSR dataset [38]. It can be clearly seen that NICE achieves significantly better results than WRPN, especially for low-weight bitwidths.

Table 3. PSNR [dB] results on joint denoising and demosaicing for different bitwidths.

Method	Bits (w = 32, a = 32)	Bits (w = 4, a = 8)	Bits (w = 4, a = 6)	Bits (w = 4, a = 5)	Bits (w = 3, a = 6)
NICE (Ours)	39.696	39.456	39.332	39.167	38.973
WRPN (our experiments)	39.696	38.086	37.496	36.258	36.002

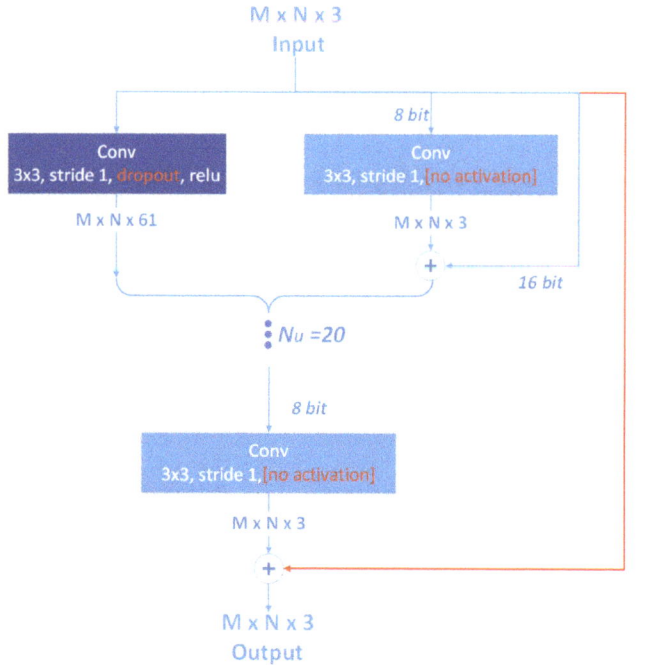

Figure 3. Model used in denoising/demosaicing experiment.

4.4. Ablation Study

In order to show the importance of each part of our NICE method, we used ResNet-18 on ImageNet. Table 4 reports the accuracy for various combinations of the NICE components. Notice that for high bitwidths, i.e., 5,5, the noise addition and gradual training contribute to the accuracy more than the clamp learning. This happens since (i) the noise distribution is indeed uniform in this case, as we show in Figure 1 and (ii) the relatively high number of activation quantization levels almost negates the effect of clamping. For low bitwidths, i.e., 3,3, we observe the opposite. The uniform noise assumption is no longer accurate. Moreover, due to the small number of bits, clamping the range of values becomes more significant.

Table 4. Ablation study of NICE scheme. Accuracy (% top-1) for ResNet-18 on ImageNet for different setups

Noise with Gradual Training	Activation Clamping Learning	Accuracy on 5,5 [w,a]	Accuracy on 3,3 [w,a]
-	-	69.72	66.51
-	✓	69.9	67.2
✓	-	70.25	66.7
✓	✓	70.3	67.68

5. Hardware Implementation

5.1. Optimizing Quantization Flow for Hardware Inference

Our quantization scheme can fit an FPGA implementation well for several reasons. Firstly, uniform quantization of both the weights and activation induces uniform steps between each quantized bin. This means that we can avoid the use of a resource costly codebook (look-up table) with the size $B_a \times B_w \times B_a$, for each layer. This also saves calculation time.

Secondly, our method enables having an integer-only arithmetic. In order to achieve that, we start, following (5), by representing each activation and network parameter in the form of $X = N \times S$, where N is the integer code and S is a pre-calculated scale. We then reformulate the scaling factors S into the form $\hat{S} = q \times 2^p$, where $q \in \mathbb{N}, p \in \mathbb{Z}$. Practically, we found that it is sufficient to constrain these values to $q \in [1, 256]$ and $p \in [-32, 0]$ without an accuracy drop. This representation allows the replacement of hardware costly floating-point operations by a combination of cheap shift operations and integer arithmetics.

5.2. Hardware Flow

In the hardware implementation, for both the regression and the classification tasks, we adopt the PipeCNN [39] implementation released by the authors. (https://github.com/doonny/PipeCNN access on 12 August 2021) In this implementation, the FPGA is programmed with an image containing data moving, convolution, and a pooling kernel. Layers are calculated sequentially. Figure 4 illustrates the flow of feature maps in the residual block from a previous layer to the next one. Sa_i, Sw_i are the activations and weights scale factors of layer i, respectively. All these factors are calculated offline and are loaded to the memory along with the rest of the parameters. Note that we use the FPGA for inference only.

We compiled the OpenCL kernel to Intel's Arria 10 FPGA and ran it with the regression architecture in Figure 3. Weights were quantized to 4 bits, activations to 8 bits, and biases, and the input image to 16 bits. The resource utilization amounts to 222 K LUTs, 650 DSP Blocks, and 35.3 Mb of on-chip RAM. With a maximum clock frequency of 240 MHz, the processing of a single image takes 250 ms. In terms of power, the FPGA requires 30 W, while an NVIDIA Titan X GPU requires 160 W. From standard hardware design practices, we can project that a dedicated ASIC manufactured using a similar process would be much more efficient by at least one order of magnitude.

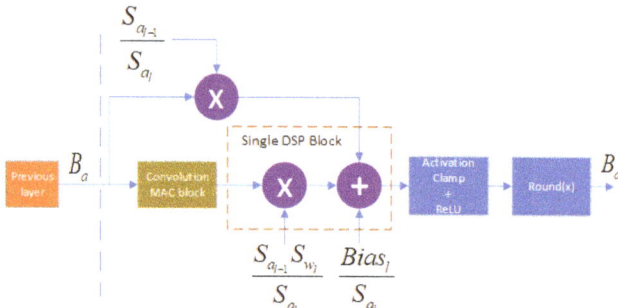

Figure 4. Residual block in hardware.

6. Conclusions

We introduced NICE —a training scheme for quantized neural networks. The scheme is based on using uniform quantized parameters, additive uniform noise injection, and learning the quantization clamping range. The scheme is amenable to efficient training by backpropagation in full precision arithmetic. One advantage of NICE is the ease of its implementation on existing networks. In particular, it does not require changes in the architecture of the network, such as increasing the number of filters as required by some previous works. Moreover, NICE can be used for various types of tasks such as classification and regression.

We report state-of-the-art results on ImageNet for a range of bitwidths and network architectures. Our solution outperforms current works on both the 4,4 and 5,5 setups, for all tested architectures, including non-uniform solutions such as [18]. It shows comparable results in the 3,3 setup.

We showed that quantization error for 4 and 5 bits distributes uniformly, which explains the larger success of our method in these bitwidths compared to the case of 3 bits. This implies that the results for less than 4 bits may be further improved by adding non-uniform noise to the parameters. However, the 4 bit quantization is of special interest since, being a power of 2, it is considered more hardware friendly, and INT4 matrix multiplications are supported by Tensor Cores in recently announced inference-oriented Nvidia's Tesla GPUs.

Author Contributions: Conceptualization, C.B., E.Z., T.R., N.L., Y.C. and E.S.; Methodology, C.B., E.Z., T.R., N.L., Y.C. and E.S.; Software, validation and formal analysis, C.B., E.Z., T.R., N.L., Y.C. and E.S.; Writing, C.B., E.Z., T.R., N.L., R.G.; Resources, A.M.B., A.M. and R.G. Project administration, A.M.B., A.M. and R.G. Funding acquisition, A.M.B., A.M. and R.G. All authors have read and agreed to the published version of the manuscript.

Funding: This research received no external funding.

Institutional Review Board Statement: Not applicable.

Informed Consent Statement: Not applicable.

Data Availability Statement: Image classification datasets (CIFAR-10 and ImageNet) are available in torchvision.datasets. MSR dataset is available in Microsoft Demosaicing Dataset (folder MSR-Demosaicing).

Conflicts of Interest: The authors declare no conflict of interest.

References

1. Hinton, G.; Deng, L.; Yu, D.; Dahl, G.E.; Mohamed, A.R.; Jaitly, N.; Senior, A.; Vanhoucke, V.; Nguyen, P.; Sainath, T.N.; et al. Deep Neural Networks for Acoustic Modeling in Speech Recognition: The Shared Views of Four Research Groups. *IEEE Signal Process. Mag.* **2012**, *29*, 82–97. [CrossRef]
2. Lai, S.; Xu, L.; Liu, K.; Zhao, J. Recurrent Convolutional Neural Networks for Text Classification. In Proceedings of the Twenty-Ninth AAAI Conference on Artificial Intelligence, AAAI'15, Austin, TX, USA, 25–30 January 2015; pp. 2267–2273.
3. Chen, L.; Papandreou, G.; Kokkinos, I.; Murphy, K.; Yuille, A.L. DeepLab: Semantic Image Segmentation with Deep Convolutional Nets, Atrous Convolution, and Fully Connected CRFs. *IEEE Trans. Pattern Anal. Mach. Intell.* **2018**, *40*, 834–848. [CrossRef] [PubMed]
4. He, K.; Zhang, X.; Ren, S.; Sun, J. Deep Residual Learning for Image Recognition. In Proceedings of the IEEE Conference on Computer Vision and Pattern Recognition (CVPR), Las Vegas, NV, USA, 27–30 June 2016.
5. He, Y.; Zhang, X.; Sun, J. Channel Pruning for Accelerating Very Deep Neural Networks. In Proceedings of the IEEE International Conference on Computer Vision (ICCV), Venice, Italy, 22–29 October 2017.
6. Molchanov, D.; Ashukha, A.; Vetrov, D. Variational Dropout Sparsifies Deep Neural Networks. In Proceedings of the International Conference on Machine Learning, Sydney, Australia, 6–11 August 2017.
7. Yu, X.; Liu, T.; Wang, X.; Tao, D. On Compressing Deep Models by Low Rank and Sparse Decomposition. In Proceedings of the IEEE Conference on Computer Vision and Pattern Recognition (CVPR), Honolulu, HI, USA, 21–26 July 2017.
8. Denton, E.; Zaremba, W.; Bruna, J.; LeCun, Y.; Fergus, R. Exploiting Linear Structure Within Convolutional Networks for Efficient Evaluation. *arXiv* **2014**, arXiv:1404.0736.
9. Iandola, F.N.; Moskewicz, M.W.; Ashraf, K.; Han, S.; Dally, W.J.; Keutzer, K. SqueezeNet: AlexNet-level accuracy with 50x fewer parameters and <1 MB model size. *arXiv* **2016**, arXiv:1602.07360.
10. Wu, B.; Wan, A.; Yue, X.; Jin, P.H.; Zhao, S.; Golmant, N.; Gholaminejad, A.; Gonzalez, J.; Keutzer, K. Shift: A Zero FLOP, Zero Parameter Alternative to Spatial Convolutions. *arXiv* **2017**, arXiv:1711.08141.
11. Gupta, S.; Agrawal, A.; Gopalakrishnan, K.; Narayanan, P. Deep learning with limited numerical precision. In Proceedings of the 32nd International Conference on Machine Learning (ICML-15), Lille, France, 6–11 July 2015; pp. 1737–1746.
12. Jacob, B.; Kligys, S.; Chen, B.; Zhu, M.; Tang, M.; Howard, A.; Adam, H.; Kalenichenko, D. Quantization and Training of Neural Networks for Efficient Integer-Arithmetic-Only Inference. In Proceedings of the IEEE Conference on Computer Vision and Pattern Recognition (CVPR), Salt Lake City, UT, USA, 18–22 June 2018.
13. Schwartz, E.; Giryes, R.; Bronstein, A.M. DeepISP: Learning End-to-End Image Processing Pipeline. *arXiv* **2018**, arXiv:1801.06724.
14. Polino, A.; Pascanu, R.; Alistarh, D. Model compression via distillation and quantization. *arXiv* **2018**, arXiv:1802.05668.
15. Rastegari, M.; Ordonez, V.; Redmon, J.; Farhadi, A. Xnor-net: Imagenet classification using binary convolutional neural networks. In *European Conference on Computer Vision*; Springer: Cham, Switzerland, 2016; pp. 525–542.
16. Hubara, I.; Courbariaux, M.; Soudry, D.; El-Yaniv, R.; Bengio, Y. Quantized Neural Networks: Training Neural Networks with Low Precision Weights and Activations. *J. Mach. Learn. Res.* **2018**, *18*, 1–30.
17. Mishra, A.; Nurvitadhi, E.; Cook, J.J.; Marr, D. WRPN: Wide Reduced-Precision Networks. *arXiv* **2017**, arXiv:1709.01134.
18. Zhang, D.; Yang, J.; Ye, D.; Hua, G. LQ-Nets: Learned Quantization for Highly Accurate and Compact Deep Neural Networks. In Proceedings of the European Conference on Computer Vision (ECCV), Munich, Germany, 8–14 September 2018.
19. Zhu, C.; Han, S.; Mao, H.; Dally, W.J. Trained ternary quantization. *arXiv* **2016**, arXiv:1612.01064.
20. Banner, R.; Hubara, I.; Hoffer, E.; Soudry, D. Scalable Methods for 8-bit Training of Neural Networks. *arXiv* **2018**, arXiv:1805.11046.
21. Zhou, A.; Yao, A.; Guo, Y.; Xu, L.; Chen, Y. Incremental Network Quantization: Towards Lossless CNNs with Low-Precision Weights. In Proceedings of the International Conference on Learning Representations, ICLR2017, Toulon, France, 24–26 April 2017.
22. Hubara, I.; Courbariaux, M.; Soudry, D.; El-Yaniv, R.; Bengio, Y. Binarized neural networks. *arXiv* **2016**, arXiv:1602.02830.
23. Zhou, S.; Wu, Y.; Ni, Z.; Zhou, X.; Wen, H.; Zou, Y. DoReFa-Net: Training low bitwidth convolutional neural networks with low bitwidth gradients. *arXiv* **2016**, arXiv:1606.06160.
24. Cai, Z.; He, X.; Sun, J.; Vasconcelos, N. Deep Learning with Low Precision by Half-wave Gaussian Quantization. In Proceedings of the IEEE Computer Society Conference on Computer Vision and Pattern Recognition, Honolulu, HI, USA, 21–26 July 2017.
25. Bengio, Y.; Léonard, N.; Courville, A.C. Estimating or Propagating Gradients Through Stochastic Neurons for Conditional Computation. *arXiv* **2013**, arXiv:1308.3432.
26. Hinton, G.; Vinyals, O.; Dean, J. Distilling the Knowledge in a Neural Network. *arXiv* **2015**, arXiv:1503.02531.
27. Mishra, A.; Marr, D. Apprentice: Using Knowledge Distillation Techniques To Improve Low-Precision Network Accuracy. *arXiv* **2017**, arXiv:1711.05852.
28. Jung, S.; Son, C.; Lee, S.; Son, J.; Kwak, Y.; Han, J.J.; Choi, C. Joint Training of Low-Precision Neural Network with Quantization Interval Parameters. *arXiv* **2018**, arXiv:1808.05779.
29. Choi, J.; Wang, Z.; Venkataramani, S.; Chuang, P.I.; Srinivasan, V.; Gopalakrishnan, K. PACT: Parameterized Clipping Activation for Quantized Neural Networks. *arXiv* **2018**, arXiv:1805.06085.
30. Dong, Y.; Ni, R.; Li, J.; Chen, Y.; Zhu, J.; Su, H. Learning Accurate Low-Bit Deep Neural Networks with Stochastic Quantization. In Proceedings of the British Machine Vision Conference (BMVC'17), London, UK, 4–7 September 2017.

31. McKinstry, J.L.; Esser, S.K.; Appuswamy, R.; Bablani, D.; Arthur, J.V.; Yildiz, I.B.; Modha, D.S. Discovering Low-Precision Networks Close to Full-Precision Networks for Efficient Embedded Inference. *arXiv* **2018**, arXiv:1809.04191.
32. Arora, S.; Ge, R.; Neyshabur, B.; Zhang, Y. Stronger generalization bounds for deep nets via a compression approach. In Proceedings of the International Conference on Machine Learning (ICML), Stockholm, Sweden, 10–15 July 2018.
33. Karbachevsky, A.; Baskin, C.; Zheltonozhskii, E.; Yermolin, Y.; Gabbay, F.; Bronstein, A.M.; Mendelson, A. Early-Stage Neural Network Hardware Performance Analysis. *Sustainability* **2021**, *13*, 717. [CrossRef]
34. Baskin, C.; Schwartz, E.; Zheltonozhskii, E.; Liss, N.; Giryes, R.; Bronstein, A.M.; Mendelson, A. UNIQ: Uniform Noise Injection for the Quantization of Neural Networks. *arXiv* **2018**, arXiv:1804.10969.
35. Sripad, A.; Snyder, D. A necessary and sufficient condition for quantization errors to be uniform and white. *IEEE Trans. Acoust. Speech, Signal Process.* **1977**, *25*, 442–448. [CrossRef]
36. Gray, R.M. Quantization noise spectra. *IEEE Trans. Inf. Theory* **1990**, *36*, 1220–1244. [CrossRef]
37. Xu, C.; Yao, J.; Lin, Z.; Ou, W.; Cao, Y.; Wang, Z.; Zha, H. Alternating Multi-bit Quantization for Recurrent Neural Networks *arXiv* **2018**, arXiv:1802.00150.
38. Khashabi, D.; Nowozin, S.; Jancsary, J.; Fitzgibbon, A.W. Joint Demosaicing and Denoising via Learned Nonparametric Random Fields. *IEEE Trans. Image Process.* **2014**, *23*, 4968–4981. [CrossRef]
39. Wang, D.; An, J.; Xu, K. PipeCNN: An OpenCL-Based FPGA Accelerator for Large-Scale Convolution Neuron Networks. *arXiv* **2016**, arXiv:1611.02450.

Article

Genetic and Swarm Algorithms for Optimizing the Control of Building HVAC Systems Using Real Data: A Comparative Study

Alberto Garces-Jimenez [1,*,†], Jose-Manuel Gomez-Pulido [2], Nuria Gallego-Salvador [2] and Alvaro-Jose Garcia-Tejedor [1,†]

1 Centro de Innovación Experimental del Conocimiento (CEIEC), Universidad Francisco de Vitoria (UFV), Carretera Pozuelo-Majadahonda, Km. 1.8, 28223 Pozuelo de Alarcón, Madrid, Spain; a.gtejedor@ceiec.es
2 Departamento de Ciencias de la Computación, Universidad de Alcala (UAH), Carretera Madrid-Barcelona, Km. 33.6, 28805 Alcalá de Henares, Madrid, Spain; jose.gomez@uah.es (J.-M.G.-P.); mnuria.gallego@edu.uah.es (N.G.-S.)
* Correspondence: alberto.garces@ufv.es or albertogarces0@gmail.com; Tel.: +34-696-018-282
† These authors contributed equally.

Abstract: Buildings consume a considerable amount of electrical energy, the Heating, Ventilation, and Air Conditioning (HVAC) system being the most demanding. Saving energy and maintaining comfort still challenge scientists as they conflict. The control of HVAC systems can be improved by modeling their behavior, which is nonlinear, complex, and dynamic and works in uncertain contexts. Scientific literature shows that Soft Computing techniques require fewer computing resources but at the expense of some controlled accuracy loss. Metaheuristics-search-based algorithms show positive results, although further research will be necessary to resolve new challenging multi-objective optimization problems. This article compares the performance of selected genetic and swarm-intelligence-based algorithms with the aim of discerning their capabilities in the field of smart buildings. MOGA, NSGA-II/III, OMOPSO, SMPSO, and Random Search, as benchmarking, are compared in hypervolume, generational distance, ε-indicator, and execution time. Real data from the Building Management System of Teatro Real de Madrid have been used to train a data model used for the multiple objective calculations. The novelty brought by the analysis of the different proposed dynamic optimization algorithms in the transient time of an HVAC system also includes the addition, to the conventional optimization objectives of comfort and energy efficiency, of the coefficient of performance, and of the rate of change in ambient temperature, aiming to extend the equipment lifecycle and minimize the overshooting effect when passing to the steady state. The optimization works impressively well in energy savings, although the results must be balanced with other real considerations, such as realistic constraints on chillers' operational capacity. The intuitive visualization of the performance of the two families of algorithms in a real multi-HVAC system increases the novelty of this proposal.

Keywords: multi-objective optimization; genetic algorithms; evolutionary computation; swarm intelligence; Heating, Ventilation and Air Conditioning (HVAC); metaheuristics search; bio-inspired algorithms; smart building; soft computing

1. Introduction

Global energy consumption has been growing at 1.4% annually over the last 10 years [1], and 94% of it is produced with combustion [2]. Greenhouse gas emissions produce adverse effects on the environment and society and cannot be completely replaced. Buildings consume on average 40% of the electrical energy in European Union cities and 32% in world cities [3], where the Heating, Ventilation, and Air Conditioning (HVAC) system requires 32.7% of the supplied electricity and up to 40.3% in public buildings [4].

Advanced control systems improve energy management by adapting fast to unforeseen events or predicting system behavior. There are some examples of this, such as the application of neural networks with genetic algorithms in building management systems, reaching savings of 27% [5]. Other studies improve the cost by 19.7%, adding an optimization module in the ambient controller [6]. Some researchers have proven that it is possible to save 30% on cold days, embedding a machine-learning-based MPC controller [7]. On the other hand, the faster the controller reaches the goals, the better the energy efficiency is obtained; for example, Adaptive LAMDA-PI (Learning Algorithm for Multivariable Data Analysis—Proportional Integral) controllers improve the Integral Absolute Error (IAE) of the response time by above 140% compared with conventional PI and Fuzzy-PI controllers [8]. Optimization works are embedded in different tasks or problems of the HVAC systems in both design and operations. They are used to adjust Proportional, Integral, and Derivative (PID) controllers to improve the logic of Model-Predicting Controllers (MPCs) or to enhance the supervision tasks in the Building Management Systems (BMSs) or Multi-Agent Controllers (MACs) [9]. There is a significant interest in embedding advanced, Artificial Intelligence (AI)-based control architectures in the BMS [10] that provide acceptable results in uncertain contexts and complex systems, while allowing the adoption of multi-objective optimization policies. There are two visible advanced control strategies: (1) predicting the system behavior with machine-learning-based simulations to obtain the optimal sequence of instructions or (2) adapting the system parameters in case of context perturbations, so that it quickly returns to the zero-error state, such as with fuzzy logic control. Artificial Intelligence (AI), together with other technologies, such as Big Data, Internet of Things (IoT), or Cloud Computing, enhances the ubiquity, accessibility, mobility, knowledge extraction, and autonomy for the new software tasks. The traditional multi-objective problem in operations is to improve the energy efficiency and maintain comfort for the users, i.e., the ideal temperature, humidity, or Indoor Environmental Quality (IEQ) that mutually conflicts. Comfort, health, or maintenance add other objectives to the optimization problem, such as the CO_2 concentration, reducing the efficiency of the optimization with fewer objectives [11].

Zadeh conceptually grouped under the umbrella of "soft computing" (SC) technologies that overperformed traditional deterministic approaches [12], at the expense of losing accuracy and generalization. Thus, SC is tolerant to imprecision and uncertain approximation and today are widely used for complex problems where moderate precision and generalization capability are acceptable, given their high-resolution speed. SC covers three main fields: (1) Machine Learning (ML), (2) metaheuristics-based optimization, and (3) Fuzzy Logic (FL) for decision-making. Metaheuristics-based optimization [11] offers good tradeoffs between consumed resources and accuracy for achieving global goals but brings challenges to face, such as algorithm convergence, stability, parameter tuning, a mathematical framework, benchmarking, generalization, and performance assessment [13]. SC also offers fitness estimation for optimization with data-based models that require fewer computer resources [14].

Digital transformation and the social trend towards standardization allow for sharing the functionality among different fields, requiring testing their approaches and convenience for specific applications. This conceptual 'liquidity' brings new challenges for optimization, such as the smart city, smart districts, and smart building, which leads to the scaling of the control and supervision capabilities to upper layers (e.g., ISA 95 and IEC 62264 L2), but constraining the lower layers. More conflicting objectives, such as the Coefficient of Performance (COP), allow for the monitoring of subtle equipment degradations, achieving considerable savings in the life cycle of the installations [15]. The system management is susceptible to becoming autonomous with the self-optimization organic function.

Thus, society, while aiming to enhance people's wealth and comfort, is forced to save energy and reduce costs. Multi-objective optimization strategies can be applied at several levels in building systems, especially HVAC, that can run at bare equipment control, at subsystems management, or at a superuser level integrating systems, buildings, blocks, or districts. In this scenario, there is a greenfield to explore, including among others autonomic building management architectures that automatically adapt their decisions to contextual changes and continuously improve with the experience. The proposed study demonstrates different multi-objective optimization techniques under this scenario that include conventional conflicting goals of comfort, observable with the ambient temperature, and energy-saving, quantifiable with the subsystems consumption, and add two new objectives: (1) the maximization of the absolute value of COP, allowing for optimal performance in saving energy and, at the same time, an enhancement of the lifecycle of the equipment, something rarely explored in operations before [16]; (2) the minimization of the rate of change in ambient temperature, which allows the system to enter into a steady-state mode since startup at nearly critical damping. The possibility for the system to automatically select the most appropriate algorithm is also proposed for the next research outcome. Although it was expected that the addition of conflicting objectives could reduce the efficiency of the optimization, the results show evidence of a wide field to be explored.

This comparative study shows the pros and cons of using different population-based multi-objective optimization algorithms for an HVAC control system. Current practices limit operation to ensure the comfort of building inhabitants dodging other objectives such as energy savings. The study will cover (1) Swarm Intelligence (SI) algorithms and (2) Genetic Algorithms (GAs) and will use real data from the HVAC system of Teatro Real de Madrid (Opera House). The individuals in the decision space are mapped in the objective space with cost functions empirically obtained with ML's Random Forest Regressors (RFRs) to assess their dominance. The RFRs have been trained with a selection of data obtained from a historic database kindly provided by the Board of Teatro Real. The selected GAs are the Multi-Objective Genetic Algorithm (MOGA) and the Non-dominated Sorting Genetic Algorithm version 2 and 3 (NSGA-II and NSGA-III), and the selected SI-based algorithms are Optimized Multi-objective Particle Swarm Optimization (OMOPSO) and Speed-constrained Multi-objective Particle Swarm Optimization (SMPSO). In the experiment, the Strength Pareto Evolutionary Algorithm Version 2 (SPEA2) was discarded, as its execution time was excessive compared to the others. Random Search (RS) results are exhibited as a reference point.

The paper is organized as follows: In Related Work, the authors bring to light significant research related to this study. Materials and Methods explain how the experiment was built and the metrics for comparing the algorithms. The Results section visualizes and discusses the outcomes. Finally, the Conclusions section compares the obtained results with other studies, outlines the novelty, and proposes possible future research lines for this work.

2. Related Work
2.1. Towards a Clear Ontology

It is common for recent literature about SC and multi-objective optimization to take for granted the approach followed in this work, due to the absence of effective classification and, therefore, the formation of an adequate body of knowledge. Although it is beyond the scope of this study, it is prudent to indicate some examples of confusing terms and try to position them.

Non-preference multi-objective optimization, i.e., those finishing with a set of non-dominant solutions, is sometimes classified as a subset of 'a posteriori' decision-making, and sometimes they are synonyms. It is often associated with multimodal optimization, although only the latter also includes local search. It is also difficult to differentiate Evolutionary Computation from GAs. While sharing a similar process, a GA includes mating and crossover to improve the search. For some articles, they are synonyms and come grouped either as evolutionary or genetic. They are sometimes considered a subset of different approaches, such as bio-inspired algorithms.

Particle Swarm Optimization (PSO) can be classified itself [17] or together with GAs [18] under Multi-Objective Evolutionary Algorithms (MOEA). MOGA is sometimes considered a separated GA [19] or the family of multi-objective GAs [20]. MOEA and MOGA may include the whole metaheuristics-based search family or only those based on the population approach.

The new algorithms based on the observation of nature can be named bio-inspired, bio-search heuristics, or metaphor-based metaheuristics, among others, exchanging their different inspirations, be they biological, chemical, or physical. GAs or SI-based algorithms can be found included in the bio-inspired family, excluding the evolutionary algorithms [19].

With regard to the optimization performance, it is possible to mislead concepts such as 'convergence' that could mean either to end the search at any point (lumps) or end it in true global optima. The diversity feature sometimes indicates the uniformity in the distribution of the solutions, how they spread, or both.

The classification by Ahma et al. [9] and Oliva et al. [21] with Zadec's original SC definition [12] supports the position of this study. At the top, algorithms are split up into stochastic techniques and intelligent agents (deterministic). Stochastic techniques then split into population-based and single individual algorithms (trajectory metaheuristics) that include Simulated Annealing (SA) and Tabu Search (TS). Population-based algorithms then split into SI and evolutionary algorithms. This study compares GAs (part of evolutionary algorithms) and SI-based particle swarms, PSO.

2.2. Research Interest

According to Wang, G. [22], at an early stage, optimization methods diversified in different fields of study: (1) linear or nonlinear programming, (2) constraints, (3) single- or multi-objective optimization, and (4) dynamic programming. The first generation introduced the iteration and gradients. The second generation brought the metaheuristics-based search for global multi-objectives that reduced computing resources and allowed for parallel computation. Soft Computing (SC) AI approaches support surrogate-based or metamodel generation, replacing computer-aided simulation software with ML models. The next-generation links and hybridizes the above approaches.

Nabaei et al. [23] provide a good reference for research interests in SI algorithms and GAs over time. GAs have been interesting since before 2000 with a peak from 2006 to 2010. PSO algorithms started to become comparable in 2006 and 2010, but there were much fewer articles published than there were regarding GAs, half of them spanning from 2011 to 2018. Another comprehensive study by Shaikh et al. [11] illustrates the research interest for the optimization in building HVAC systems in which GA articles are 24% of the total and MOGA represent 3%. PSO is present in 5%, and MOPSO in 7%. Scheduling Optimization, Hooke and Jeeves, and Linear Quadratic shares range between 3% and 6%.

Optimization can be used for designing systems or in real-time operations [24]. A GA is used for both design and operations, and so is NSGA-II, but only in a third of the articles reviewed. There are more articles about PSO in operations than in design, but a Differential Evolutionary (DE) algorithm is only used in designing, and the number of articles about the combination of these algorithms is similar to the number of articles related to NSGA-II.

2.3. Genetic and Swarm Intelligence Outcomes

Algorithms based on metaheuristics are good options for characterizing the behavior of complex, dynamic, and nonlinear systems [25].

A GA puts together a set of individuals (chromosomes) 'coded' with genes (variables), marking them with fitness functions. It then uses a selection strategy to obtain a new population ready for the next iteration. Mutation and crossover operators regulate the speed and variety of chromosome changes in the GA. While the crossover 'exploits' the search, the mutation widens the explored space. One key point is the adjustment of the parameters to the specific problem. The mutation operator can generate solutions with polynomial or uniform probability distributions. The non-uniform probability prevents the population from decaying in the early stages of the evolution by generating distant solutions with a random probability. Simulated Binary Crossover (SBX) generates offspring from two parents attending to their probability distributions.

GAs discovers the optimal set in three different ways [26]. (1) The first approach is known as Pareto-based dominance with a two-level ranking scheme, one to obtain the dominance and diversity assessment and the other, containing such metrics as the total nondominated vectors generation, the hypervolume, the generational distance or spacing, and the error rate [27], to determine the convergence to local or global minima [28]. NSGA-II and SPEA2 make use of these principles. (2) The second approach uses unary or binary indicators to check their performance, for example, with the coefficient of determination, the R2-like S-Metric Selection Evolutionary Multi-Objective Optimization Algorithm (SMS-EMOA) that maximizes the hypervolume (HV). (3) The third approach is based on decomposition that splits the overall problem into smaller problems for the search. There is not a common procedure for these algorithms. Splitting up complicated Pareto Fronts (PFs) to apply a local search and Tchebycheff's scalarization is one of these methods, as well as the Multi-Objective Evolutionary Algorithm based on decomposition (MOEA/D) and NSGA-III.

The advantages of GAs are that they (1) have simple fitness arrangement schemes; (2) do not need derivatives or gradients; (3) are relatively robust; (4) are easy to parallelize. However, although they require less information about the problem, (1) designing an objective function, (2) getting a representation, and (3) adjusting the operators can be a difficult task. In addition, they are computationally expensive compared with others. NSGA and NSGA-II perform niching, decide deterministically the tournaments, and avoid chaotic perturbations of the population composition with updated fitness sharing. However, the niching function is too complex and scales poorly as the number of objectives increases [24].

SI-based optimization is also population-based, where its individuals are bio-inspired on natural ecosystem metaphors, such as ants, bees, or particles [29]. Swarm algorithms still generate some skepticism because of the mentioned metaphoric ornaments describing their operators [30].

In the case of PSO, the particles move around in the decision space with simple mathematical equations that yield their position and velocity. Each particle's best-known local position and velocity determine its movement towards the optimum. PSO (1) is easy to adjust; (2) can be implemented and provide fast speed results; (3) is capable of finding the global optimal solutions in most cases. However, (1) strict convergence cannot be assured; (2) they are relatively weak in terms of local search abilities; (3) in multi-modal problems, they are prone to obtain local optima [23].

2.4. Research Activity

There are two schools of thought for improving the efficiency of population-based optimization. One focuses on balancing the explore and exploit strategy with many variations, such as the elitist strategy found in some GAs. The other seeks simplification, as decisions cannot be well understood, especially for large search spaces, discontinuities, noise, or algorithms with time-varying parameters, such as PSO. The revision of the research activity is guided by the following goals:

- to find which metaheuristic among some GAs and SI algorithms performs better and discover possible ways for the system to automate the decision among them;
- to study multi-objective optimization in the transient time at the startup of HVAC systems;
- to include new optimization objectives for enhancing the lifecycle of the equipment and specifically to facilitate the transition to the steady state.

Sharif et al. [31] included the assessment of the lifecycle cost (LCC) in addition to the energy consumption and environmental impact as a new optimization objective in the passive and active building design with a GA. They managed conflicting objectives such as renovating the envelope (passive structure) or the systems (active structure). Lee [32] also combined a GA with Computational Fluid Dynamics (CFD) for the building geometry (passive design) and the HVAC system (active design), having the temperature, energy consumption, and the Index of Air Quality (IAQ) as objectives.

Gagnon et al. [33] compared the computational resources spent in sequential and holistic approaches of a net-zero building design, using NSGA-II to optimize the carbon footprint, lifecycle cost, and thermal comfort. The experiment proved that the holistic approach achieved 59% of the optimal solutions in 100 h, and the sequential approach achieved 41% in 765 h.

The work of Haniff et al. [34] is representative of introducing minor changes to an algorithm that improves the addressed problem. They modified the Global PSO so that it can outperform the optimization of the energy consumption and the temperature, considering the weather forecast, an estimation of the characteristics of the building, and the Predicted Mean Vote (PMV) for Air Conditioning scheduling.

Cai et al. [35] proposed hybridizing a multi-objective evolutionary algorithm with a quantum-behaved PSO after dividing the problem into subproblems with Tchebycheff's decomposition: Decomposition-based Multi-Objective Binary Quantum-behaved Particle Swarm Optimization (MOMBQPSO/D). The algorithm minimizes the temperature mean and deviation in area-to-point heat conduction.

Zhai et al. [36] enhanced MOGA for the secondary cooling process in continuous casting by dynamically tuning the mutation and crossover operators with the probability method. They compared it with MOPSO and MOGA and showed a 10% water reduction.

Oliva et al. [21] reviewed different metaheuristics-based algorithms applied to the estimation of solar cell parameters. They outlined the advantages and disadvantages of the GA, Harmony Search (HS), Artificial Bee Colony (ABC), SA, Cat Swarm Optimization, Differential Evolutionary, PSO, Advanced Bee Swarm Optimization, Whale Optimization Algorithm (WOA), Gravitational Search Algorithm, Flower Pollination Algorithm, Shuffled Complex Evolution, and Wind-Driven Optimization. They concluded that WOA performs better than the others regarding the accuracy and convergence speed and avoided local minima trapping.

Aguilar et al. [37] proposed a new flexible architecture for Building Management Systems (BMSs), with an Autonomic Cycle of Data Analysis Tasks (ACODAT) that makes use of banks of optimization algorithms for HVAC system control and hinted at its use for supervisory and self-optimization tasks. In fact, in a later study, they developed a Fault Detect and Diagnosis (FDD) system optimized with MOPSO, also capable of long-term equipment degradation, using the COP [15].

Awan et al. [17] analyzed the design of a solar tower plant using fuzzy goals with PSO, showing significant improvements in most of the design parameters (solar multiple, tower height, and others).

Afzal et al. [38] compared the results of applying Fuzzy Logic (FL) in both a GA and PSO to optimize the Nusselt number, friction coefficient, and maximum temperature of a battery thermal management, observing that GAs provide better results, though they are less widespread than PSO.

Suthar et al. [39] compared NSGA-II, NSGA-III, and MOPSO, applying the Technique for the Order of Preference by the Similarity to Ideal Solution (TOPSIS) for tuning the parameters of a 2 Degree-of-Freedom (DoF) controller: the setpoint track, flow variation, and input fluid. The performance was measured with IAE, ISE, ITAE errors, and the execution time, and the step function reaction was analyzed.

Waseem Ahmad et al. [9] assessed several optimization methods and indicated that GAs perform global searches well but show poor convergence. Swarm-based algorithms are good for local searches but are slower than genetic algorithms for global searches. However, Ant Colony Optimization (ACO) is faster at searching compared to others and at converging compared to simple genetic algorithms. In an HVAC system's control, the most studied multi-objective optimization techniques are GAs, in 29% of the related literature, and MOPSO, in 10%. MOGA also stands out among them.

Behrooz et al. [40] confirmed that GAs provide optimization for comfort and energy savings because of their good behavior with nonlinear systems but are challenged with variable context information and perturbances [41]. They are sometimes combined with fuzzy control [8].

Previous and current research does not fully cover the topics addressed in this article, which constitutes a novelty. Most of the studies demonstrate GA and SI optimization in HVAC systems in both design and operations, but few compare them. Some research deals with dynamic adaptation, such as dynamic PID tunning, but none of them include optimization of the COP to enlarge the lifecycle and the rate of change in ambient temperature at the end of the transient state to moderate the damping to a steady state. Table 1 shows all cited works related to this section.

Table 1. This research's topics addressed in the cited articles.

Topics Addressed	References
HVAC system applied research	[9,17,32,34–37,39,40]
Improving operations	[9,15,34–40]
New optimization objectives	[15,31–33,35,37,39]
Genetic and swarm comparison	[9,21,36,38]
Algorithm improvements	[32,34,35,38,39]
Dynamic objectives	[35,39,41]

3. Materials and Methods

3.1. Teatro Real: The Opera House of Madrid

The case study is the HVAC system of the emblematic Opera House of Madrid (Spain), known as Teatro Real. The building has a floor size of 65,000 m^2 (700,000 ft^2) in 10 levels above the ground and 6 underneath. The 1430 m^2 (15,400 ft^2) stage includes the most advanced scenic technology and hosts opera and concerts for 1746 seated people in the stalls, the boxes, the balcony, and the paradise areas. The building has 11 lounges, four rehearsal rooms, and seven studios, and the scenic 'box' is surrounded by offices, warehouses, and technical premises. Figure 1 is a recent photo of the building.

Figure 1. Main façade of the Opera House of Madrid. Courtesy of Fundacion del Teatro Real.

The Opera House is open from September to July and closed in August every year. Madrid climate changes abruptly with cold winters, with an average of 0 °C (32 °F), and hot summers, with an average of 35 °C (95 °F), requiring heating and cooling. Teatro Real is also used out of the shows for rehearsals, celebrations, and product launches, making the HVAC operation a complex task.

The HVAC system of Teatro Real is an iconic example of a heterogenous HVAC system built with several refurbishments, allocating two 195 kW water–air heat pumps for both heating and cooling, and two 350 kW water–water chillers for extra cooling, managed with the same BMS. There is also a boiler and an ice accumulator that are falling into disuse.

The database provided by the Administration of Teatro Real contains historical data registered in the BMS between 1 January 2016 and 4 June 2018.

3.2. Selection of the Optimization Algorithms

The selection of the multi-objective optimization algorithms for HVAC analyzed in this study is based on the observations of Ekici et al.'s comprehensive review [42]. The initial selection of evolutionary algorithms is MOGA, NSGA-II, NSGA-III, and SPEA2.

3.2.1. The Multi-Objective Genetic Algorithm (MOGA)

Fonseca et al. [27] proposed in 1993 to compute the fitness of each individual as a weighted sum of the objective functions with random weights to obtain the probability to either select or discard it. MOGA yields interesting results, but it is not yet widely spread in real building HVAC systems.

3.2.2. The Non-Dominated Sorting Genetic Algorithm Version 2 (NSGA-II)

Deb et al. [43] proposed in 2002 to sort the individuals into categories based on non-dominance. Thus, the non-dominated individuals are in the first category. The individuals dominated by others in upper levels belong to the second and next categories. Figure 2 shows how the algorithm works.

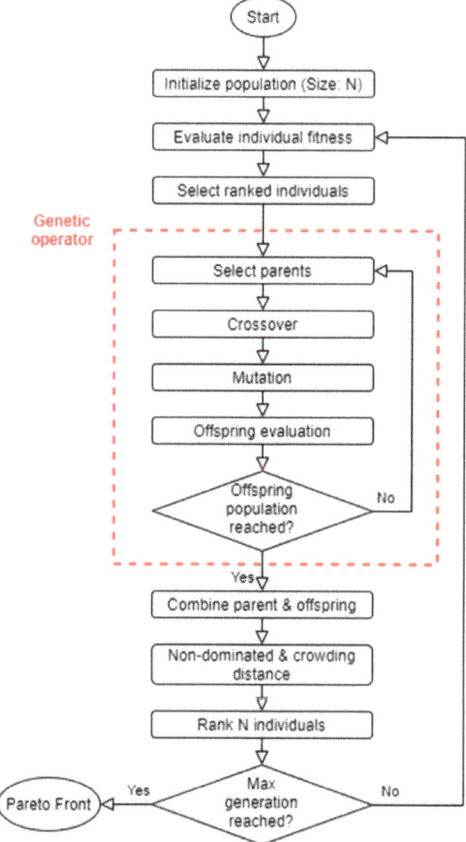

Figure 2. NSGA-II algorithm flowchart.

At the end of each iteration, the algorithm computes the distances among the individuals, known as crowding distance, for ranking.

3.2.3. The Non-Dominated Sorting Genetic Algorithm Version 3 (NSGA-III)

NSGA-III is a variant of NSGA-II that Deb et al. proposed later in 2014 [44] with an adaptive selection of the operator and a set of pre-specified (or manually) points of reference that generate a hyper-plane that improves the diversity of the population. It is conceived for improving performance when the number of objectives is larger.

3.2.4. The Strength Pareto Evolutionary Algorithm Version 2 (SPEA2)

Zitzler et al. [45] proposed in 2001 a fitness function to sort the individuals by identifying how many were dominated by a given solution and how many dominate it. The density is estimated with the k-Nearest Neighbor (k-NN) technique that prunes the elitist set (non-dominated) so that the algorithm delivers the desired number of solutions. Figure 3 shows how SPEA2 works.

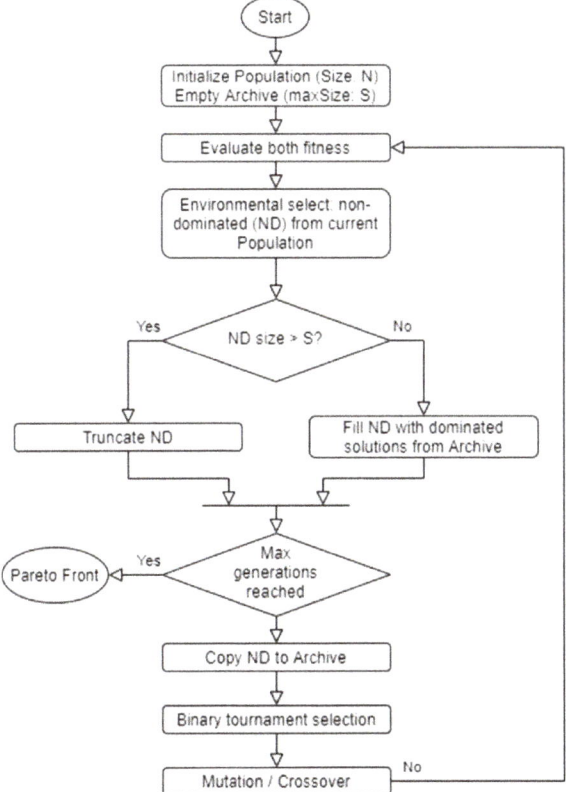

Figure 3. SPEA2 algorithm flowchart.

The other side of this analysis considers the SI-based algorithms, OMOPSO and SMPSO.

3.2.5. Optimized Multi-Objective Particle Swarm Optimization (OMOPSO)

OMOPSO is one of the MOPSO versions proposed by Reyes-Sierra et al. [46] in 2006 that uses Pareto's non-dominance to identify the leaders and the crowding distance to regulate the maximum number of them. Each iteration proclaims a leader, modifying the speed of the rest to head for it. The leaders of the current generation are set apart from the global leaders. The algorithm splits the population into groups with different mutation operators. Figure 4 shows how these algorithms work.

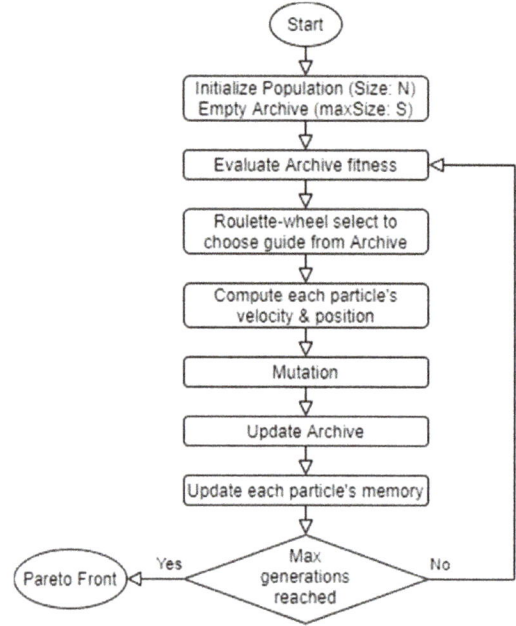

Figure 4. OMOPSO algorithm flowchart.

3.2.6. Speed-Constrained Multi-Objective Particle Swarm Optimization (SMPSO)

SMPSO, proposed by Nebro et al. in 2009, is another version of MOPSO [47] that includes a speed constraint mechanism for each individual, being good when individuals are excessively accelerated. The optimization is no-preference, bringing an important Degree of Freedom (DoF) for making tactic and strategic decisions. The result consists of "nondominated" solutions located in the hyper-plane of optimum values or the Pareto Front (PF). Thus, for instance, the operation can take optimal values increasing the ventilation to reduce the risk of transmission of disease, e.g., COVID-19, or aiming toward maximum comfort, allowing the manager or the system to pick up the best value of the PF to accomplish the goal.

In any case, diversity is preserved by either the density estimation or truncation. Fitness with the k-NN of the ith individual, F(i), is computed as

$$F(i) = R(i) + D(i)$$

When F(i) < 1, the individual is non-dominated. R(i) is the raw fitness, obtained from

$$R(i) = \sum_{j \in (\text{Population} + \text{Archive}), j \succ i} S(j)$$

where S(j) is the strength value, representing the number of solutions in both Population and Archive, when i dominates:

$$S(i) = \{j \ / \ j \in (\text{Population} + \text{Archive}) \wedge i \succ j\}$$

D(i) is the density that allows the discrimination between individuals with identical fitness values, and it is obtained from

$$D(i) = \frac{1}{\sigma_i^k + 2}$$

$$k = \sqrt{|Population| + |Archive|}$$

where σ_i^k is the distance in the objective space to the kth nearest neighbor in both Population and Archive. In the case of truncation,

$$i \text{ is removed, if } i \prec j, \forall j$$

The performances of these metaheuristics are compared with Random Search acting as a baseline, for not having any specific speeding up mechanism for exploring and exploiting the decision space.

3.3. Selection of Metrics

The "no free lunch" theorem is applicable for assessing the optimization [9], as the improvements on one feature reduce the effectiveness on another. The algorithm performance is a balance between the achievement of solutions with values close to the PF and the runtime resources required. This proves the algorithms empirically. Riquelme et al. [48] identified up to 54 metrics to prove (1) the cardinality or the number of solutions in the approximation set; (2) the accuracy, convergence, or distance to the PF; and (3) the diversity, which measures the distribution of the fitness values and how they spread. Another classification of metrics is given by the generic definition of Zitzler et al. [49], being unary if only one approximation set is received and binary if two are received. This analysis takes the top three metrics in the ranking and the one that records the runtime [48]:

- Hypervolume (HV), S metric, or Lebesgue measure: a unary metric that obtains the total space covered by the found solutions or approximation set using a reference point [11]. It considers accuracy, cardinality, and diversity.
- Generational Distance (GD): the average Euclidean distance between the approximation set with the nearest member of the ideal PF [50]. It only considers the accuracy.
- ε-Indicator (EI): a binary indicator that gives a factor by which an approximation set is worse than another considering all objectives.
- Execution Time (ET) or runtime: the time consumed by the optimization algorithm to fully complete the task.

3.4. Auxiliary Tools

The simulation was coded in Python, using basic NumPy, Pandas, and Datetime libraries for managing vectors, matrices, and time series. The simulation module, RFR, is implemented with Scikit recommended for machine learning [51]. The optimization is built with the JMetalPy framework [52], well proved for solving multi-objective optimization problems with metaheuristics [41]. The visualization of the obtained results is built with Matplotlib.

4. Problem Formulation

The HVAC system of Teatro Real is set to follow the mechanical and comfort setpoints required for a near event. The time spent to climatize and several HVAC parameters are those that the chiller's manufacturer initially recommended just after installation. The BMS sends commands to the HVAC system to start/stop the chillers in a certain sequence to ensure that, at the time of the event, the comfort parameters will be appropriate.

The proposed control loop for the multi-HVAC system performance optimization is depicted in Figure 5.

The Control Module, with the same functions as today, initiates the process by requesting the Optimization Module for instructions to improve its operation. The Optimization Module, which performs a metaheuristic search in the space of possible solutions, returns the best candidate obtained with the algorithm used in each model run (either with GA or SI). The fitness functions of the candidates are evaluated by the Simulation Module that receives every individual of the population and performs the simulation of the HVAC behavior (non-linear system) [53], as defined by the candidate control parameters. The simulation is carried out with an ML algorithm, specifically a Random Forest Regressor (RFR), previously trained with historical data from the database of Teatro Real, by minimizing the Mean Squared Error (MSE) and maximizing the coefficient of determination (R^2). The RFR also requests contextual information to compute the simulation, which is provided by external sources. Finally, the Control Module translates the optimal recommendations into instructions for the actuators.

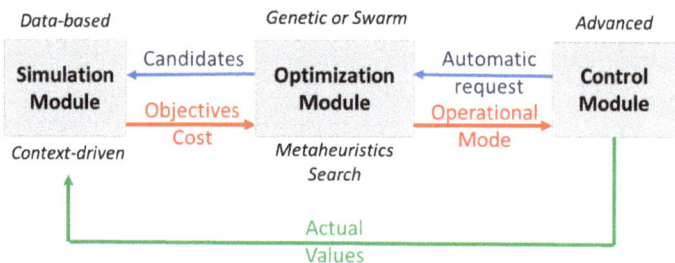

Figure 5. Advanced control optimized with a predicting context-driven model.

Each experiment carried out in this study executes one control cycle (petition) and addresses the optimization, without delving into the control stage. Inspired by the ACO-DAT management architecture for HVAC systems [37], an autonomous cycle updates the model offline, maintaining its accuracy in real operational conditions, as shown by the green arrow in Figure 5.

The primary objectives are to maximize comfort and minimize the consumed energy.

$$\text{Comfort} = |T_0 - T_r|$$

$$E = \sum_1^n E_i$$

where T_0 is the setpoint temperature, and T_r is the indoor room temperature, both in °C. The maximum comfort for the optimization is therefore 0. The consumed electrical energy, E, is the sum of the consumed energy in kW.h in each chiller group, the multi-HVAC concept [37]. N is the number of chiller groups. The energy of one chiller group, E_i, is

$$E_i = E_{\text{chiller, i}} + E_{\text{CT,i}} + E_{\text{cwp,i}} + E_{\text{wpp,i}}$$

where $E_{\text{chiller,i}}$ is the energy consumed in the chiller machine, $E_{\text{CT,i}}$ is that in the cooling tower, $E_{\text{cwp,i}}$ is that in the cooling water pump, and $E_{\text{wpp,I}}$ is that in the chilled water primary pump.

This study includes two new objectives in the optimization as a novelty. The first one is the Coefficient of Performance, COP. The higher the COP is, the better the performance of the equipment, resulting in better energy efficiency and lower maintenance costs:

$$\text{COP} = \frac{W}{P}$$

The COP is the engineering ratio of the supplied thermal power, W, to the consumed electric power, P. The optimization of the COP brings two important advantages for the

HVAC system. HVAC equipment is designed to work at maximum performance, and in this regime, the system obtains its best energy efficiency. With the appropriate autonomous cycle of data tasks [8], the supervisory system detects the degradation of the system, providing predictive maintenance [15].

The second novel objective is the rate of change in the ambient temperature, $\dot{T}_r$, that is the rate at which the temperature varies when it reaches the setpoint. This objective leads the system to rapidly reach the steady state with convenient damp that minimizes the overshooting:

$$\dot{T}_r = \frac{dT_r}{dt}$$

This parameter is important at sudden startups when there is a transient time before the steady state [16]. The lower the slope of the derivative is, the less impact on overshooting and steady noise there is on the next control phase.

The optimization requires Comfort, E, and $\dot{T}_r$ to be minimized and COP to be maximized. The decision space is formed with the chillers' capacities, C_i [%], the setpoint, T_0 [°C], and the schedule or the date and time at which the system is expected to reach the setpoint, t_{start}. The indoor ambient temperature when the system starts, $T_r(t=0)$ [°C], the number of occupants, N, and the outdoor ambient temperature, OAT [°C], are the contextual information that determines the system. This study uses the capacity of the chillers as actuators on the subsystems, and this is justified with this simplified model:

$$P_i = P_{max} \, C_i$$

where P_i is the electrical power actually supplied from the ith chiller, and P_{max} is the maximum power of the chiller. The chiller's thermal power is generated according to the machine performance that is added to the other chillers, W_{HVAC}.

$$W_i = COP_i \, P_i$$

$$W_{HVAC} = \sum_1^n W_i$$

Thermal power conditions indoors compensate for the outdoor weather conditions and the corporal temperature of the occupants:

$$W = W_{HVAC} + W_{SUN} + W_{OCC}$$

The thermal energy, Q, is then obtained from the power, and T_r is obtained from ΔT_r, the indoor temperature variance.

$$Q = \int_0^{t_{end}} W \, dt$$

$$Q = C_e \, m \, \Delta T_r$$

Figure 6 shows the model with the inputs required, grouped in controllable and control variables, and the outputs, differentiating the normal optimization objectives of the thermal inertia for the next control plan [37].

An individual in the population consists of a sequence of four operational modes of the chillers based on their capacities, C_i [%], at certain times, t_i, before the event starts at t_{end} [37]. Each operational mode is a 5-tuple consisting of the proposed capacities for the four chillers ranging from 0% to 100% and the time that they start. Thus, a single individual contains four of these 5-tuples. The RFR performs a simulation for each 5-tuple, chaining them according to their start-up time. The last 5-tuple indicates the operational values applied to the chillers until the system reaches the steady state, t_{end}.

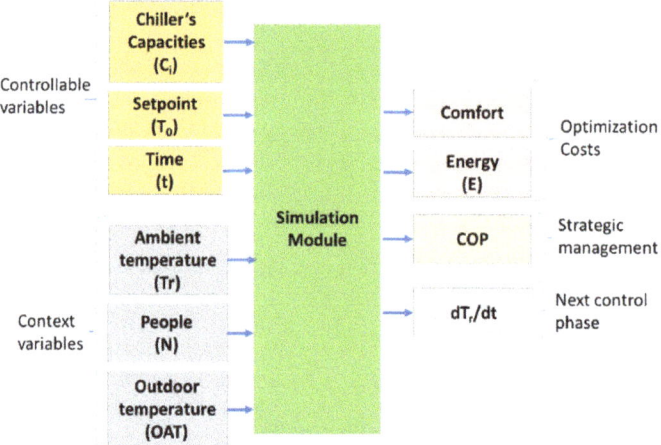

Figure 6. Simulation module's functionality to compute the cost functions for the optimization.

The multi-objective optimization problem would be formally defined as follows:
1. Find the vector $\bar{x}$ in the decision space:

$$\bar{x} = \begin{bmatrix} t_1 & t_2 & t_3 & t_4 \\ C_1^1 & C_1^2 & C_1^3 & C_1^4 \\ C_2^1 & C_2^2 & C_2^3 & C_2^4 \\ C_3^1 & C_3^2 & C_3^3 & C_3^4 \\ C_4^1 & C_4^2 & C_4^3 & C_4^4 \end{bmatrix}$$

t_i, where i = 1, 2, 3, and 4, represents the starting dates and times to configure the capacities of every subsystem, while C_j^i, where j = 1, 2, 3, and 4, represents the capacities of the chiller j during the period that starts at i and ends at i + 1. The last period is between t_4 and t_{end}.

2. $\bar{x}$ will satisfy these inequality constraints at the following point:

$$\left| C_j^i \right| \leq 100$$

$$t_{i+1} \geq t_i$$

3. $\bar{x}$ will optimize the vector function $\bar{f}(\bar{x})$ in the objective space:

$$\bar{f} = \begin{bmatrix} \text{Comfort}(\bar{x}) \\ E(\bar{x}) \\ \text{COP}(\bar{x}) \\ \frac{dT_r}{dt}(\bar{x}) \end{bmatrix}$$

Comfort and COP must be maximized, while consumed energy, E, and the rate of change of ambient temperature, $\dot{T}_r$ must be minimized.

5. Results

5.1. Dataset

The BMS is connected to 1824 digital and analog sensors, prompting the ambient and return temperatures, frozen water flow rates, valve states, chiller's performance, secondary circuit values, air flow rate, fan speeds, pumps rotational speeds, controller status, etc., and allows the operator to send instructions to the actuators from the centralized platform. However, the historical data only keeps 169 variables: outdoor temperature, room temperatures, electrical supplied power, thermal energy generated by each of the four HVAC subsystems, and their COPs grouped in several tables with different sampling rates (10 min, 15 min, 1 h, daily). Usable records are from January 2016 to June 2018. The data have been cleaned to improve the accuracy by removing nonessential fields, records with outliers, nulls, and/or zeros, getting 9898 (80%) registers for training and 2475 (20%) for validation.

The Department of Engineering prepares the work order, based on the HVAC operational mode (HOM) for the field operators based on the events schedule and the weather forecasts, and consisting of pre-programmed routines. This is, however, inefficient because the complexity of the system operation reduces all possible variations to a small set of HOMs, based on the primitive recommendations of the installers. The occupancy of the building can reach up to 1700 during performances, while the number of people on labor days is around 600.

5.2. Data Model

The multi-objective estimation is computed with the RFR with good accuracy and speed balance. The model simulates the outputs in intervals of 15 min, which is a tradeoff between the system inertia and the discretization of the system dynamics. The model receives the time required for starting up the HVAC system, t_0, the time of the venue or the moment in which the room temperature, t_{end}, must reach the setpoint, T_0, the room temperature at the beginning, $T_r(t = t_0)$, the number of people, N, and the outdoor temperature forecast, which is a vector of temperatures from t_0 to t_{end} every 15 min. Table 2 represents an example where the temperature at 17 °C must reach the setpoint, 23.5 °C, in an hour.

Table 2. Control request & context data.

Feature	Value
HVAC startup time, t_0	18:30
Event start time, t_{end}	19:30
Setpoint, T_0	23.5 °C
Ambient Temperature, $T_r(t_0)$	17 °C
Outdoors Temperature vector, $\overline{OAT}$	[14 14.5 14.7 14.6]

The model also requires the outdoor temperature from the weather forecast. The optimization algorithm then releases the proposed individual for fitness.

In addition, the simulation receives the set of HOMs searched by the algorithms that will work in each interval. The algorithm is a sequence of HOMs proposed for the slots in the interval from t_0 to t_{end}, consisting of the power capacities of each chiller. Following the example, Table 3 shows one of these candidate solutions.

Table 3. Individual consisting of a sequence of four operational modes.

Time	C_1	C_2	C_3	C_4
18:30	−30	−21	0	−3
18:45	−27	−20	−10	−4
19:00	−32	−15	0	−10
19:15	−28	−20	0	−10

A negative capacity indicates that the chiller is cooling, while a positive one indicates that it is heating. Real implementations will impose restrictions that are not considered here, such as smoothing the capacity transitions from one slot to another or preparing the chiller for cooling or heating modes. Table 4 depicts the result of the optimization for this example.

Table 4. Model prediction applying optimal operational modes.

Feature	Value
Comfort	0.12 °C
Consumed Energy, E	1300 kW.h
Coef. of Performance, COP	4.5 kW/kW
Ambient Temp. Change Rate, $\frac{dT_r}{dt}$	0.15 °C/min
Time before performance	15 min

5.3. Algorithm Analysis

The analysis compares the performance and execution time (ET) of the algorithms. They start with the same expected number of solutions, i.e., the population size for the GA and the swarm size for the SI algorithm. The experiment involved population/swarm sizes ranging from 100 to 350 in steps of 50. The mutation probabilities were the same, and the SBX crossover probabilities and distribution index were the same for the GAs. The mutation scheme followed a polynomial probability distribution, except for OMOPSO, which combined uniform and nonuniform distributions with the same perturbation index, 0.5.

The algorithms stopped after 5000 iterations, and GAs stopped earlier if they were triggered with the dominance threshold. In order to obtain stable results, the algorithms were proved 10 times to determine the average of the obtained values. Figures 7–9 represent the objective space for the variables Comfort, Consumed Energy, and COP in 2D diagrams.

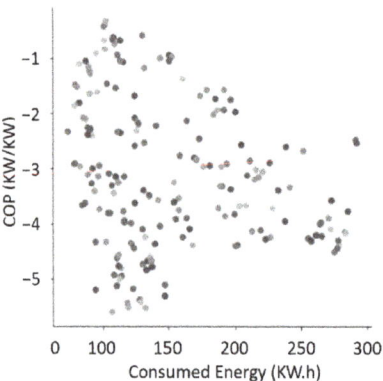

Figure 7. 2D objective space of COP [kW/kW] vs. the chiller's Consumed Energy [kW.h].

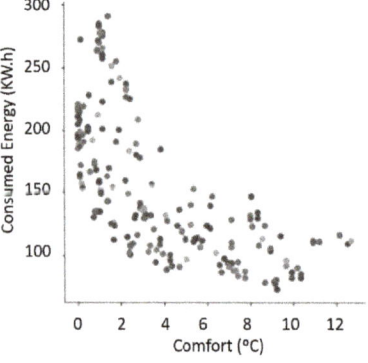

Figure 8. 2D objective space of Consumed Energy [kW.h] vs. Comfort [°C].

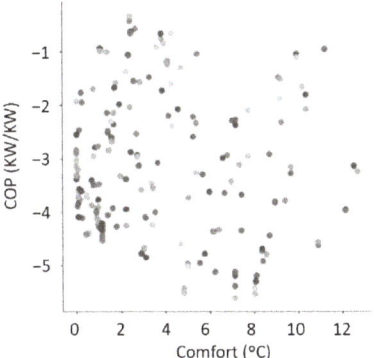

Figure 9. 2D objective space of COP [kW/kW] vs. Comfort [°C].

Metrics used in the comparison were computed with the JMetal framework, and the ETs were recorded. The SPEA2 algorithm was dropped from the analysis, as it takes 22-fold more runtime than MOGA [45]. Figure 10 shows the obtained ET values.

Figure 10. Average execution time for each algorithm.

The GAs ran faster than the SI-based algorithms. MOGA improved the Random Search by 13%, and OMOPSO improved it by 9%. It was observed that NSGA-III takes more time than NSGA-II to execute. This is because of the extra computation required for the adaptive operator and the generation of hyperplanes. On the other hand, the speed constraint mechanism seemed to increase the ET of the SMPSO, compared with OMOPSO. All outperformed RS.

GD showed how close the fitness of the set of solutions was from the ideal PF, and this is depicted in Figure 11.

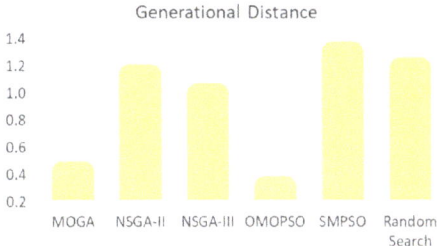

Figure 11. Average GD to the Pareto's ideal front by each algorithm.

An approximate PF was constructed running the NSGA-II 20,000 times, simulating a limit behavior. The accuracy of OMOPSO and MOGA with 75% and 65% improvements compared to Random Search was observed. It was unavoidable that the quasi-ideal PF construction was insufficient for the rest of the algorithms. HV and EI are shown in Figures 12 and 13.

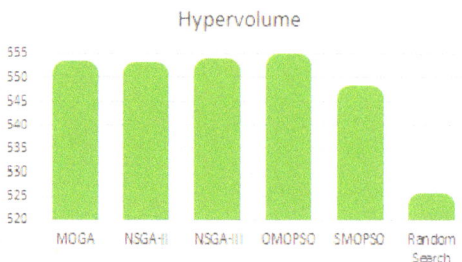

Figure 12. Average hypervolume of each algorithm.

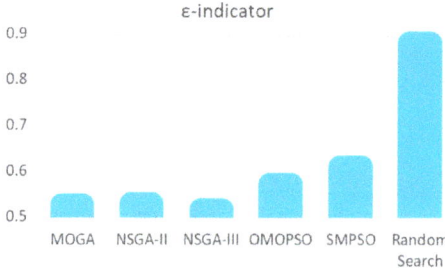

Figure 13. Average ε-Indicator of each algorithm.

In both metrics, it is possible to identify the significant improvements of all the algorithms compared with Random Search. The ε-Indicator show NSGA-III and MOGA as the best algorithms, outperforming Random Search by 42% and 40%, while the SI algorithms were worse (31–35%). The HV does not show significant differences among algorithms but shows an improvement of 5% on average.

5.4. Visualization

Regarding the question of whether an algorithm outperforms another with a combination of any quality measures, such as those seen above, Zitzler came to the conclusion that there was no such combination, but it could be seen as the equivalence to the concept of dominating [54]. Thus, Figures 14–16 show 2D maps formed with the metrics of this study, those closer to the bottom left corner being the most appropriate. The best algorithms are found in the lower-left corner in all cases. The charts also show the distance among them, presenting an intuitive method with which to make decisions as to which performs better. Figure 14 shows the behavior of the algorithms when setting the priority in ET and GD.

This case yields the selection of either MOGA or OMOPSO algorithms as the best for optimization accuracy. Both metrics penalize SMPSO, which obtains a GD even worse than RS. Figure 15 prioritizes the HV (the inverse in this case for obtaining a homogeneous visualization) with the ET.

In this case, SMPSO still performs worse than the others in terms of accuracy, but much better than RS, likely due to diversity. All the rest behave similarly, the GA family standing out. Figure 16 prioritizes the ε-Indicator and ET.

ε-Indicator also measures the cardinality and maintains SMPSO at the back, followed by OMOPSO, while GAs shows better behavior.

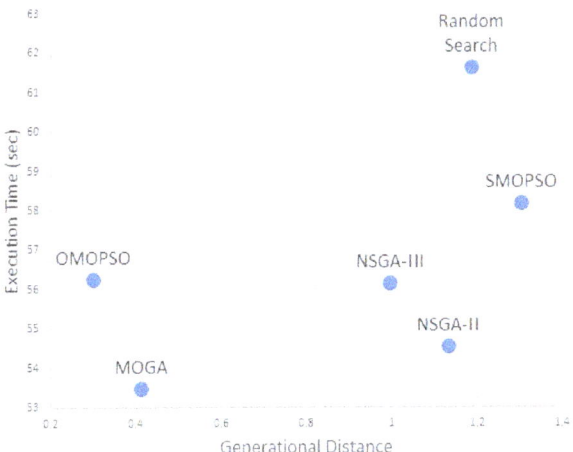

Figure 14. Plot chart mapping the studied algorithms according to the ET and the GD.

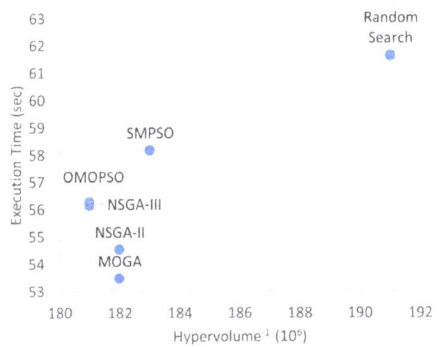

Figure 15. Plot chart mapping the studied algorithms according to the ET and the HV^{-1}.

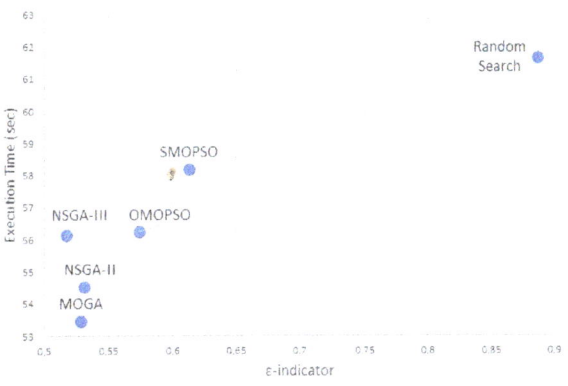

Figure 16. Plot chart mapping the studied algorithms according to the ET and the ε-Indicator.

5.5. Energy Efficiency Improvements

To complete the experiment, four differentiated events available in the historical data of the building were randomly selected to compare the performance of the HVAC equipment in terms of energy efficiency, with the results that would have been obtained by applying the proposed optimization. This indicates what can be expected from this approach. The events are defined in Table 5.

Table 5. Model prediction applying optimal HOMs.

Dataset Cases	Event 1	Event 2	Event 3	Event 4
Date	27 September 2017	30 December 2017	29 July 2017	16 February 2018
Outdoors Temperature	35.87 °C	15.12 °C	35.51 °C	14.99 °C
Simulation Start Time	18:00	16:30	18:00	16:30
Performance Schedule	21:00	19:30	21:00	19:30
Ambient Temp. (start)	25.85 °C	23.03 °C	26.03 °C	22.82 °C
Ambient Temp. (end)	25.54 °C	23.41 °C	25.66 °C	23.26 °C

To illustrate the example, a second decision-making process with a weighted sum was set to select one of the solutions with values in the PF. Weights slightly favored Consumed Energy savings over the others. Table 6 shows the results.

Table 6. Results obtained with MOGA optimization and comparison with real data.

Objective Space	Historical Data	MOGA	Improvement
Event 1			
Consumed Energy	1138.69 KW.h	154.63 KW.h	86%
Comfort	2.04 °C	0.13 °C	<0.5 °C
COP	3.29	3.25	>3.00
Event 2			
Consumed Energy	365.25 KW.h	132.51 KW.h	64%
Comfort	−0.08 °C	0.27 °C	<0.5 °C
COP	4.31	3.42	>3.00
Event 3			
Consumed Energy	931.02 KW.h	126.85 KW.h	86%
Comfort	2.16 °C	0.20 °C	<0.5 °C
COP	3.59	4.69	>3.00
Event 4			
Consumed Energy	338.04 KW.h	132.44 KW.h	61%
Comfort	−0.23 °C	0.40 °C	<0.5 °C
COP	3.64	3.69	>3.00

The right column shows the theoretical energy savings in each case with the optimized HOMs compared with what was actually recorded in the dataset. This column also stresses the achievements in comfort with expected deviations of less than 0.5 °C and HVAC subsystems working with COPs above 3.00, which is considered a good value. These impressive results of 60–80% in energy savings, preserving the comfort and the system performance, must be adjusted with further research considering real restrictions, but they hint toward a promising line of research.

5.6. Comparison with Other Works

Several authors have proposed comparisons between NSGA-II and MOPSO, which may contribute to the comprehension of the results. Keshavarz et al. [55] compared NSGA-II and MOPSO for the stochastic optimization of an inventory control system, showing that NSGA-II has better performance in spacing and in the number of Pareto optimal solutions, while MOPSO better spreads the fitness of the solution set and consumes fewer computational resources. Niyomubyeyi et al. [56] studied optimization in evacuation planning, obtaining better convergence and spread with MOPSO, but the algorithm execution took five times longer than NSGA-II. Saldanha et al. [18] obtained similar results in convergence and spread for MOPSO and NSGA-II, although MOPSO yielded better results in spacing. Elgammal et al. [57] studied the integration of hybrid wind photovoltaic and fuel cells, obtaining similar system operating costs with both, but in this case, the MOPSO execution time was shorter than NSGA-II.

6. Conclusions

This study shows the performance of several genetic and SI-based algorithms when optimizing the control of a building HVAC system. The study works with the real historical data of a complex and singular building by adapting the control logic to the available sensed measures and individual chiller actuators. The results yield that simple MOGA and NSGA-II/III run faster than MOPSOs, confirming the pure Random Search algorithm as the slowest. The best convergence is obtained with OMOPSO according to GD and HV.

The achievement on energy consumption is impressive, as shown with several events randomly selected from the data, reaching savings from 60% to 80%. These results will be proved for generalization purposes with further research that will include the new model's restrictions.

This study is the first to take two new objectives into the optimization problem: the HVAC subsystem's performance, COP, and the rate of change in ambient temperature at the end of the system startup stage. The first objective brings the possibility of advanced supervisory policies that improve the maintenance of the equipment and extend its lifecycle. The minimization of the second allows for a smooth transition to the permanent stage of the HVAC operation, reducing the overshoots or the underdamping effect of the room temperature values. In the following works, the dominance variation produced when adding new conflicting objectives and how this affects control system decision-making will be analyzed.

The proposed simple visualization of the algorithms not only allows for an intuitive understanding of which algorithm performs better but also opens the possibility of the automatic real-time instantiation of the most convenient algorithm from a bank of optimizers according to given contextual information. This is important because there are no rigid rules, but rather, existing or new strategies, such as running out of time, operations when the building is closed, etc.

The article also claims for consensus in optimization with a body of knowledge that integrates the contribution of the different disciplines that theorize or are applicable to the case.

This study requires generalization to demonstrate its scope with other different buildings, HVAC systems, and overall different variables extracted from the control logic. It is also of interest to work on parameter tuning to characterize the inherent "no free lunch" theorem.

The use of real data has made the study more reliable. The singularity of the building and the heterogeneous equipment that forms the HVAC system represents a demanding test for this research.

This research will contribute to the development of the smart city with autonomic management systems capable of learning from experience and improving with the context using AI to overcome the complexity of the managed systems and changing the user's requirements.

Author Contributions: Conceptualization, N.G.-S., J.-M.G.-P. and A.G.-J.; methodology, A.-J.G.-T.; software, A.G.-J.; validation, A.G.-J. and N.G.-S.; formal analysis, A.G.-J.; investigation, A.G.-J. and A.-J.G.-T.; resources, N.G.-S, A.-J.G.-T. and J.-M.G.-P.; data curation, A.G.-J.; writing—original draft preparation, A.G.-J.; writing—review and editing, A.G.-J., A.-J.G.-T. and J.-M.G.-P.; visualization, A.G.-J.; supervision, A.G.-J., A.-J.G.-T. and J.-M.G.-P.; project administration, A.-J.G.-T. and A.G.-J.; funding acquisition, A.-J.G.-T. All authors have read and agreed to the published version of the manuscript.

Funding: This work was partially supported by the Vice Rectorate for Research of the Universidad Francisco de Vitoria with grant code MOGA-TR, Reference: UFV2020-34.

Acknowledgments: This work was possible thanks to the contribution of the Management Committee of Teatro Real de Madrid by providing the database of the HVAC BMS System. The authors wish to thank Raul Jiménez-Juarez for his help in preparing the code as part of his end-of-degree project in the Universidad Francisco de Vitoria.

Conflicts of Interest: The authors declare no conflict of interest. The funders had no role in the design of the study, in the collection, analyses, or interpretation of data, in the writing of the manuscript, or in the decision to publish the results.

References

1. Plecher, H. Global Gross Domestic Product (GDP) at Current Prices from 2014 to 2024 (in Billion U.S. Dollars). Statista. 2019. Available online: https://www.statista.com/statistics/268750/global-gross-domestic-product-gdp/ (accessed on 12 May 2020).
2. Ritchie, H.; Roser, M. Energy. Our World in Data. 2019. Available online: https://ourworldindata.org/energy (accessed on 12 May 2020).
3. Mauro, G.M.; Hamdy, M.; Vanoli, G.P.; Bianco, N.; Hensen, J.L. A new methodology for investigating the cost-optimality of energy retrofitting a building category. *Energy Build.* **2015**, *107*, 456–478. [CrossRef]
4. Minoli, D.; Sohraby, K.; Occhiogrosso, B. IoT considerations, requirements, and architectures for smart buildings—Energy optimization and next-generation building management systems. *IEEE Internet Things J.* **2017**, *4*, 269–283. [CrossRef]
5. Reynolds, J.; Rezgui, Y.; Kwan, A.; Piriou, S. A zone-level, building energy optimisation combining an artificial neural network, a genetic algorithm, and model predictive control. *Energy* **2018**, *151*, 729–739. [CrossRef]
6. Wang, F.; Yoshida, H.; Ono, E. Methodology for optimizing the operation of heating/cooling plants with multi-heat-source equipments. *Energy Build.* **2009**, *41*, 416–425. [CrossRef]
7. Aste, N.; Manfren, M.; Marenzi, G. Building Automation and Control Systems and performance optimization: A framework for analysis. *Renew. Sustain. Energy Rev.* **2017**, *75*, 313–330. [CrossRef]
8. Escobar, L.M.; Aguilar, J.; Garcés-Jiménez, A.; De Mesa, J.A.G.; Gomez-Pulido, J.M. Advanced Fuzzy-Logic-Based Context-Driven Control for HVAC Management Systems in Buildings. *IEEE Access* **2020**, *8*, 16111–16126. [CrossRef]
9. Ahmad, M.W.; Mourshed, M.; Yuce, B.; Rezgui, Y. Computational Intelligence Techniques for HVAC systems: A review. In *Building Simulation*; Tsinghua University Press: Beijing, China, 2016; Volume 9, pp. 359–398.
10. Mallawaarachchi, V. Introduction to genetic algorithms-including example code. *Towards Data Science.* 2017. Available online: https://towardsdatascience.com/introduction-to-genetic-algorithms-including-example-code-e396e98d8bf3 (accessed on 12 May 2020).
11. Shaikh, P.H.; Nor, N.B.M.; Nallagownden, P.; Elamvazuthi, I.; Ibrahim, T. A review on optimized control systems for building energy and comfort management of smart sustainable buildings. *Renew. Sustain. Energy Rev.* **2014**, *34*, 409–429. [CrossRef]
12. Zadeh, L.A. Fuzzy Logic, Neural Networks and Soft Computing. In *Safety Evaluation Based on Identification Approaches Related to Time-Variant and Nonlinear Structures*; Springer: Berlin/Heidelberg, Germany, 1993; pp. 320–321.
13. Yang, X.S. Nature-inspired optimization algorithms: Challenges and open problems. *J. Comput. Sci.* **2020**, *46*, 101104. [CrossRef]
14. Afroz, Z.; Shafiullah, G.M.; Urmee, T.; Higgins, G. Modeling techniques used in building HVAC control systems: A review. *Renew. Sustain. Energy Rev.* **2018**, *83*, 64–84. [CrossRef]
15. Aguilar, J.; Ardila, D.; Avendaño, A.; Macias, F.; White, C.; Gomez-Pulido, J.; Gutiérrez de Mesa, J.A.; Garces-Jimenez, A. An Autonomic Cycle of Data Analysis Tasks for the Supervision of HVAC Systems of Smart Building. *Energies* **2020**, *13*, 3103. [CrossRef]
16. Aguilar, J.; Garces-Jimenez, A.; Gómez-Pulido, J.M.; R-Moreno, M.D.; Gutiérrez de Mesa, J.A.; Gallego-Salvador, N. Autonomic Management of a Building's multi-HVAC System Start-Up. *IEEE Access* **2021**, *9*, 70502–70515. [CrossRef]
17. Awan, A.B.; Chandra Mouli, K.V.V.; Zubair, M. Performance enhancement of solar tower power plant: A multi-objective optimization approach. *Energy Convers. Manag.* **2020**, *225*, 113378. [CrossRef]
18. Saldanha, W.H.; Soares, G.L.; Machado-Coelho, T.M.; dos Santos, E.D.; Ekel, P.I. Choosing the best evolutionary algorithm to optimize the multiobjective shell-and-tube heat exchanger design problem using PROMETHEE. *Appl. Therm. Eng.* **2017**, *127*, 1049–1061. [CrossRef]
19. Kuan, Y.N.; Ong, H.S. Optimization of heating, Ventilating and Air Conditioning (HVAC) systems: A review. *Univ. Tenaga Nasional.* **2018**, *13*, 9049–9906. Available online: http://dspace.uniten.edu.my/jspui/handle/123456789/11701 (accessed on 12 May 2020).
20. Nasruddin, S.; Satrio, P.; Mahlia, T.M.I.; Giannetti, N.; Saito, K. Optimization of HVAC system energy consumption in a building using artificial neural network and multi-objective genetic algorithm. *Sustain. Energy Technol. Assess.* **2019**, *35*, 48–57. [CrossRef]
21. Oliva, D.; Elaziz, M.A.; Elsheikh, A.H.; Ewees, A.A. A review on meta-heuristics methods for estimating parameters of solar cells. *J. Power Sources* **2019**, *435*, 126683. [CrossRef]
22. Wang, G.G.; Shan, S. Review of metamodeling techniques in support of engineering design optimization. *J. Mech. Des.* **2007**, *129*, 370–380. [CrossRef]

23. Nabaei, A.; Hamian, M.; Parsaei, M.R.; Safdari, R.; Samad-Soltani, T.; Zarrabi, H.; Ghassemi, A. Topologies and performance of intelligent algorithms: A comprehensive review. *Artif. Intell. Rev.* **2018**, *49*, 79–103. [CrossRef]
24. Gao, L.; Hwang, Y.; Cao, T. An overview of optimization technologies applied in combined cooling, heating and power systems. *Renew. Sustain. Energy Rev.* **2019**, *114*, 109344. [CrossRef]
25. Castillo-Martinez, A.; Ramon Almagro, J.; Gutierrez-Escolar, A.; Del Corte, A.; Castillo-Sequera, J.L.; Gómez-Pulido, J.M.; Gutiérrez-Martínez, J.M. Particle swarm optimization for outdoor lighting design. *Energies* **2017**, *10*, 141. [CrossRef]
26. Wang, H.; Deutz, A.H.; Bäck, T.; Emmerich, M. Hypervolume Indicator Gradient Ascent Multi-Objective Optimization. In *EMO*; Volume 10173 of Lecture Notes in Computer Science; Springer: Berlin/Heidelberg, Germany, 2017; pp. 654–669.
27. Fonseca, C.M.; Fleming, P.J. Genetic Algorithms for Multiobjective Optimization: Formulation, Discussion and Generalization. In *ICGA*; Morgan Kaufmann: Burlington, MA, USA, 1993; pp. 416–423.
28. Zhang, Q.; Li, H. MOEA/D: A multiobjective evolutionary algorithm based on decomposition. *IEEE Trans. Evol. Comput.* **2007**, *11*, 712–731. [CrossRef]
29. Kumar, D.; Kumar, S.; Bansal, R.; Singla, P. A survey to nature inspired soft computing. IGI Global. *Int. J. Inf. Syst. Modeling Des.* **2017**, *8*, 112–133. [CrossRef]
30. Robic, T.; Filipic, B. DEMO: Differential Evolution for Multiobjective Optimization. In *EMO*; Volume 3410 of Lecture Notes in Computer Science; Springer: Berlin/Heidelberg, Germany, 2005; pp. 520–533.
31. Sharif, S.A.; Hammad, A. Simulation-Based Multi-Objective Optimization of institutional building renovation considering energy consumption, Life-Cycle Cost and Life-Cycle Assessment. *J. Build. Eng.* **2019**, *21*, 429–445. [CrossRef]
32. Lee, J. Multi-objective optimization case study with active and passive design in building engineering. *Struct. Multidiscip. Optim.* **2019**, *59*, 507–519. [CrossRef]
33. Gagnon, R.; Gosselin, L.; Armand Decker, S. Performance of a sequential versus holistic building design approach using multi-objective optimization. *J. Build. Eng.* **2019**, *26*, 100883. [CrossRef]
34. Haniff, M.F.; Selamat, H.; Khamis, N.; Alimin, A.J. Optimized scheduling for an air-conditioning system based on indoor thermal comfort using the multi-objective improved global particle swarm optimization. *Energy Effic.* **2019**, *12*, 1183–1201. [CrossRef]
35. Cai, H.; Guo, K.; Liu, H.; Xiang, W.; Liu, C. Multiobjective optimization of area-to-point heat conduction structure using binary quantum-behaved PSO and Tchebycheff decomposition method. *Can. J. Chem. Eng.* **2020**, *99*, 1211–1227. [CrossRef]
36. Zhai, Y.-Y.; Li, Y.; Ao, Z.-G. Optimization of Continuous Casting Secondary Cooling Based on an Enhanced Multi-objective Genetic Algorithm. *Northeast. Univ. Dongbei Daxue Xuebao J. Northeast. Univ.* **2019**, *40*, 658–662.
37. Aguilar, J.; Garcès-Jimènez, A.; Gallego-Salvador, N.; Gutièrrez de Mesa, J.A.; Gomez-Pulido, J.M.; Garcìa-Tejedor, À.J. Autonomic management architecture for multi-HVAC systems in smart buildings. *IEEE Access* **2019**, *7*, 123402–123415. [CrossRef]
38. Afzal, A.; Ramis, M.K. Multi-objective optimization of thermal performance in battery system using genetic and particle swarm algorithm combined with fuzzy logics. *J. Energy Storage* **2020**, *32*, 101815. [CrossRef]
39. Suthar, H.A.; Gadit, J.J. Multiobjective optimization of 2DOF controller using Evolutionary and Swarm intelligence enhanced with TOPSIS. *Heliyon* **2019**, *5*, e01410. [CrossRef]
40. Behrooz, F.; Mariun, N.; Marhaban, M.H.; Mohd Radzi, M.A.; Ramli, A.R. Review of control techniques for HVAC systems—Nonlinearity approaches based on Fuzzy cognitive maps. *Energies* **2018**, *11*, 495. [CrossRef]
41. Emmerich, M.T.; Deutz, A.H. A tutorial on multiobjective optimization: Fundamentals and evolutionary methods. *Nat. Comput.* **2018**, *17*, 585–609. [CrossRef]
42. Ekici, B.; Cubukcuoglu, C.; Turrin, M.; Sariyildiz, I.S. Performative computational architecture using swarm and evolutionary optimisation: A review. *Build. Environ.* **2019**, *147*, 356–371. [CrossRef]
43. Deb, K.; Pratap, A.; Agarwal, S.; Meyarivan, T. A fast and elitist multiobjective genetic algorithm: NSGA-II. *IEEE Trans. Evol. Comput.* **2002**, *6*, 182–197. [CrossRef]
44. Deb, K.; Jain, H. An evolutionary many-objective optimization algorithm using reference-point-based nondominated sorting approach, part I: Solving problems with box constraints. *IEEE Trans. Evolut. Comput.* **2014**, *18*, 577–601. [CrossRef]
45. Zitzler, E.; Laumanns, M.; Thiele, L. *SPEA2: Improving the Strength Pareto Evolutionary Algorithm*; TIK—Report 103; Eidgenössische Technische Hochschule Zürich, Institut für Technische Informatik und Kommunikationsnetze: Zürich, Switzerland, 2001.
46. Reyes-Sierra, M.; Coello, C.C. Multi-objective particle swarm optimizers: A survey of the state-of-the-art. *Int. J. Comput. Intel. Res.* **2006**, *2*, 287–308.
47. Nebro, A.J.; Durillo, J.J.; Garcia-Nieto, J.; Coello, C.C.; Luna, F.; Alba, E. SMPSO: A new PSO-based metaheuristic for multi-objective optimization. In Proceedings of the 2009 IEEE Symposium on Computational Intelligence in Multi-Criteria Decision-Making (MCDM), Nashville, TN, USA, 30 March–2 April 2009; pp. 66–73.
48. Riquelme, N.; Von Lücken, C.; Baran, B. Performance metrics in multi-objective optimization. In Proceedings of the 2015 Latin American Computing Conference (CLEI), Arequipa, Peru, 19–23 October 2015; pp. 1–11.
49. Zitzler, E.; Thiele, L.; Laumanns, M.; Fonseca, C.M.; Da Fonseca, V.G. Performance assessment of multiobjective optimizers: An analysis and review. *IEEE Trans. Evol. Comput.* **2003**, *7*, 117–132. [CrossRef]
50. Van Veldhuizen, D.A. *Multiobjective Evolutionary Algorithms: Classifications, Analyses, and New Innovations (No. AFIT/DS/ENG/99-01)*; Air Force Institute of Technology: Wright-Patterson AFB, OH, USA, 1999.
51. Géron, A. *Hands-On Machine Learning with Scikit-Learn, Keras, and TensorFlow: Concepts, Tools, and Techniques to Build Intelligent Systems*; O'Reilly Media: Sebastopol, CA, USA, 2019.

52. Barba-Gonzaléz, C.; García-Nieto, J.; Nebro, A.J.; Aldana-Montes, J.F. Multi-objective big data optimization with jmetal and spark. In Proceedings of the International Conference on Evolutionary Multi-Criterion Optimization, East Lansing, MI, USA, 10–13 March 2017; pp. 16–30.
53. Zadeh, L.A. Fuzzy Logic, Neural Networks, and Soft Computing. In *Fuzzy Sets, Fuzzy Logic, and Fuzzy Systems: Selected Papers by Lotfi a Zadeh*; World Scientific: Singapore, 1996; pp. 775–782.
54. Zitzler, E.; Laumanns, M.; Thiele, L.; Fonseca, C.M.; da Fonseca, V.G. Why quality assessment of multiobjective optimizers is difficult. In Proceedings of the 4th Annual Conference on Genetic and Evolutionary Computation, New York, NY, USA, 9–13 July 2002; pp. 666–674.
55. Keshavarz, M.; Pasandideh, S.H.R. Multi-objective optimisation of continuous review inventory system under mixture of lost sales and backorders within different constraints. Inderscience Publishers. *Int. J. Logist. Syst. Manag.* **2018**, *29*, 327–348.
56. Niyomubyeyi, O.; Sicuaio, T.E.; Díaz González, J.I.; Pilesjö, P.; Mansourian, A. A Comparative Study of Four Metaheuristic Algorithms, AMOSA, MOABC, MSPSO, and NSGA-II for Evacuation Planning. *Algorithms* **2020**, *13*, 16. [CrossRef]
57. Elgammal, A.; El-Naggar, M. Energy management in smart grids for the integration of hybrid wind–PV–FC–battery renewable energy resources using multi-objective particle swarm optimisation (MOPSO). *J. Eng.* **2018**, *11*, 1806–1816. [CrossRef]

Article

Effect of Initial Configuration of Weights on Training and Function of Artificial Neural Networks

Ricardo J. Jesus [1,2], Mário L. Antunes [1,3,*], Rui A. da Costa [4], Sergey N. Dorogovtsev [4], José F. F. Mendes [4] and Rui L. Aguiar [1,3]

[1] Departamento de Eletrónica, Telecomunicações e Informática, Campus Universitario de Santiago, Universidade de Aveiro, 3810-193 Aveiro, Portugal; ricardojesus@ua.pt (R.J.J.); ruilaa@ua.pt or ruilaa@av.it.pt (R.L.A.)
[2] EPCC, The University of Edinburgh, Edinburgh EH8 9YL, UK
[3] Instituto de Telecomunicações, Campus Universitario de Santiago, Universidade de Aveiro, 3810-193 Aveiro, Portugal
[4] Departamento de Física & I3N, Campus Universitario de Santiago, Universidade de Aveiro, 3810-193 Aveiro, Portugal; americo.costa@ua.pt (R.A.d.C.); sdorogov@ua.pt (S.N.D.); jfmendes@ua.pt (J.F.F.M.)
* Correspondence: mario.antunes@ua.pt or mario.antunes@av.it.pt

Citation: Jesus, R.J.; Antunes, M.L.; da Costa, R.A.; Dorogovtsev, S.N.; Mendes, J.F.F.; Aguiar, R.L. Effect of Initial Configuration of Weights on Training and Function of Artificial Neural Networks. *Mathematics* **2021**, *9*, 2246. https://doi.org/10.3390/math9182246

Academic Editor: Freddy Gabbay

Received: 20 August 2021
Accepted: 7 September 2021
Published: 13 September 2021

Publisher's Note: MDPI stays neutral with regard to jurisdictional claims in published maps and institutional affiliations.

Copyright: © 2021 by the authors. Licensee MDPI, Basel, Switzerland. This article is an open access article distributed under the terms and conditions of the Creative Commons Attribution (CC BY) license (https://creativecommons.org/licenses/by/4.0/).

Abstract: The function and performance of neural networks are largely determined by the evolution of their weights and biases in the process of training, starting from the initial configuration of these parameters to one of the local minima of the loss function. We perform the quantitative statistical characterization of the deviation of the weights of two-hidden-layer feedforward ReLU networks of various sizes trained via Stochastic Gradient Descent (SGD) from their initial random configuration. We compare the evolution of the distribution function of this deviation with the evolution of the loss during training. We observed that successful training via SGD leaves the network in the close neighborhood of the initial configuration of its weights. For each initial weight of a link we measured the distribution function of the deviation from this value after training and found how the moments of this distribution and its peak depend on the initial weight. We explored the evolution of these deviations during training and observed an abrupt increase within the overfitting region. This jump occurs simultaneously with a similarly abrupt increase recorded in the evolution of the loss function. Our results suggest that SGD's ability to efficiently find local minima is restricted to the vicinity of the random initial configuration of weights.

Keywords: training; evolution of weights; deep learning; neural networks; artificial intelligence

1. Introduction

Training of neural networks is based on the progressive correction of their weights and biases (model parameters) performed by such algorithms as gradient descent, which compare actual outputs with the desired ones for a large set of input samples [1]. Consequently, the understanding of the internal operation of neural networks should be intrinsically based on the detailed knowledge of the evolution of their weights in the process of training, starting from their initial configuration. Recently, Li and Liang [2] revealed that, during training, weights in neural networks only slightly deviate from their initial values in most practical scenarios. In this paper, we explore in detail how training changes the initial configuration of weights, and the relations between those changes and the effectiveness of the networks' function. We track the evolution of the weights of networks consisting of two Rectified Linear Unit (ReLU) hidden layers trained on three different classification tasks with Stochastic Gradient Descent (SGD), and measure the dependence of the distribution of deviations from an initial weight on this initial value. In all of our experiments, we observe no inconsistencies in the results of the three tasks.

By experimenting with networks of different sizes, we have observed that, to reach an arbitrarily chosen loss value, the weights of larger networks tend to deviate less from their initial values than those of smaller networks. This suggests that larger networks tend to converge to minima which are closer to their initialization. On the other hand, we observe that for a certain range of network sizes, the deviations from initial weights abruptly increase at some moment during their training within the overfitting regime.

This effect is illustrated in Figure 1 by the persistence and disappearance of an initialization mask (the letters are stamped to a network's initial configuration of weights by creating a bitmap of the same shape as the matrix of weights of the layer being marked, rasterizing the letter to the bitmap, and using the resulting binary mask to set to zero the weights laying outside the mark's area.) in panels (a) and (b), respectively, for two network sizes. We find that the sharp increase on the deviations of the weights closely correlates with the crossover between two regimes of the network—trainability and untrainability—occurring in the course of the training.

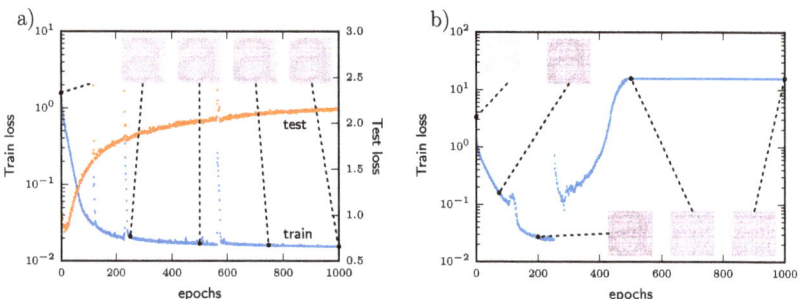

Figure 1. Train and test loss of networks consisting of two equally sized hidden layers of nodes, trained on HASYv2. Some of the weights connecting the two hidden layers were initially set to zero so that the weight matrix of these layers resembles the letter a (at initialization). The evolution of the weight matrix is shown in the subplots. (**a**) Loss of a stable learner network with 512 nodes in each hidden layer. (**b**) Training loss of an unstable network with 256 nodes in each hidden layer, illustrating the effect of crossing over from trainability to untrainability regimes on the network's weights (disappearance of the initialization mask). A single curve is shown for clarity, but networks of the same width show the same behavior. Experiments with other symbols (that serve as mask for the initialization) exhibit similar behavior.

The main contributions of this work are the following: (I) a quantitative statistical characterization of the deviation of the weights of two-hidden-layer ReLU network of various sizes, from their initial random configuration, and (II) we show a correlation between the magnitude of deviations of weights and the successful training of a network. Recent works [2–4] showed that in highly over-parametrized networks, the training process implies a fine-tuning of the initial configuration of weights, significantly adjusting only a small portion of them. Our quantitative statistical characterization describes this phenomenon in greater detail and empirically verifies the small deviations that occur when the training process is successful. Furthermore, our analysis allows us to draw some insights regarding the training process of neural network and pave the way for future research.

Our paper is organized as follows. In Section 2, we summarize some background topics on neural networks' initializations, and review a series of recent papers pertaining to ours. Section 3 presents the problem formulation, experimental settings and datasets used in this paper. In Section 4, we explore the shape of the distribution of the deviations of weights from their initial values and its dependence on the initial weights. We continue these studies in Section 5 by experimenting with networks of different widths and find that, whenever a network's training is successful, the network does not travel far from its initial configuration. Finally, Section 6 provides concluding remarks and points out directions for future research.

2. Background and Related Work

2.1. Previous Works

It is widely known that a neural network's initialization is instrumental in its training [5–8]. The works of Glorot and Bengio [7], Chapelle and Erhan [9] and Krizhevsky et al. [10], for instance, showed that deep networks initialized with random weights and optimized with methods as simple as Stochastic Gradient Descent could, surprisingly, be trained successfully. In fact, by combining momentum with a well-chosen random initialization strategy, Sutskever et al. [11] managed to achieve performance comparable to that of Hessian-free methods.

There are many methods to randomly initialize a network. Usually, they consist of drawing the initial weights of the network from uniform or Gaussian distributions centered at zero, and setting the biases to zero or some other small constant. While the choice of the distribution (uniform or Gaussian) does not seem to be particularly important [12], the scale of the distribution from which the initial weights are drawn does. The most common initialization strategies—those of Glorot and Bengio [7], He et al. [8], and LeCun et al. [5]—define rules based on the network's architecture for choosing the variance that the distribution of initial weights should have. These and other frequently used initialization strategies are mainly heuristic, seeking to achieve some desired properties at least during the first few iterations of training. However, it is generally unclear which properties are kept during training or how they vanish [12] (Section 8.4). Moreover, it is also not clear why some initializations are better from the point of view of optimization (i.e., achieve lower training loss), but are simultaneously worse from the point of view of generalization.

Frankle and Carbin [13] recently observed that randomly initialized dense neural networks typically contain subnetworks (called winning tickets) that are capable of matching the test accuracy of the original network when trained for the same amount of time in isolation. Based on this observation, they formulate the Lottery Ticket Hypothesis, which essentially states that this effect is general and manifests with high probability in this kind of network. Notably, these subnetworks are part of the network's initialization, as opposed to an organization that emerges throughout training. The subsequent works of Zhou et al. [14] and Ramanujan et al. [15] corroborate the Lottery Ticket Hypothesis and propose that winning tickets may not even require training to achieve quality comparable to that of the trained networks.

In their recent paper, Li and Liang [2] established that two-layer over-parameterized ReLU networks, optimized with SGD on data drawn from a mixture of well-separated distributions, probably converge to a minimum close to their random initializations. Around the same time, Jacot et al. [3] proposed the neural tangent kernel (NTK), a kernel that characterizes the dynamics of the training process of neural networks in the so-called infinite-width limit. These works instigated a series of theoretical breakthroughs, such as the proof that SGD can find global minima under conditions commonly found in practice (e.g., over-parameterization) [16–23], and that, in the infinite-width limit, neural networks remain in an $O(1/\sqrt{n})$ neighborhood of their random initialization (n being the width of the hidden layers) [24,25]. Lee et al. [4] make a similar claim about the distance a network may deviates from its linearized version. Chizat et al. [26], however, argue that such wide networks operate in a regime of "lazy training" that appears to be incompatible with the many successes neural networks are known for in difficult, high dimensional tasks.

2.2. Our Contribution

From distinct perspectives, these previous works have shown that, in highly over-parametrized networks, the training process consists of a fine-tuning of the initial configuration of weights, adjusting significantly just a small portion of them (the ones belonging to the winning tickets). Furthermore, Frankle et al. [27] recently showed that the winning ticket's weights are highly correlated with each other.

The previous investigations on the role of the initial weights configuration focus on networks with potentially infinite width, in which, as our results also show, the persistence of the initial configuration is more noticeable. In contrast, we explore a wide range of network sizes from untrainable to trainable by varying the number of units in the hidden layers. This approach allows us to explore the limits of trainability, and characterize the trainable–untrainable network transition that occurs at a certain threshold of the width of hidden layers.

A few recent works [28,29] indicated the existence of 'phase transitions' from narrow to wide networks associated to qualitative changes in the set of loss minima in the configuration space. These results resonate with ours, although neither relations to trainability nor the role of the initial configuration of weights were explored.

On one hand, we observe that, when the networks are trainable (large networks), they always converge to a minima in the vicinity of the initial weight configuration. On the other hand, when the network is untrainable (small networks) the weight configuration drifts away from the initial configuration. Moreover, in our simulations, we found an intermediate size range for which the networks train reasonably well for a while, decreasing the loss consistently, but, later in the overfitting region, their loss abruptly increases dramatically (due to overshooting). Past this point of divergence, the loss can no longer be reduced by more training. The behavior of these ultimately untrainable networks further emphasizes the connection between trainability (ability to reduce train loss) and proximity to the initial configuration: the distance to the initial configuration remains small in the first stage of training, while the loss is reduced, and later increases abruptly, simultaneously with the loss.

We hypothesize that networks initialized with random weights and trained with SGD can only find good minima in the vicinity of the initial configuration of weights. This kind of training procedure is unable to effectively explore more than a relatively small region of the configuration space around the initial point.

3. Problem Formulation

Our aim in this work is to contribute to the conceptual understanding of the influence that random initializations have in the solutions of feedforward neural networks trained with SGD. In order to avoid undesirable effects, specific to particular architectures, training methods, etc., we set up our experiments with very simple, vanilla, settings.

We trained feedforward neural networks with two layers of hidden nodes (three layers of links) with all-to-all connectivity between adjacent layers. In our experiments, we vary the widths (i.e., numbers of nodes) of the two hidden layers between 10 and 1000 nodes simultaneously, always keeping them equal to each other. The number of nodes of the input and output layers is determined by the dataset, specifically by the number of pixels of the input images and the number of classes, respectively. This architecture is largely based on the multilayer perceptron created by Keras for MNIST (https://raw.githubusercontent.com/ShawDa/Keras-examples/master/mnist_mlp.py, accessed on 8 September 2021).

Let us denote the weight of the link connecting nodes i in a given layer and node j in the next layer by w_{ij}. The output of a node j in the hidden and output layers, denoted by o_j, is determined by an activation function of the weighted sum of the outputs of the previous layer added by the node's bias, b_j, as $b_j + \sum_i w_{ij} o_i \equiv x_j$. The nodes of two hidden layers employ the Rectified Linear Unit (ReLU) activation function

$$f(x_j) = \begin{cases} x_j \text{ if } x_j \geq 0, \\ 0 \text{ if } x_j < 0. \end{cases} \quad (1)$$

The ReLU is a piecewise linear function that will output the input directly if it is positive; otherwise, it will output zero. It has become the default activation function for many types of neural networks because a model that uses it is easier to train and often achieves better performance.

The nodes in the output layer employ the softmax activation function

$$f(x_j) = \frac{e^{x_j}}{\sum_{k=1}^{K} e^{x_k}} \qquad (2)$$

where K is the number of elements in the input vector (i.e., the number of classes of the dataset). The softmax activation function is a generalization of the logistic function to multiple dimensions. It is used in multinomial logistic regression and is often used as the last activation function of a neural network to normalize the output of a network to a probability distribution over predicted output classes.

Unless otherwise stated, the biases of the networks are initialized at zero and the weights are initialized with Glorot's uniform initialization [7]:

$$w_{ij} \sim U\left(-\frac{\sqrt{6}}{\sqrt{m+n}}, \frac{\sqrt{6}}{\sqrt{m+n}}\right), \qquad (3)$$

where $U(\alpha, \beta)$ is the uniform distribution in the interval (α, β), and m and n are the number of units of the layers that weight w_{ij} connects. In some of our experiments, we apply various masks to these uniformly distributed weights, setting to zero all weights w_{ij} not covered by a mask (see Figure 1). The loss function to be minimized is the categorical cross-entropy, i.e.,

$$L = -\sum_{i=1}^{C} y_i \ln o_i, \qquad (4)$$

where C is the number of output classes, $y_i \in \{0, 1\}$ the i-th target output, and o_i the i-th output of the network. The neural networks were optimized with Stochastic Gradient Descent with a learning rate of 0.1 and in mini-batches of size 128. The networks were defined and trained in Keras [30] using its TensorFlow [31] back-end.

Throughout this paper, we use three datasets to train our networks: MNIST, Fashion MNIST, and HASYv2. Figure 2 displays samples of them. These are some of the most standard datasets used in research papers on supervised machine learning.

MNIST (http://yann.lecun.com/exdb/mnist/, accessed on 5 September 2021) [32] is a database of gray-scale handwritten digits. It consists of 6.0000×10^4 training and 1.0000×10^4 test images of size 28×28, each showing one of the numerals 0 to 9. It was chosen due to its popularity and widespread familiarity.

Fashion MNIST (https://github.com/zalandoresearch/fashion-mnist, accessed on 5 September 2021) [33] is a dataset intended to be a drop-in replacement for the original MNIST dataset for machine learning experiments. It features 10 classes of clothing categories (e.g., coat, shirt, etc.) and it is otherwise very similar to MNIST. It also consists of 28×28 gray-scale images, 6.0000×10^4 samples for training, and 1.0000×10^4 for testing.

HASYv2 (https://github.com/MartinThoma/HASY, accessed on 5 September 2021) [34] is a dataset of 32×32 binary images of handwritten symbols (mostly LATEX symbols, such as α, σ, $\int$, etc.). It mainly differentiates from the previous two datasets in that it has many more classes (369) and is much larger (containing around 150,000 train and 17,000 test images).

In this paper, the number of epochs elapsed on the process of training is denoted by t. We typically trained networks for very long periods (up to $t = 1000$), and, consequently, for most of their training, the networks were in the overfitting regime. However, since we are studying the training process of these network and making no claims concerning the networks' ability to generalize on different data, overfitting does not affect our conclusions. In fact, our results are usually even stronger prior to overfitting. For similar reasons, we will be considering only the loss function of the networks (and not other metrics such as their accuracy), since it is the loss function that the networks are optimizing.

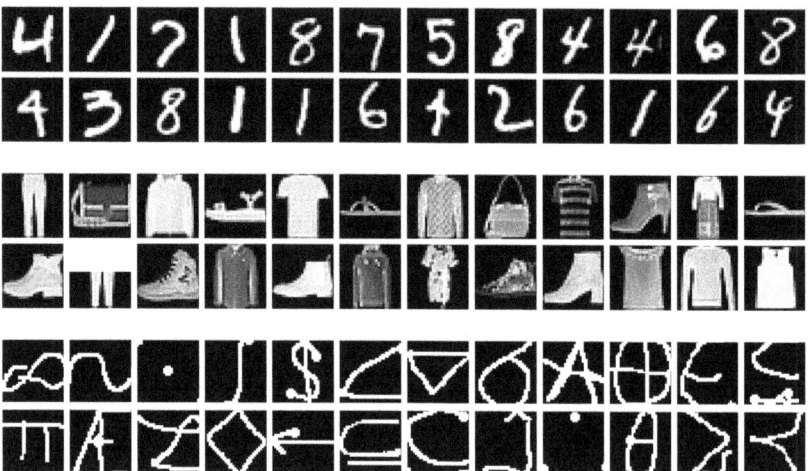

Figure 2. Samples of the datasets used in our experiments. Top: MNIST. Middle: Fashion MNIST. Bottom: HASYv2 (colors reversed).

4. Statistics of Deviations of Weights from Initial Values

To illustrate the reduced scale of the deviations of weights during the training, let us mark a network's initial configuration of weights using a mask in the shape of a letter, and observe how the marking evolves as the network is trained. Naturally, if the mark is still visible after the weights undergo a large number of updates and the networks converge, it indicates that the training process does not shift the majority of the weights of a network far from their initial states.

Figure 1a shows typical results of training a large network whose initial configuration is marked with the letter a. One can see that the letter is clearly visible after training for as many as 1000 epochs. In fact, one observes the initial mark during all of the network's training, without any sign that it will disappear. Even more surprisingly, these marks do not affect the quality of training. Independently of the shape marked (or whether there is a mark or not) the network trains to approximately the same loss across different realizations of initial weights. This demonstrates that randomly initialized networks preserve features of their initial configuration along their whole training—features that are ultimately transferred into the networks' final applications.

Figure 1b demonstrates an opposite effect for midsize networks that cross over between the regimes of trainability and untrainability. As it illustrates, the initial configuration of weights of these unstable networks tend to be progressively lost, suffering the largest changes when the networks diverge (loss function sharply increases at some moment).

By inspecting the distribution of the final (i.e., last, after training) values of the weights of the network of Figure 1a versus their initial values, portrayed in Figure 3, we see that weights that start with larger absolute values are more likely to suffer larger updates (in the direction that their sign points to). This trend can be observed in the plot by the tilt of the interquartile range (yellow region in the middle) with respect to the median (dotted line). The figure demonstrates that initially large weights in absolute value have a tendency to become even larger, keeping their original sign; it also shows the maximum concentration of weights near the line $w_f = w_i$, indicating that most weights either change very little or nothing at all throughout training.

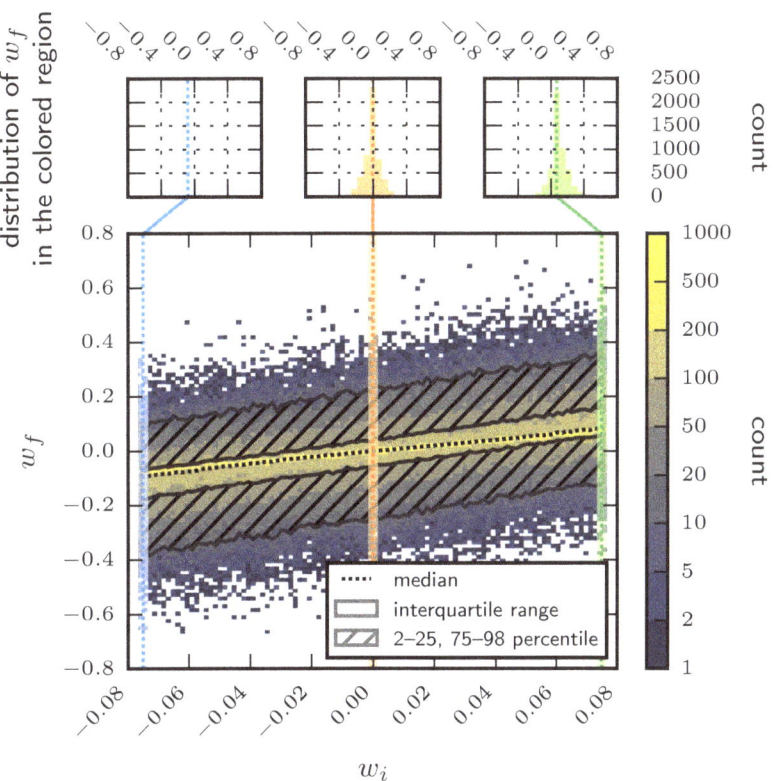

Figure 3. Distribution of the final values of the weights of the network of Figure 1a, trained for 1000 epochs on HASYv2, as function of their initial value. The peak of the distribution is at $w_f = w_i$, which is extremely close to the median. The skewness of the distribution for large absolute values of w_i is evidenced in the histograms at the top.

This effect may be explained by the presence of a winning ticket in the network's initialization. Our results suggest that the role of the over-parametrized initial configuration is decisive in successful training: when we reduce the level of over-parametrization to a point where the initial configuration stops containing such winning tickets, the network becomes untrainable by SGD.

The skewness in the distribution of the final weights can be explained by the randomness of the initial configuration, which initializes certain groups of weights with more appropriate values than the others, which makes them better suited for certain features of the dataset. This subset of weights does not need to be particularly good, but as long as it provides slightly better or more consistent outputs than the rest of the weights, the learning process may favor their training, improving them further than the rest. Over the course of many epochs, the small preference that the learning algorithm keeps giving them adds up and causes these weights to become the best recognizers for the features that they initially, by chance, happened to be better at. In this hypothesis, it is highly likely that weights with larger initial values are more prone to be deemed more important by the learning algorithm, which will try to amplify their 'signal'. This effect has several bearings with, for instance, the real-life effect of the month of birth in sports [35].

5. Evolution of Deviations of Weights and Trainability

One may understand the relationship between the success of training and the fine-tuning process observed in Section 4, during which a large fraction of the weights of a network suffer very tiny updates (and many are not even changed at all), in the following way. We suggest that the neural networks typically trained are so over-parameterized that, when initialized at random, their initial configuration has a high probability of being close to a proper minimum (i.e., a global minimum where the training loss approaches zero). Hence, to reach such a minimum, the network needs to adjust its weights only slightly, which causes its final configuration of weights to have strong traces of the initial configuration (in agreement with our observations).

This hypothesis raises the question of what happens when we train networks that have a small number of parameters. At some point, do they simply start to train worse? Or do they stop training at all? It turns out that, during the course of their training, neural networks cross over between two regimes—trainability and untrainability. The trainability region may be further split into two distinct regimes: a regime of *high trainability* where training drives the networks towards global minima (with zero training loss), and a regime of *low trainability* where the networks converge to sub-optimal minima of significantly higher loss. Only high trainability allows a network to train successfully (i.e., to be trainable), since in the remaining two regimes, of untrainability and low trainability, either the networks do not learn at all, or they learn but very poorly. Figure 4 illustrates these three regimes. Note that we use the term trainability/untrainability referring to the regimes of the training process, in which loss and deviations of weights are, respectively, small/large. We reserve the terms trainable-untrainable to refer to the capability of a network to keep a low train loss after infinite training, which depends essentially on the network's architecture.

We measure the dependence of the time at which these crossovers happen on the network size and build a diagram showing the network's training regime for each network width and training time. This diagram, Figure 4c, resembles a phase diagram, although the variable t, the training time, is not a control parameter, but it is rather the measure of the duration of the 'relaxation' process that SGD training represents. One may speak about a phase transition in these systems only in respect of their stationary state, that is, the regime in which they finally end up, after being trained for a very long (infinite) time. Figure 4c shows three characteristic instants (times) of training for each network width: (i) the time at which the minimum of the test loss occurs, (ii) the time of the minimum of the train loss, (iii) the time at which the loss abruptly increases ('diverges'). Each of these times differ for different runs, and, for some widths, these fluctuations are strong or they even diverge. The points in this plot are the average values over ten independent runs. By the error bars, we show the scale of the fluctuations between different runs. Notice that the times (ii) and (iii) approach infinity as we approach the threshold of about 300 nodes from below (which is specific to the network's architecture and dataset). Therefore, wide networks ($\gtrsim$300 nodes in each hidden layer) never cross over to the untrainability regime; wide networks should stabilize in the trainability regime as $t \to \infty$. The untrainability region of the diagram exists only for widths smaller than the threshold, which is in the neighborhood of 300 nodes. Networks with such widths initially demonstrate a consistent decrease in the train loss. However, at some later moment, during the training process, the systems abruptly cross over from the trainability regime, with small deviations of weights from their initial values and decreasing train loss, to the untrainability regime, with large loss and large deviations of weights.

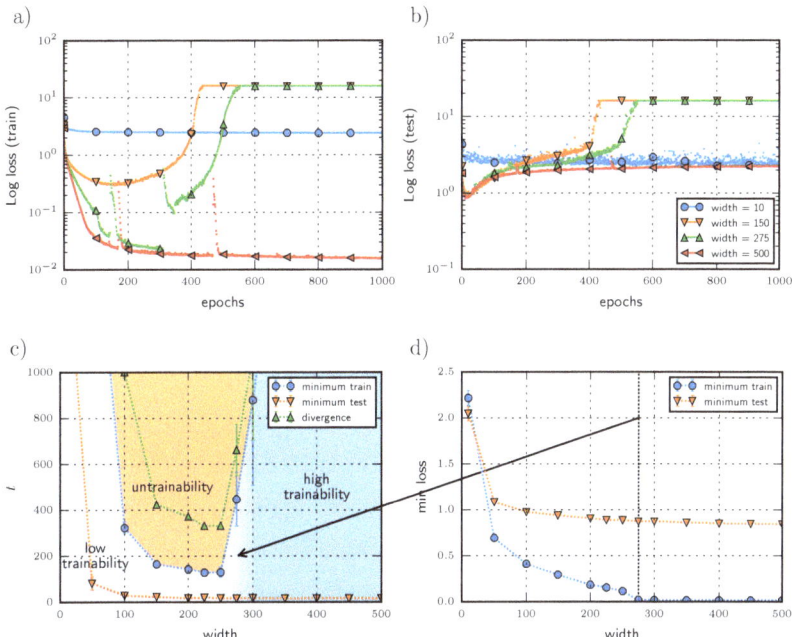

Figure 4. The regimes of a neural network over the course of its training: Evolution of (**a**) train and (**b**) test loss functions of networks of various widths. Panels (**a**,**b**) show single typical training runs of the full set which we explored. (**c**) Average times (in epochs) taken by the networks to reach the minima of the train loss and of the test loss functions, and to diverge (i.e., reach the plateau of the train loss). For each network width, we calculate the averages and standard deviations of these times (represented by error bars) over ten independent runs trained on HASYv2. (**d**) Average values of minimum loss at the train and test sets reached during individual runs (different runs reach their minima at different times). These averages were measured over the same ten independent realizations as in panel (**c**).

By gradually reducing the width, and looking at the trainability regime in the limit of infinite training time, we find a phase transition from trainable to untrainable networks. In the diagram of Figure 4c, this transition corresponds to a horizontal line at $t = \infty$, or, equivalently, to the projection of the diagram on the horizontal axis (notice that the border between regimes is concave).

The phase diagram in Figure 4 (the bottom left panel) suggests the existence of three different classes of networks: weak, strong, and, in between them, unstable learners. Weak learners are small networks that, throughout their training, do not tend to diverge, but only train to poor minima. They mostly operate in a regime of low trainability, since they can be trained, but only to ill-suited solutions. On the other side of the spectrum of trainability are large networks. These are strong learners, as they train to very small loss values and they are not prone to diverge (they operate mostly in the regime of high trainability). In between these two classes are unstable learners, which are midsize networks that train to progressively better minima (as their size increases), but that, at some point, in the course of their training, are likely to diverge and become untrainable (i.e., they tend to show a crossover from the regime of trainability to the one of untrainability).

Remarkably, we observe the different regimes of operation of a network not only in the behavior of its loss function, but also in the distance it travels from its initial configuration of weights. We have already demonstrated in Figure 1 how the mark of the initial configuration of weights of a network persists in large networks (i.e., strong learners that over the course of their training were always on the regime of high trainability), and vanishes for midsize networks that ultimately cross over to the regime of untrainability. In Appendix A we supply detailed description of the evolution of the statistics of weights during the training of the networks used to draw Figure 4. Figures A1 and A2 show that, as the network width is reduced, the highly structured correlation between initial and final weight, illustrated by the coincidence of the median with the line $w_f = w_i$ (see Figure 3), remains in effect in all layers of weights until the trainability threshold. Below that point the structure of the correlations eventually breaks down, given enough training time. The reliability of the observation of this breakdown in Figures A1 and A2, for widths below ~300, is reinforced by the robust fitting method based in cumulative distributions that is explained in Appendix B.

To quantitatively describe how distant a network becomes from its initial configuration of weights we consider the root mean square deviation (RMSD) of its system of weights at time t with respect to its initial configuration, i.e.,

$$\text{RMSD}(t) \equiv \sqrt{\frac{1}{m} \sum_{j=1}^{m} \left[w_j(t) - w_j(0) \right]^2}, \tag{5}$$

where m is the number of weights of the network (which depends on its width), and $w_j(t)$ is the weight of edge j at time t.

Figure 5 plots, for three different datasets, the evolution of the loss function of networks of various widths alongside the deviation of their configuration of weights from its initial state. These plots evidence the existence of a link between the distance a network travels away from its initialization and the regime in which it is operating, which we describe below.

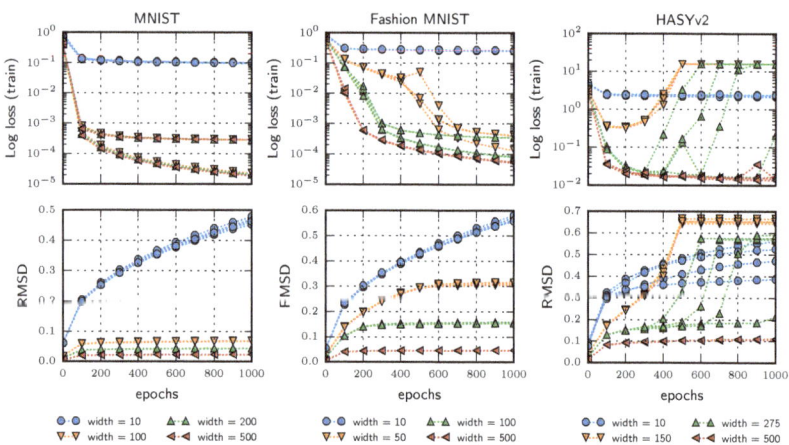

Figure 5. Top: Temporal evolution of the loss function of networks of various widths. Bottom: Evolution of the root mean square deviation (RMSD) between the initial configuration of weights of a network and its current configuration. From left to right: networks trained on the MNIST, Fashion MNIST, and HASYv2 datasets. Five independent test runs are plotted (individually) for each value of width and each dataset.

For all the datasets considered, the blue circles (●) show the training of networks that are weak learners—hence, they only achieve very high losses and are continuously operating in a regime of low trainability. These networks experience very large deviations on their configuration of weights, getting further and further away from their initial state. In contrast, the red left triangles (◀) show the training of large networks that are strong learners (in fact, for MNIST all the networks marked with triangles are strong learners; in our experiments we could not identify unstable learners on this dataset). These networks are always operating in the regime of high trainability, and over the course of their training they deviate very slightly from their initial configuration (compare with the results of Li and Liang [2]). Finally, for the Fashion MNIST and HASYv2 datasets, orange down (▼) and green up (▲) triangles show unstable networks of different widths (the former being smaller than the latter). While on the regime of trainability, these networks deviate much further from their initial configuration than strong learners (but less than weak learners). However, as they diverge and cross over into the untrainability regime (which could only be observed on networks training with the HASYv2 dataset), the RMSD suffers a sharp increase and reaches a plateau. These observations highlight the persistent coupling between the network's trainability (measured as train loss) and the distance it travels away from the initial configuration (measured as RMSD), as well as their dependence of the network's width.

To complete the description of the behavior of these networks on the different regimes, Figure 6 plots, for networks of different widths, the time at which they reach a loss below a certain value θ, and the RMSD between their configuration of weights at that time and the initial one. It shows that networks that are small and are operating under the low trainability regime fail to reach even moderate losses (e.g., on Fashion MNIST, no network of width 10 reaches a loss of 0.1, whereas networks of width 100 reach losses that are three orders of magnitude smaller). Moreover, even when they reach these loss values, they take a significantly longer time to do so, as the plots for MNIST demonstrate. Finally, the figure also shows that, as the networks grow in size, the displacement each weight has to undergo to allow the network to reach a particular loss decreases, meaning that the networks are progressively converging to minima that are closer to their initialization. We can treat this displacement as a measure of the work the optimization algorithm performs during the training of a network to make it reach a particular quality (i.e., value of loss). Then one can say that using larger networks eases training by decreasing the average work the optimizer has to spend with each weight.

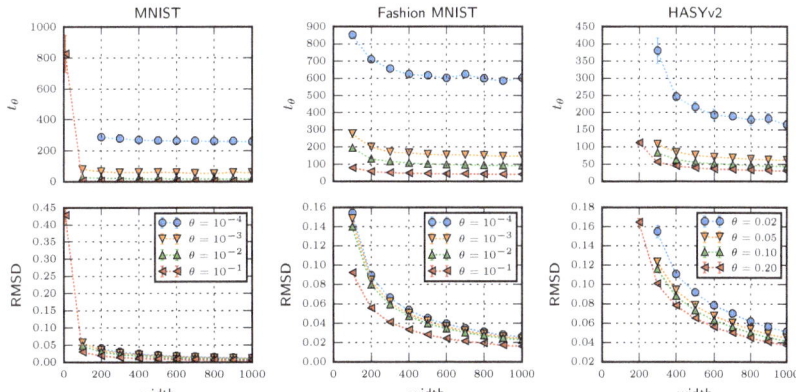

Figure 6. Top: Time t_θ (in epochs) at which the networks first reach a given value θ. Bottom: Root mean square deviation (RMSD) between the initial configuration of weights of a network and its configuration of weights at time t_θ. From left to right: networks trained on the MNIST, Fashion MNIST, and HASYv2 datasets. Each point in the figure is the average of five independent test runs. The absence of a point in a plot indicates that the network does not reach this loss during the entire period of training.

6. Conclusions

In this paper, we explored the effects of the initial configuration of weights on the training process and function of shallow feedforward neural networks. We performed a statistical characterization of the deviation of the weights of two-hidden-layer networks of various sizes trained via Stochastic Gradient Descent from their initial random configuration. Our analysis has shown that there is a strong correlation between the successful training of a shallow feedforward network and the magnitude of the weights' deviations from their initial values. Furthermore, we were able to observe that the initial configuration of weights typically leaves recognizable traces on the final configuration after training, which provides evidence that the learning process is based on fine-tuning the weights of the network.

We investigated the conditions under which a network travels far from its initial configurations. We observed that a neural network learns in one of two major regimes: trainability and untrainability. Moreover, its size (number of parameters) largely determines the quality of its training process and its chance to enter the untrainability regime. By comparing the evolution of the distribution function of the deviations of the weights with the evolution of the loss function during training, we have shown that a network only travels far away from its initial configuration of weights if it is either (i) a poor learner (which means that it never reaches a good minimum) or (ii) when it crosses over from trainability to untrainability regimes. In the alternative (good) case, in which the network is a strong learner and it does not become untrainable, the network always converges to the neighbourhood of its initial configuration (keeping extensive traces of its initialization); in all of our experiments, we never observed a network converging to a good minimum outside the vicinity of the initial configuration. The results and analysis of our simulations point out that the typical black-box model, used in most applications of neural networks, hides the trainability capacity of the networks. For a set of three typical classification problems, these results indicate a range of network sizes where the training process is successful. Our conclusions are consistent with recent finds, specifically the Lottery Ticket Hypothesis [13].

Finally, it is important to mention that most of our analysis was conducted when overfitting was already taking place. At shorter times, the deviations of weights from their initial values are even smaller, and our conclusions remain valid. Our conclusions were based on the analysis of the loss function of the networks since this was the actual function that the networks were optimizing. However, we argue that equivalent results can be obtained by using the accuracy or other similar metric. Our analysis was conducted on a specific set of conditions. To fully generalize our findings, different initialization methods and datasets should be considered in order to fully validate the hypothesis stated in Section 2.2. The generalization of the results is outside of the scope of this paper and is intended as future work.

Author Contributions: Conceptualization, R.J.J., M.L.A., R.A.d.C., S.N.D., J.F.F.M. and R.L.A.; methodology, R.J.J., M.L.A., R.A.d.C., S.N.D. and R.L.A.; software, R.J.J.; validation, R.J.J., M.L.A., R.A.d.C. and R.L.A.; formal analysis, R.A.d.C. and S.N.D.; investigation, R.J.J., R.A.d.C., M.L.A.; writing—original draft preparation, R.J.J., M.L.A. and R.A.d.C.; writing—review and editing, S.N.D., J.F.F.M. and R.L.A. All authors have read and agreed to the published version of the manuscript.

Funding: This work was developed within the scope of the project i3N, UIDB/50025/2020 and UIDP/50025/2020, financed by national funds through the FCT/MEC. RAC acknowledges the FCT Grant No. CEECIND/04697/2017. The work was also supported under the project YBN2020075021- Evolution of mobile telecommunications beyond 5G inside IT-Aveiro; by FCT/MCTES through national funds and when applicable co-funded EU funds under the project UIDB/50008/2020- UIDP/50008/2020; under the project PTDC/EEI-TEL/30685/2017 and by the Integrated Programme of SR&TD "SOCA" (Ref. CENTRO-01-0145-FEDER-000010), co-funded by Centro 2020 program, Portugal 2020, European Union, through the European Regional Development Fund.

Institutional Review Board Statement: Not applicable.

Informed Consent Statement: Not applicable.

Data Availability Statement: Not applicable.

Conflicts of Interest: The authors declare no conflict of interest.

Appendix A. Statistics of Weights across the Trainable–Untrainable Transition

The distribution of deviations from the initial weight is qualitatively similar across the trainable phase. The proximity between the line $w_f = w_i$ and the median of the distributions of w_f for fixed w_i, observed in Figure 3, is a distinctive feature of that kind of distribution. Figure A1 shows the slope obtained by fitting a straight line to the peaks (or mode) of the distributions of trained weights w_f for fixed w_i. In large networks, the fitted slope of the peaks, c_{opt}, is very close 1 in all layers (even for very large training times), independently of the width. Below the width threshold for the network to be trainable, of about 300 nodes per hidden layer, the slope c_{opt} shows significant deviations from 1, and its value strongly fluctuates among realizations of the initialization and training. (The borders of the shaded area in the plots of Figures A1 and A2 represent the standard deviation measured in ten independent realizations.) To emphasize the coupling between trainability and proximity to the initial configuration of weights, we used data from the same ten realizations to plot Figures 4, A1 and A2. Combined, these figures show the simultaneousness of the abrupt increase in the loss and of the deviation from the initial configuration.

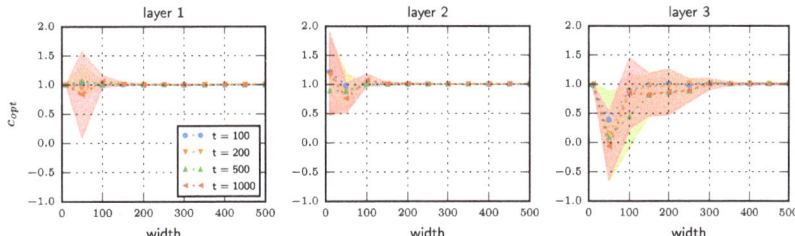

Figure A1. Fitting of the line across the maxima of the distribution of final weights in the networks of Figure 4. The parameter c_{opt} is the slope obtained by fitting a straight line to the peaks of the distributions of trained weights w_f for fixed w_i. These results are averages measured in ten independent realizations, and the shaded areas represents the dispersion (standard deviation across realizations).

For the sake of completeness, we perform linear fittings also to the mean trained weight as a function of the initial weight. Figure A2 shows the results of these fittings: a_0 and a_1 denote the constant and the slope, respectively. Similarly to the c_{opt}, while above a width of about 300 nodes the values of a_0 and a_1 are stable, below the threshold they suffer an abrupt change at some moment of training. The dispersion of the trained weights around their initial value, measured by the standard deviation, is also shown in Figure A2 for the set of weights that are initialized with the value $w_i = 0$, displaying the same transition at a width of about 300. We observed that the distribution of trained weights for other $w_i \neq 0$ behaves similarly with the variation of the network's width. Notice that, since the weights are initialized from a continuous distribution, we are able to measure the mode (peaks), average, and standard deviation of weights as functions of the initial value w_i by applying the special procedure described in Appendix B, which is less affected by the presence of noise (fluctuations) in the data that the standard binning methods.

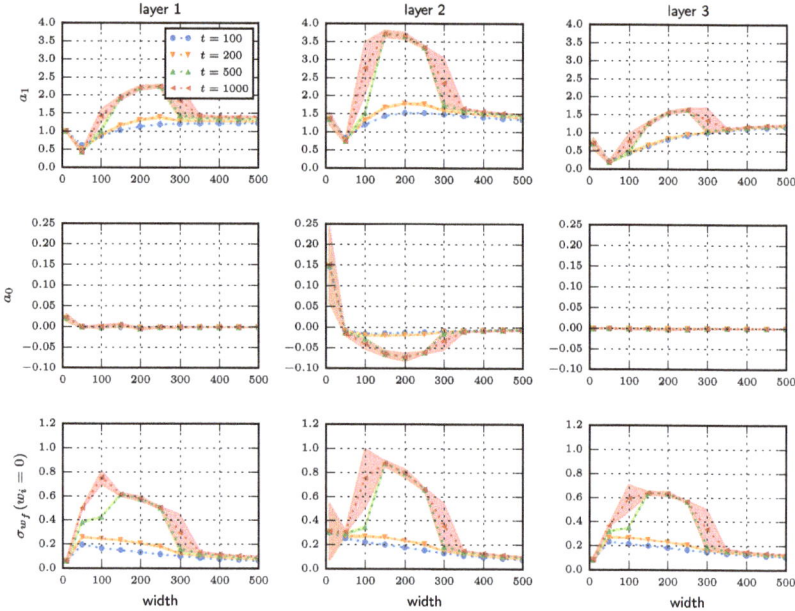

Figure A2. Mean and standard deviation of the trained weights in the networks of Figure 4. Top and middle rows: results of fitting a straight line $a_0 + a_1 w_i$ to the mean of the trained weights $w_f(wi)$. Bottom row: standard deviation of the distribution of trained weights for the set of weights that are initialized with the value $w_i = 0$. These results are averages measured in ten independent realizations, and the shaded areas represents the dispersion (standard deviation across realizations).

Appendix B. Fitting the Statistics of Weights in a Single Realization

This appendix briefly describes the methods used in this work to characterize the statistics of the displacements of weights produced by training with SGD. The problem is that we cannot directly obtain the distributions $P_{w_i}(w_f)$ of the final weights w_f for each value of the initial weight w_i, because the w_is are drawn from a continuous (uniform) distribution, see Equation (3). In practice, for a single realization of training, we have a set of points (w_i, w_f), one for each link, in the continuous plane, as shown in Figure 3. In this situation, calculating the mean $\langle w_f \rangle(w_i)$ and standard deviation $\sigma_{w_f}(w_i)$ of the distribution $P_{w_i}(w_f)$ may follow one of two approaches: either using a binning procedure or cumulative distributions. In our analysis, we employed the latter, which is less affected by random fluctuations than the standard binning methods.

We assume a linear fit $\langle w_f \rangle(w_i) = a_0 + a_1 w_i$, and obtain the constants a_1 and a_0 as follows. Let us define the function

$$W_f(w) = \int_{\min(w_i)}^{w} \langle w_f \rangle(x) dx = C + a_0 w + \frac{a_1}{2} w^2, \tag{A1}$$

where C is a constant resulting from the lower limit of the integral. For one given realization, we can estimate this function from the following cumulative sum

$$W_f(w) \approx \frac{\max(w_i) - \min(w_i)}{N} \sum_{j: w_j(0) \leq w} w_j(t_f), \tag{A2}$$

where $w_j(t_f)$ denotes the value of the weight of link j at time t_f, N is the number of links of the network, and $\min(w_i)/\max(w_i)$ is the minimum/maximum value of the initialization weights. (The sum in the right-hand side of Equation (A2) runs over all links whose initial weight is not larger than w.) Finally, we fit a second degree polynomial to $W_f(w)$, and get the constants a_1 and a_0 from its coefficients.

We use the same 'cumulative-based' approach to find the second moment of $P_{w_i}(w_f)$, denoted by $\langle w_f^2 \rangle(w_i)$. In this case, we assume the polynomial $\langle w_f^2 \rangle(w_i) = b_0 + b_1 w_i + b_2 w_i^2$. We again define the cumulative $W_f^{(2)}(w) = \int_{\min(w_i)}^{w} \langle w_f^2 \rangle(x) dx$. Similarly to $W_f(w)$, we estimate $W_f^{(2)}(w)$ as

$$W_f^{(2)}(w) \approx \frac{\max(w_i) - \min(w_i)}{N} \sum_{j: w_j(0) \leq w} w_j^2(t_f), \tag{A3}$$

and fit a third degree polynomial to get the coefficients b_0, b_1, and b_2. Then, we calculate $\sigma_{w_f}(w_i)$ as

$$\sigma_{w_f}(w_i) = \sqrt{\langle w_f^2 \rangle(w_i) - \left[\langle w_f \rangle(w_i)\right]^2}. \tag{A4}$$

The method for fitting the peak (or the mode) of the distribution $P_{w_i}(w_f)$ is also based on a cumulative distribution. In our experiments we observe that, in the trainability regime, the peak of the distribution of w_f as a function of w_i is indistinguishable from a straight line, see Figure 3. Accordingly, we define

$$N_c(b) = \left|\left\{(w_i, w_f): w_f \leq b + c w_i\right\}\right|, \tag{A5}$$

which is a function that counts the number of points (w_i, w_f) below or at the line $b + cw_i$. Then, we fit the peak of $P_{w_i}(w_f)$ by optimizing the expression

$$\max_c \max \left(\frac{d}{db} N_c(b)\right), \tag{A6}$$

In other words, we look for the slope that causes the largest rate of change in the function $N_c(b)$. This slope, c_{opt}, is the slope of the linear function that best aligns with the peak of the distribution $P_{w_i}(w_f)$.

References

1. LeCun, Y.; Bengio, Y.; Hinton, G. Deep learning. *Nature* **2015**, *521*, 436. [CrossRef] [PubMed]
2. Li, Y.; Liang, Y. Learning overparameterized neural networks via stochastic gradient descent on structured data. In *Advances in Neural Information Processing Systems 31*; Curran Associates Inc.: Red Hook, NY, USA, 2018; pp. 8157–8166.
3. Jacot, A.; Gabriel, F.; Hongler, C. Neural tangent kernel: Convergence and generalization in neural networks. In *Advances in Neural Information Processing Systems 31*; Curran Associates Inc.: Red Hook, NY, USA, 2018; pp. 8571–8580.
4. Lee, J.; Xiao, L.; Schoenholz, S.; Bahri, Y.; Novak, R.; Sohl-Dickstein, J.; Pennington, J. Wide neural networks of any depth evolve as linear models under gradient descent. In *Advances in Neural Information Processing Systems 32*; Curran Associates Inc.: Red Hook, NY, USA, 2019; pp. 8572–8583.
5. LeCun, Y.; Bottou, L.; Orr, G.B.; Müller, K.R. Efficient BackProp. In *Neural Networks: Tricks of the Trade*; Springer: Berlin/Heidelberg, Germany, 1998; pp. 9–50.
6. Yam, J.Y.; Chow, T.W. A weight initialization method for improving training speed in feedforward neural network. *Neurocomputing* **2000**, *30*, 219–232. [CrossRef]
7. Glorot, X.; Bengio, Y. Understanding the difficulty of training deep feedforward neural networks. In Proceedings of the Thirteenth International Conference on Artificial Intelligence and Statistics, Sardinia, Italy, 13–15 May 2010; pp. 249–256.
8. He, K.; Zhang, X.; Ren, S.; Sun, J. Delving deep into rectifiers: Surpassing human-level performance on imagenet classification. In Proceedings of the IEEE International Conference on Computer Vision, Santiago, Chile, 7–13 December 2015; pp. 1026–1034.
9. Chapelle, O.; Erhan, D. Improved preconditioner for hessian free optimization. In *NIPS Workshop on Deep Learning and Unsupervised Feature Learning*; Sierra Nevada, Spain, 2011. Available online: https://citeseerx.ist.psu.edu/viewdoc/download?doi=10.1.1.297.3089&rep=rep1&type=pdf (accessed on 5 September 2021).
10. Krizhevsky, A.; Sutskever, I.; Hinton, G.E. Imagenet classification with deep convolutional neural networks. In *Advances in Neural Information Processing Systems*; Curran Associates Inc.: Red Hook, NY, USA, 2012; pp. 1097–1105.

11. Sutskever, I.; Martens, J.; Dahl, G.; Hinton, G. On the importance of initialization and momentum in deep learning. In Proceedings of the International Conference on Machine Learning, Atlanta, GA, USA, 16–21 June 2013; pp. 1139–1147.
12. Goodfellow, I.; Bengio, Y.; Courville, A. *Deep Learning*; MIT Press: Cambridge, MA, USA, 2016.
13. Frankle, J.; Carbin, M. The lottery ticket hypothesis: Finding sparse, trainable neural networks. In Proceedings of the International Conference on Learning Representations, New Orleans, LA, USA, 6–9 May 2019.
14. Zhou, H.; Lan, J.; Liu, R.; Yosinski, J. Deconstructing lottery tickets: Zeros, signs, and the supermask. In Proceedings of the Advances in Neural Information Processing Systems 32: Annual Conference on Neural Information Processing Systems 2019, NeurIPS 2019, Vancouver, BC, Canada, 8–14 December 2019; pp. 3592–3602.
15. Ramanujan, V.; Wortsman, M.; Kembhavi, A.; Farhadi, A.; Rastegari, M. What is Hidden in a Randomly Weighted Neural Network? *arXiv* **2019**, arXiv:1911.13299.
16. Du, S.; Lee, J.; Li, H.; Wang, L.; Zhai, X. Gradient descent finds global minima of deep neural networks. In Proceedings of the 36th International Conference on Machine Learning, PMLR, Long Beach, CA, USA, 9–15 June 2019; Volume 97, pp. 1675–1685.
17. Du, S.S.; Zhai, X.; Póczos, B.; Singh, A. Gradient descent provably optimizes over-parameterized neural networks. In Proceedings of the 7th International Conference on Learning Representations, ICLR 2019, New Orleans, LA, USA, 6–9 May 2019.
18. Allen-Zhu, Z.; Li, Y.; Liang, Y. Learning and generalization in overparameterized neural networks, going beyond two layers. In *Advances in Neural Information Processing Systems 32*; Curran Associates Inc.: Red Hook, NY, USA, 2019; pp. 6158–6169.
19. Allen-Zhu, Z.; Li, Y.; Song, Z. A convergence theory for deep learning via over-parameterization. In Proceedings of the 36th International Conference on Machine Learning, Long Beach, CA, USA, 10–15 June 2019; Chaudhuri, K., Salakhutdinov, R., Eds.; PMLR: Long Beach, CA, USA, 2019; Volume 97, pp. 242–252.
20. Allen-Zhu, Z.; Li, Y.; Song, Z. On the convergence rate of training recurrent neural networks. In *Advances in Neural Information Processing Systems 32*; Curran Associates, Inc.: Red Hook, NY, USA, 2019; pp. 6676–6688.
21. Oymak, S.; Soltanolkotabi, M. Overparameterized nonlinear learning: Gradient descent takes the shortest path? In Proceedings of the 36th International Conference on Machine Learning, Long Beach, CA, USA, 10–15 June 2019; PMLR: Long Beach, CA, USA, 2019; Volume 97, pp. 4951–4960.
22. Oymak, S.; Soltanolkotabi, M. Toward Moderate Overparameterization: Global Convergence Guarantees for Training Shallow Neural Networks. *IEEE J. Sel. Areas Inf. Theory* **2020**, *1*, 84–105. [CrossRef]
23. Zou, D.; Cao, Y.; Zhou, D.; Gu, Q. Gradient descent optimizes over-parameterized deep ReLU networks. *Mach. Learn.* **2020**, *109*, 467–492. [CrossRef]
24. Arora, S.; Du, S.; Hu, W.; Li, Z.; Wang, R. Fine-grained analysis of optimization and generalization for overparameterized two-layer neural networks. In Proceedings of the 36th International Conference on Machine Learning, Long Beach, CA, USA, 10–15 June 2019; PMLR: Long Beach, CA, USA, 2019; Volume 97, pp. 322–332.
25. Arora, S.; Du, S.S.; Hu, W.; Li, Z.; Salakhutdinov, R.R.; Wang, R. On exact computation with an infinitely wide neural net. In *Advances in Neural Information Processing Systems*; Wallach, H., Larochelle, H., Beygelzimer, A., d'Alché-Buc, F., Fox, E., Garnett, R., Eds.; Curran Associates, Inc.: Red Hook, NY, USA, 2019; Volume 32, pp. 8141–8150.
26. Chizat, L.; Oyallon, E.; Bach, F. On lazy training in differentiable programming. In *Advances in Neural Information Processing Systems 32*; Curran Associates, Inc.: Red Hook, NY, USA, 2019; pp. 2937–2947.
27. Frankle, J.; Schwab, D.J.; Morcos, A.S. The early phase of neural network training. In Proceedings of the International Conference on Learning Representations, Addis Ababa, Ethiopia, 26–30 April 2020.
28. Lu, Z.; Pu, H.; Wang, F.; Hu, Z.; Wang, L. The expressive power of neural networks: A view from the width. In *Advances in Neural Information Processing Systems*; Curran Associates, Inc.: Red Hook, NY, USA, 2017; pp. 6231–6239.
29. Li, D.; Ding, T.; Sun, R. On the benefit of width for neural networks: Disappearance of bad basins. *arXiv* **2018**, arXiv:1812.11039.
30. Chollet, F. Keras. 2015. Available online: https://keras.io (accessed on 5 September 2021).
31. Abadi, M.; Agarwal, A.; Barham, P.; Brevdo, E.; Chen, Z.; Citro, C.; Corrado, G.S.; Davis, A.; Dean, J.; Devin, M.; et al. TensorFlow: Large-Scale Machine Learning on Heterogeneous Systems, 2015. Available online: tensorflow.org (accessed on 5 September 2021).
32. LeCun, Y.; Bottou, L.; Bengio, Y.; Haffner, P. Gradient-based learning applied to document recognition. *Proc. IEEE* **1998**, *86*, 2278–2324. [CrossRef]
33. Xiao, H.; Rasul, K.; Vollgraf, R. Fashion-MNIST: A Novel Image Dataset for Benchmarking Machine Learning Algorithms. *arXiv* **2017**, arXiv:1708.07747.
34. Thoma, M. The hasyv2 dataset. *arXiv* **2017**, arXiv:1701.08380.
35. Helsen, W.F.; Van Winckel, J.; Williams, A.M. The relative age effect in youth soccer across Europe. *J. Sport. Sci.* **2005**, *23*, 629–636. [CrossRef] [PubMed]

Article

Compression of Neural Networks for Specialized Tasks via Value Locality

Freddy Gabbay [1,*] and Gil Shomron [2]

1. Computer Science Department, Ruppin Academic Center, Emek Hefer 4025000, Israel
2. Faculty of Electrical and Computer Engineering, The Technion—Israel Institute of Technology, Haifa 3200000, Israel; gilsho@campus.technion.ac.il
* Correspondence: freddyg@ruppin.ac.il

Abstract: Convolutional Neural Networks (CNNs) are broadly used in numerous applications such as computer vision and image classification. Although CNN models deliver state-of-the-art accuracy, they require heavy computational resources that are not always affordable or available on every platform. Limited performance, system cost, and energy consumption, such as in edge devices, argue for the optimization of computations in neural networks. Toward this end, we propose herein the value-locality-based compression (VELCRO) algorithm for neural networks. VELCRO is a method to compress general-purpose neural networks that are deployed for a small subset of focused specialized tasks. Although this study focuses on CNNs, VELCRO can be used to compress any deep neural network. VELCRO relies on the property of value locality, which suggests that activation functions exhibit values in proximity through the inference process when the network is used for specialized tasks. VELCRO consists of two stages: a preprocessing stage that identifies output elements of the activation function with a high degree of value locality, and a compression stage that replaces these elements with their corresponding average arithmetic values. As a result, VELCRO not only saves the computation of the replaced activations but also avoids processing their corresponding output feature map elements. Unlike common neural network compression algorithms, which require computationally intensive training processes, VELCRO introduces significantly fewer computational requirements. An analysis of our experiments indicates that, when CNNs are used for specialized tasks, they introduce a high degree of value locality relative to the general-purpose case. In addition, the experimental results show that without any training process, VELCRO produces a compression-saving ratio in the range 13.5–30.0% with no degradation in accuracy. Finally, the experimental results indicate that, when VELCRO is used with a relatively low compression target, it significantly improves the accuracy by 2–20% for specialized CNN tasks.

Keywords: machine learning; deep neural networks; convolutional neural network; deep compression

Citation: Gabbay, F.; Shomron, G. Compression of Neural Networks for Specialized Tasks via Value Locality. *Mathematics* **2021**, *9*, 2612. https://doi.org/10.3390/math9202612

Academic Editor: Oliviu Matei

Received: 30 September 2021
Accepted: 15 October 2021
Published: 16 October 2021

Publisher's Note: MDPI stays neutral with regard to jurisdictional claims in published maps and institutional affiliations.

Copyright: © 2021 by the authors. Licensee MDPI, Basel, Switzerland. This article is an open access article distributed under the terms and conditions of the Creative Commons Attribution (CC BY) license (https://creativecommons.org/licenses/by/4.0/).

1. Introduction

Convolutional Neural Networks (CNNs) are broadly employed by numerous computer vision applications such as autonomous systems, healthcare, retail, and security. Over time, the processing requirements and complexity of CNN models have significantly increased. For example, AlexNet [1], which was introduced in 2012, has eight layers, whereas ResNet-101 [2], which was released in 2015, uses 101 layers and requires an approximately sevenfold-greater computational throughput [3]. The increasing model complexity in conjunction with large datasets used for model training has endowed CNNs with phenomenal performance for various computer vision tasks [4]. Typically, large complex networks can further extend their capacity to learn complex image features and properties. The growing model size of CNNs and the requirement of significant processing power have become major deployment challenges for migrating CNN models into mobile, Internet of Things, and edge applications. Such applications incur limited computational and memory resources, energy constraints, and system cost and, in many cases, cannot rely on cloud computational resources due to privacy, online communication network availability, and real-time considerations.

The compression of CNN models without excessive performance loss significantly facilitates their deployment by a variety of edge systems. Such compression has the potential to reduce computational requirements, save energy, reduce memory bandwidth and storage requirements, and shorten inference time. Various techniques have been suggested to compress CNN models, one of the most common of which is pruning [5–7], which exploits the tendency to over-parameterize CNNs [8]. Pruning trades off degradation in model prediction accuracy for model size by removing weights, Output Feature Maps (OFMs), or filters that make minor or no contribution to the inference of a network. Quantization [9–12] is another common technique that attempts to further compress network size by reducing the number of bits used to represent weights, filters, and OFMs with only a minor impact on accuracy. These methods and other compression approaches are discussed in more detail in Section 2.

This paper focuses on machine learning models that are used for specialized tasks. A specialized neural network is typically a general-purpose model which has been adjusted and optimized to carry out a set of specific tasks. Specialized neural networks have recently become common not only for edge devices but also for datacenters [13–15]. Unlike general-purpose neural networks that are used for a diverse range of classification tasks, specialized neural networks are used for a small number of specific classification tasks. For example, a CNN model that is used to detect vehicles does not use its animal classification capabilities. A common usage of specialized CNN is as a fast filter in front of a heavy general-purpose CNN model. A typical example to such usage is related to offline video analytics [14], which is processed by a specialized CNN model, and only when the model has a low level of confidence are the corresponding frames sent to a general-purpose CNN. Another example is related to game scrapping, where specialized CNNs are used to classify video stream events in by scraping in-game text appearing in frames. A cascaded-CNN [16] is another approach that employs multiple specialized CNNs. The result of each specialized CNN is combined to produce a complete prediction map. The mixture-of-experts model [17] employs a combination of expert models, where each expert is a neural network specialized in specific tasks. Hierarchical classification is another example for specialized CNN usage. Since image categories are typically organized in hierarchical manner, hierarchical classification can be employed by performing a prediction starting from a super class and can only perform detailed classification within the super class. We introduce in this study the value-locality-based compression (VELCRO) algorithm. VELCRO is a method to compress deep neural networks that were originally trained for a large set of diverse classification tasks but are deployed for a smaller subset of specialized tasks. Although this work focuses on CNN models, VELCRO can be used to compress any deep neural network. The main principle of VELCRO is based on the property of value locality, which we introduce herein in the context of neural networks. This property suggests that, when the network is used for specialized tasks, a proximal range of values are produced by the activation functions in the inference process. VELCRO consists of two stages: a preprocessing stage, which identifies activation-function output elements with a high degree of value locality, and a compression stage, which replaces these activation elements with their corresponding arithmetic averages. As a result, VELCRO avoids not only the computation of these activation elements but also the convolution computation of their corresponding OFM elements. VELCRO also requires significantly fewer computational resources than common pruning techniques due to the avoidance of back propagation training. For our experimental analysis we use three CNN models: ResNet-18 [2], MobileNet V2 [18], and GoogLeNet [19] with the ILSVRC-2012 (ImageNet) [20] dataset to examine compression capabilities and model accuracy. Lastly, we implement VELCRO in hardware on a Field Programable Gate-Array (FPGA) and demonstrate the computational and energy savings.

The contributions of this paper are summarized as follows:

1. We introduce the notion of value locality in the context of deep neural networks used for specialized tasks.
2. We present the VELCRO algorithm, which exploits value locality to compress neural networks that are deployed for specialized tasks.
3. VELCRO introduces a fast compression process which solely employs statistics gathering through the inference process and avoids heavy computations involved in back-propagation training, which is usually used by traditional compression approaches such as pruning.
4. VELCRO can be used directly in conjunction with other compression methods such as pruning and quantization.
5. The results of our experiments indicate that
    a. VELCRO produces a compression-saving ratio of computations in the range 20.0–27.7% for ResNet-18, 25–30% for GoogLeNet, and 13.5–20% for MobileNet V2 with no impact on model accuracy;
    b. VELCRO significantly improves accuracy by 2–20% for specialized-task CNNs when given a relatively small compression-savings target.
6. We demonstrate the computational and energy savings of VELCRO by implementing the compression algorithm in hardware on FPGA. Our experimental results indicate a 13.5–30% reduction in energy consumption with VELCRO, which corresponds to the compression-saving ratio.

The remainder of this paper is organized as follows: Section 2 reviews previous work. Section 3 introduces the proposed method and algorithm. Section 4 presents the experimental results. Finally, Section 5 summarizes the conclusions and suggests future research directions.

2. Prior Works

Numerous recent studies have proposed various techniques to optimize CNN computations, reduce redundancy, and improve computational efficiency and memory storage. This section describes the following related methods: pruning, quantization, knowledge distillation, deep compression, CNN folding, ablation and CNN filters compression methods.

Pruning is one of the most common methods used for CNN optimization and was introduced in Refs. [5–7]. The concept of pruning, which is inspired by neuroscience, assumes that some network parameters are redundant and may not contribute to network performance. Various pruning techniques [5,21–25] suggest the removal of activations, weights, OFMs, or filters that make a minor or no contribution to the inference process of an already-trained network. Thereby, pruning can significantly reduce the network size and the number of computations. Traditional pruning techniques typically require fine-tuned training on the full model, which may involve significant computational overhead [26].

Pruning techniques can be classified into unstructured and structured classes. Unstructured pruning imposes no constraints on the activations or weights with respect to the network structure (i.e., individual weights or activations are removed by replacing them with zero). Structured pruning [27], in contrast, restricts the pruning process to a set of weights, channels, filters, or activations. Whereas structured pruning incurs limitations on the sparsity that can be exploited in the network due to its coarse pruning granularity, unstructured pruning uses a broader scope of the available sparsity. Conversely, unstructured pruning may involve additional overhead for representing the pruned elements and may not always fit parallel processing elements such as GPUs.

The process of pruning is typically performed by ranking the network elements in accordance with their contribution. The rank can be determined by using various functions such as the L1 or L2 norms [28–31] of weights, activations, or other metrics [32]. Activation pruning requires dynamic mechanisms to monitor activation values because activation importance may depend on the model input. For example, Ref. [33] employs reinforcement learning to prune channels, and Refs. [34,35] leverage spatial correlations of CNN OFMs to predict and prune zero-value activations. Further pruning techniques based on weight magnitudes were recently introduced in Refs. [21,36,37], which demonstrate that computation efficiency and network scale can be improved significantly. Various gradual pruning approaches [38], given memory footprints and computational bounds, were studied by examining the accuracy and size tradeoffs. The neuron importance score propagation, introduced by Ref. [39], suggests jointly pruning neurons based on a unified goal. Other approaches such as random neuron pruning and random grouping of weight connections into hash buckets were introduced in Refs. [40,41]. Pruning based on a Taylor-expansion criterion [42] focuses on transfer learning by optimizing a network trained to a large dataset of images into a smaller and more efficient network specialized in a subset of classes. Their pruning method performs an iterative backpropagation pruning by removing feature maps with the least level of importance. Ref. [42] evaluated their pruning method by using various criteria such as weight pruning, using l2 norm, and activation pruning, using mean, variance, mutual information, and Taylor-expansion criteria. Their results indicate that the importance of OFMs decreases with layer depth and that each layer has feature maps with both high and low degrees of importance. Ref. [43] introduced pruning by compression using residual connections and limited data (CURL) for residual CNN compression when relying on small datasets that represent specialized tasks.

Quantization methods attempt to reduce the number of bits used to represent the values of weights, filters, and OFMs from 32-bit floating point to 8 bit or less with a slight degradation in model accuracy while simplifying computational complexity. Employing quantization methods that use fewer than 8 bits, however, is not trivial because quantization noise excessively degrades model accuracy. Quantization-aware training uses training processes for quantization to reduce quantization noise and recover model accuracy [44–46]. This approach can be limited when the training process cannot be used due to lack of dataset availability or lack of computational resources. Various fixed-point and vector quantization methods, introduced in Refs. [9–12], present tradeoffs between network accuracy and quantization-compression ratios. A combination of pruning and quantization was introduced in Ref. [22]. Post-training quantization methods [47–50] avoid these limitations by searching for the optimal tensor-cutting values to reduce quantization noise after the network model has been trained.

Knowledge distillation is another machine learning optimization [51,52] that transfers knowledge from a large machine learning model into a smaller compact model that mimics the original model (instead of being trained on the original dataset) to perform competitively. These systems consist of three main elements: knowledge, an algorithm for knowledge distillation, and a teacher–student model. A broad survey of knowledge distillation is available in Ref. [53].

Deep compression was introduced in Ref. [22] and consists of a three-stage pipeline: pruning, trained quantization, and Huffman coding, which operate simultaneously to optimize model size. The first stage prunes the model by learning the important connections, the second stage performs weight quantization and sharing, and the last stage uses Huffman coding. Ref. [54] extends the deep compression idea and introduces the once-for-all network, which can be installed under diverse architectural constraints and configurations, such as performance, power, and cost. The once-for-all approach introduces the progressive shrinking techniques that generalize pruning. Whereas pruning shrinks the network width, progressive shrinking operates on four dimensions: image resolution, kernel size, depth, and width, thereby achieving a higher level of flexibility.

FoldedCNN [15] is another approach to optimize CNNs for specialized-inference tasks. Unlike compression techniques, FoldedCNN does not aim at compressing the CNN model but rather attempts to increase the inference throughput and hardware utilization. The FoldedCNN approach suggests CNN model transformations to increase their arithmetic intensity when processing a large batch size without increasing processing requirements.

Additional studies have attempted to understand the internal mechanisms of CNNs and their contribution to classification tasks. From various CNN models, Refs. [55,56] created visualized images based on the OFMs of different layers and units. Their results indicate that OFMs extract features that detect patterns, textures, shapes, concepts, and various elements related to the classified images. Ablation techniques were used by Ref. [57] to further quantify the contribution of OFM units to the classification task. Their results indicate that elements that are selective to certain classes may be excluded from the network without necessarily impacting the overall model performance. The impact on ablation of a subset of classes was further studied in Ref. [58], which found that single-OFM-unit ablation can significantly impact the model accuracy for a subset of classes, leading them to suggest different methods to measure the importance of internal OFM units to specific classification accuracy.

CNN filter compression techniques attempt to remove kernel and filters that have small contribution to the network performance. Removal of specific convolution filters based on their importance has been introduced in Ref. [59]. The authors suggest considering two consecutive network layers as a coupled function where the weights are used to compute the coupling factors. In addition, they suggest using the coupling factors to prune filter and maximize the variance of feature maps. Another study on convolution filters compression [60] has highlighted that certain feature maps inside and across CNN layers may have a different contribution to the accuracy of the inference process. The authors indicate that, first model layers typically extract semantic features while the deep layers may extract simple features. Thereby, understanding the importance of feature map can help the compression of the network. They investigate the relationship between input feature map and filter kernels and suggest Kernel Sparsity and Entropy (KSE) as a quantitative indicator for the feature map importance.

These recent studies [55–60] provide the motivation for the present study by suggesting that, when using the CNN model for specialized tasks, we eliminate unrelated computations and thereby compress the model, all with minimal impact on classification accuracy.

3. Method and Algorithm

Our proposed VELCRO compression algorithm relies on the fundamental property of value locality. We start our discussion by first presenting qualitative and quantitative aspects of value locality, following which we describe the VELCRO compression algorithm for specialized neural networks.

3.1. Value Locality of Specialized Convolutional Neural Networks

The principle of the method proposed to compress specialized CNNs is based on the property of value locality. Value locality suggests that, when a CNN model runs specialized tasks, the output values of the activation tensor is in proximity through inference of images. The rationale behind this theory relies on the assumption that the inferred images, which already have a certain level of similarity, exhibit common features such as patterns, textures, shapes, and concepts. As a result, the intermediate layers of the model produce similar values in the vicinity. Figure 1 explains the property of value locality by illustrating the activation-function output tensors in each convolution layer k and channel c. In this example, the set of elements $A^{(m)}[k][c][i][j]$ for images $m = 0, 1, \ldots, N-1$ in the activation tensor is populated with values in proximity through the inference between images.

For every convolution layer k, we define a variance tensor V[k], where each element V[k][c][i][j] in the variance tensor is defined as

$$\begin{aligned} V[k][c][i][j] &= \text{Var}(A[k][c][i][j]) = E(A[k][c][i][j]^2) - E(A[k][c][i][j])^2 \\ &= \tfrac{1}{N}\sum_{m=0}^{N-1} A^{(m)}[k][c][i][j]^2 - \left(\tfrac{1}{N}\sum_{m=0}^{N-1} A^{(m)}[k][c][i][j]\right)^2, \end{aligned} \quad (1)$$

where c is the channel index and i and j are the element coordinates.

We use the variance tensor as a measure to quantify the proximity of values for every activation tensor element A[k][c][i][j]. Thereby, a small value of V[k][c][i][j] suggests that the corresponding activation element has a high degree of value locality. The proposed compression algorithm leverages such activation elements for compression. Section 4 presents an experimental analysis of the distribution of the variance tensor for various specialized CNN models.

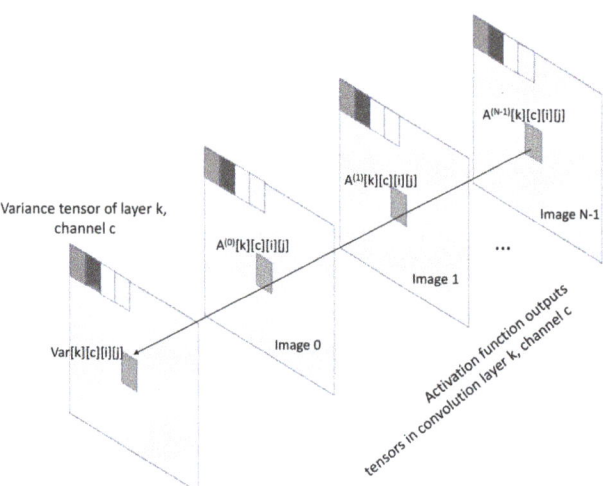

Figure 1. Value locality: The elements with coordinates i, j of the activation-function output tensor in convolutional layer k, channel c, are populated with values in proximity through the inference between images 0 to N − 1. The variance tensor V serves to measure the degree of value locality.

3.2. VELCRO Algorithm for Specialized Neural Networks

The VELCRO algorithm consists of two stages: preprocessing and compression.

1. Preprocessing stage: In this stage, VELCRO makes an inference by applying the original CNN model to a small subset of images from the specialized task preprocessing dataset. Note that the performance of the compressed model is evaluated on a validation dataset which is distinct from the preprocessing dataset. This is discussed in detail in Section 4. During this stage, the variance tensor is calculated by using Equation (1) for each activation output in each convolution layer in the CNN model. Because the preprocessing stage of VELCRO relies only on inference, it involves a significantly smaller computational overhead with respect to traditional compression methods, which employ heavy backpropagation training processes that can last from a few hours up to hundreds of hours [61].

2. Compression stage: The compression stage uses a tuple of threshold values provided by the user as a hyperparameter for the algorithm. Each threshold element in the tuple corresponds to an individual activation function in each convolution layer. The threshold value of each layer represents the percentile of elements in the variance tensor to be compressed by the algorithm. All elements in the activation tensor with a variance within the percentile threshold are replaced by the arithmetic average constant of the elements located in the same corresponding coordinates. All other activation elements remain unchanged. Replacing activation function output elements by constants avoids not only the activation function computation but also the particular convolution computation of their related OFM elements. In fact, the compression savings of each layer is determined by the corresponding threshold, so the user can determine the overall compression-saving ratio C for the model through the threshold tuple as follows:

$$C = 1 - \frac{\text{Compressed model computations}}{\text{Original model computations}} = \sum_{k=0}^{K-1} T_k c_k w_k h_k \quad (2)$$

where the tuple $T = \{T_0, T_1, \ldots, T_K\}$ contains the threshold values for the activation in each convolution layer. In addition, c_k, w_k, and h_k are the number of channels, the width, and the height of the activation function output tensor for convolution layer k, respectively.

The complete and formal definition of the algorithm is given in Algorithm 1.

A simple example that demonstrates the VELCRO algorithm is illustrated in Figure 2, which shows the activation output tensor in convolution layer k for a preprocessing dataset of N = 3 images. The dimensions of the activation tensor are $c_k = 1$, $w_k = 3$, and $h_k = 3$. The VELCRO preprocessing stage performs inference on the preprocessing dataset to create a variance tensor V[k] and an arithmetic average tensor B[k]. The hyperparameter threshold value for layer k is defined in this example as $T_k = 0.33$, which means that the three elements in the activation function output tensor with the lowest variance (highlighted in red) are replaced with their arithmetic average. The remaining elements remain unchanged. The outcome of the VELCRO compression stage is given by the compressed activation-function output tensor $\widetilde{A}[k]$, where the computation of three elements (highlighted in green) are replaced by the arithmetic averages.

Algorithm 1: VELCRO algorithm for specialized neural networks

Input: A CNN model M with K activation-function outputs (each in a different convolution layer), N preprocessing images, and a threshold tuple $T = \{T_0, T_1, \ldots, T_K\}$, where $\forall\ 0 \leq n < N\ \ 0 \leq T_n < 1$.
Output: A compressed CNN Model M_C.
Preprocessing stage
Step 1: Let A(k) be the activation-function output tensor in convolution layer k and let $A^{(m)}(k)$ be the corresponding activation-tensor values at the inference of image m, $0 \leq m < N$, where the tensors A[k] and $A^{(m)}[k]$ have dimension $c_k \times w_k \times h_k$ and c_k, w_k, and h_k are the number of channels, the width, and the height of the tensor at convolution layer k, respectively.
Step 2: For every $0 \leq k < K$, $0 \leq c < c_k$, $0 \leq i < w_k$, and $0 \leq j < h_k$:
Let tensors S and K be initialized such that S[k][c][i][j] = 0 and Q[k][c][i][j] = 0
Step 3: For each image $0 \leq m < N$:
Perform inference by model M on image m.
For every convolution layer $0 \leq k < K$:
For every $0 \leq c < c_k$, $0 \leq i < w_k$, and $0 \leq j < h_k$,
Let the tensors S and Q be
S[k][c][i][j] = S[k][c][i][j] + $A^{(m)}$[k][c][i][j]
Q[k][c][i][j] = Q[k][c][i][j] + $(A^{(m)}[k][c][i][j])^2$.
Step 4: Let B[k] be the arithmetic average tensor in convolution layer k such that each tensor element is
$B[k][c][i][j] = \frac{1}{N} S[k][c][i][j]$
For every $0 \leq c < c_k$, $0 \leq i < w_k$, and $0 \leq j < h_k$,
Step 5: Let V[k] be the variance tensor of convolution layer k such that each tensor element is
$V[k][c][i][j] = \frac{1}{N} Q[k][c][i][j] - (B[k][c][i][j])^2$
For each $0 \leq c < c_k$, $0 \leq i < w_k$, and $0 \leq j < h_k$
Compression stage:
Step 6: For each convolution layer $0 \leq k < K$:
Let p(x,Y) be the percentile function of element x in tensor Y. p returns the percentile value for x with respect to all elements in tensor Y.
Let the tensor $\widetilde{A}[k]$ be
$$\widetilde{A}[k][c][i][j] = \begin{cases} A[k][c][i][j] & p(V[k][c][i][j], A[k]) > T_k \\ B[k][c][i][j] & p(V[k][c][i][j], A[k]) \leq T_k \end{cases}$$
For each $0 \leq c < c_k$, $0 \leq i < w_k$, and $0 \leq j < h_k$
Step 7: Let the compressed CNN model M_C be such that every activation function output tensor A[k] is replaced with $\widetilde{A}[k]$ for every convolution layer $0 \leq k < K$.

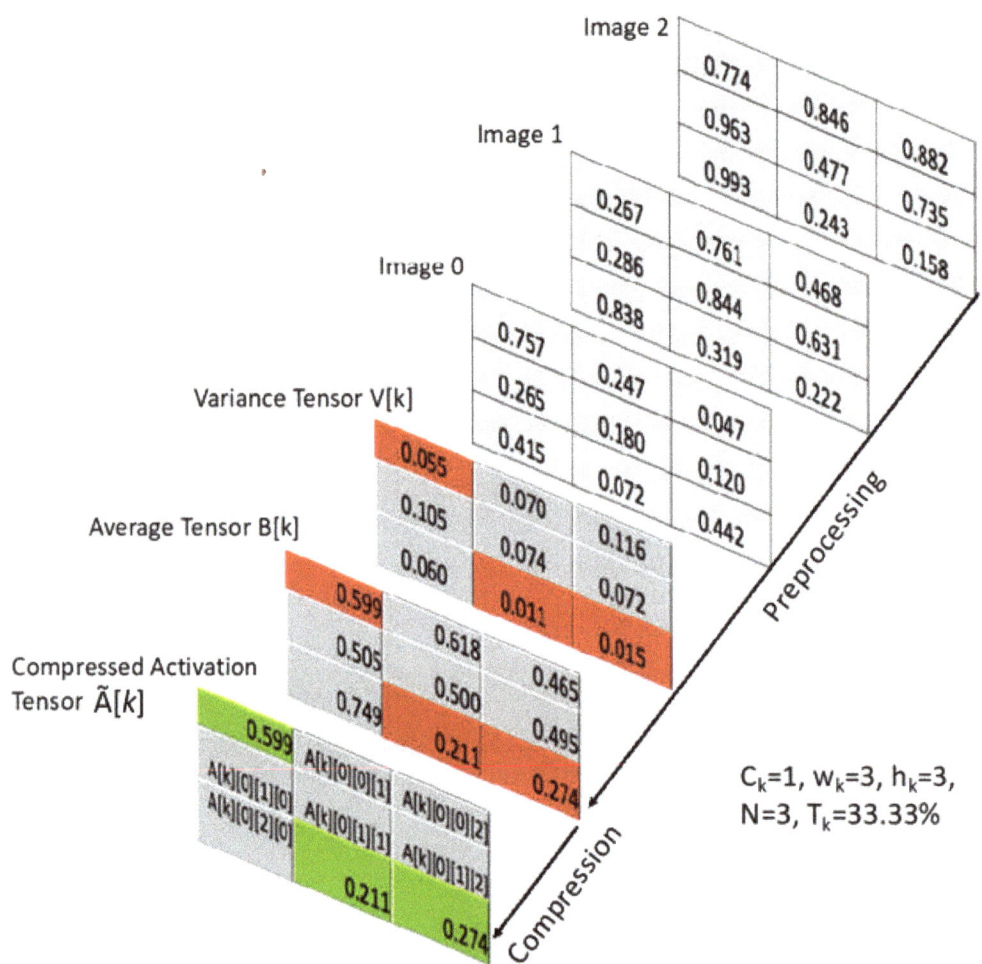

Figure 2. Example of VELCRO preprocessing and compression stages.

4. Experimental Results and Discussion

Our experimental study consists of a comprehensive analysis of both value locality and the performance of various CNN models when used for specialized tasks. In the following, we first describe the experimental environment and then introduce the value locality experimental measurements. Next, we discuss the performance of the VELCRO compression algorithm. Finally, we demonstrate the computational and energy savings of VELCRO by designing a hardware that implements the compression algorithm on FPGA.

4.1. Experimental Environment

Our experimental environment is based on PyTorch [62], the ILSVRC-2012 dataset (also known as "ImageNet") [20,60], and the ResNet-18, MobileNet V2, and GoogLeNet CNN models [18,19,55] with their PyTorch pretrained models. The VELCRO algorithm, described in Algorithm 1, has been fully implemented on the PyTorch environment. Table 1 summarizes the specialized tasks used for our experimental analysis. The experiments examine five groups of specialized tasks: the groups Cats-2, Cats-3, and Cats-4 include two, three, and four classes from the ILSVRC-2012 dataset, respectively, and the groups Dogs and Cars include four classes each. Throughout the experimental analysis, we do not modify the first layer of the model, which is a common practice that has been used by numerous studies [46].

Table 1. Specialized tasks summary.

Specialized Tasks	ILSVRC-2012 Classes
Cats-2	Egyptian cat
	Persian cat
Cats-3	Egyptian cat
	Persian cat
	Cougar
Cats-4 (Cats)	Egyptian cat
	Persian cat
	Cougar
	Tiger cat
Dogs	English setter
	Siberian husky
	English springer
	Scottish deerhound
Cars	Beach wagon
	Cab
	Convertible
	Minivan

4.2. Experimental Analysis of Value Locality

The distribution of the variance tensor elements in each layer (skipping the first layer) is a measure to quantify the proximity of the activation-function output. Figure 3 shows the distribution of the variance tensor elements for the selected activation function outputs in the convolution layers 1, 3, 7, 10, and 14 in ResNet-18. The distribution is shown for the groups of classes Cats-2, Cats-3, and Cats-4, which include two, three, and four classes of cats from the dataset, respectively. The group "all" contains a mixture of all ILSVRC-2012 dataset classes and represents the case when the CNN model is used for general tasks. When the CNN model is used for specialized tasks (Cats-1, -2, and -3), the distribution of the variance tensor elements clearly shifts toward zero with respect to the distribution when the model is used for general tasks (all), which indicates that the CNN model produces values of closer proximity (i.e., a higher degree of value locality) for specialized tasks. Another important outcome made apparent in Figure 3 is that the three groups of specialized tasks behave similarly regardless of the number of classes. The distribution of variance tensor elements in all ReseNet-18 layers is presented in Figure A1 (Appendix A) and behaves similarly to the distribution presented herein.

Figure 4 illustrates the same experimental analysis but for the GoogLeNet CNN model for selected layers 1, 6, 12, 21, 32, 38, 47, 51, and 56. The variance tensor elements of GoogLeNet behave very similarly to those of ResNet-18. When the model is used for specialized tasks, the variance distribution shifts left with respect to the general-purpose use, indicating a higher degree of value locality. The distribution in all GoogLeNet layers is presented in Figure A2 (Appendix A).

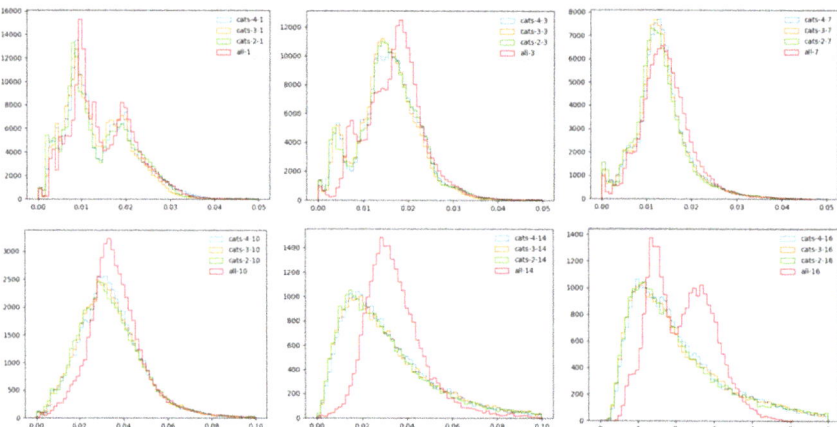

Figure 3. Distribution of ResNet-18 variance tensor elements in layers 1, 3, 7, 10, 14, and 16 for specialized tasks: all ImageNet classes, Cats-2, Cats-3, and Cats-4.

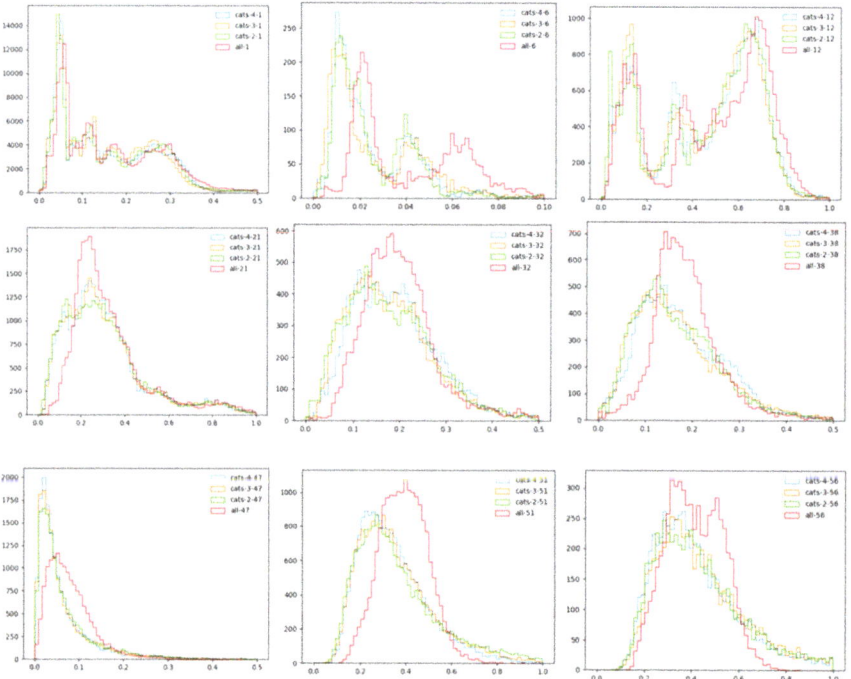

Figure 4. Distribution of GoogLeNet variance tensor elements in layers 1, 6, 12, 21, 32, 38, 47, 51, and 56 for specialized tasks: all ImageNet classes, Cats-2, Cats-3, and Cats-4.

Figure 5 presents a similar experimental analysis for MobileNet V2 layers 1, 6, 12, 19, 28, 30, and 35, and the distribution in all MobileNet V2 layers is presented in Figure A3 (Appendix A). The results indicate that a lower degree of value locality occurs relative to ResNet-18 and GoogLeNet when MobileNet V2 is used for specialized tasks. The results indicate that the shift of the variance tensor elements distribution is smaller with respect to the other CNN models. These observations reflect the highly compact nature of the

MobileNet V2 network with respect to ResNet-18 and GoogLeNet, which results in a lower potential for leveraging value locality for the former.

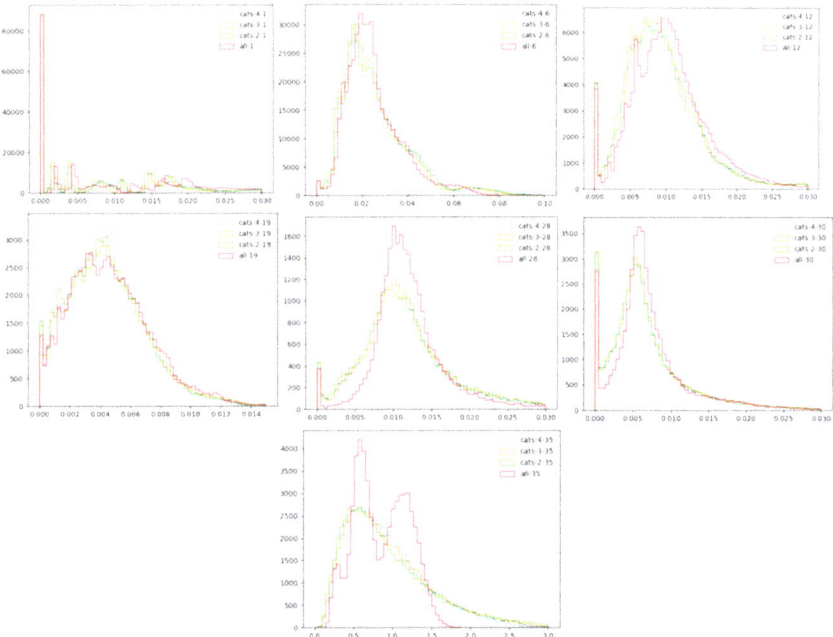

Figure 5. Distribution of MobileNet-V2 variance tensor elements in layers 1, 6, 12, 19, 28, 30, and 35 for specialized tasks: all ImageNet classes, Cats-2, Cats-3, and Cats-4.

Figures 6–8 extend our experimental analysis for additional groups of specialized tasks, Dog and Cars, each of which includes four classes from the ILSVRC-2012 dataset. Note that the Cats group corresponds to the group Cats-4. The results further confirm those shown in Figures 3–5. In all the examined CNN models and in the majority of activation-function outputs in all convolution layers, the distribution of variance tensor elements for the specialized tasks clearly shifts toward zero relative to the distribution when the model is used for general tasks (all). Like the results presented in Figure 5, we also observe that MobileNet V2 can leverage value locality but in a smaller magnitude with respect to ResNet-18 and GoogLeNet.

These experimental results support our expectations that CNN models that are used for specialized tasks exhibit a high degree of value locality. Figures A4–A6 (Appendix A) show the experimental results for all layers of all models. The complete experimental results for all layers behave similarly to the distribution presented in Figures 6–8.

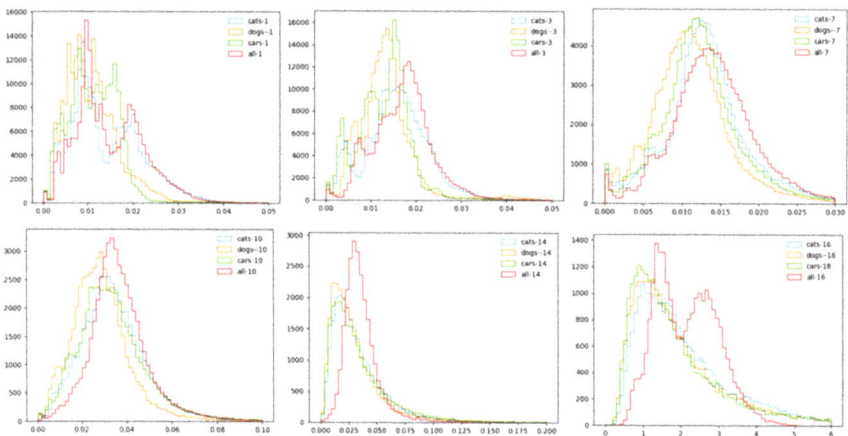

Figure 6. Distribution of ResNet-18 variance tensor elements in layers 1, 3, 7, 10, 14, and 16 for specialized tasks: Cats, Dogs, Cars, and all ImageNet classes.

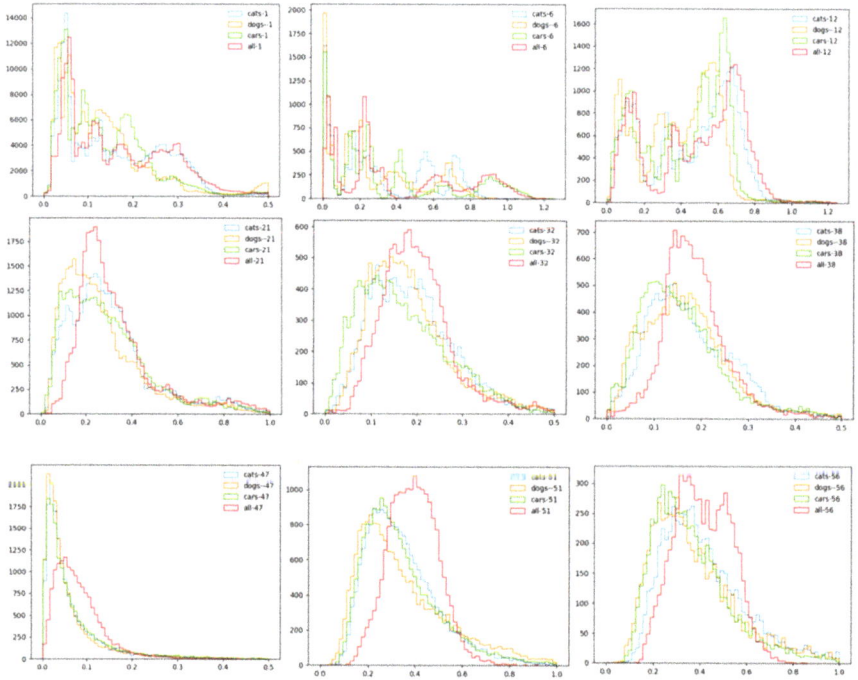

Figure 7. Distribution of GoogLeNet variance tensor elements in layers 1, 6, 12, 21, 32, 38, 47, 51, and 56 for specialized tasks: Cats, Dogs, Cars, and all ImageNet classes.

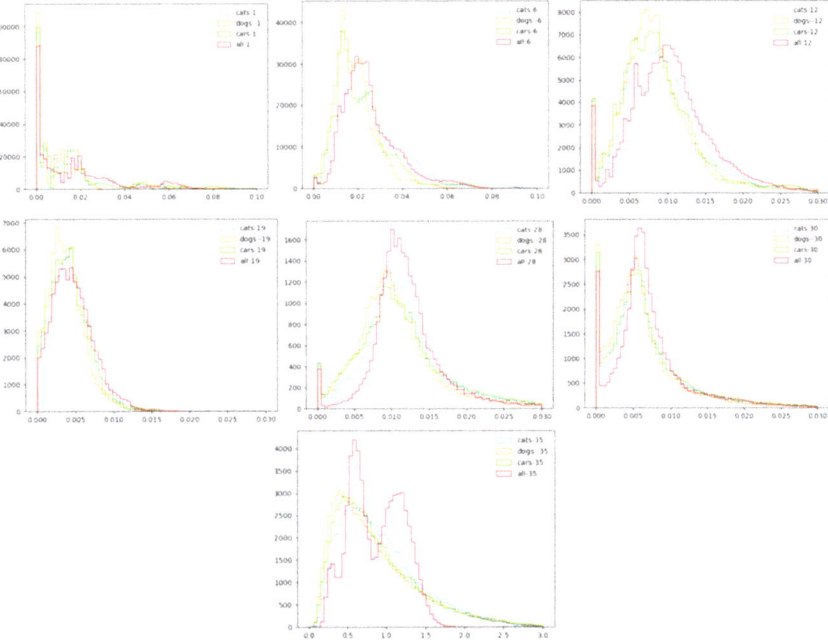

Figure 8. Distribution of MobileNet V2 variance tensor elements in layers 1, 6, 12, 19, 28, 30, and 35 for specialized tasks: Cats, Dogs, Cars, and all ImageNet classes.

4.3. Performance of Compression Algorithm

As part of our experimental analysis, we examine the compression-saving ratio of the VELCRO algorithm on three groups of specialized tasks: cats, cars, and dogs (see Table 1). Only a very small subset (<2%) of images from the preprocessing dataset has been used for the preprocessing stage of the algorithm, while the remaining images have used for the validation of the compressed model. This approach is essential in order to perform an unbiased evaluation of the model performance and preserved the generalization property of the model. Figure 9a–c present the top-1 prediction accuracy versus the compression-saving ratio for cars, dogs, and cats, respectively. The experimental analysis is applied to the ResNet-18, GoogLeNet, and MobileNet V2 CNN models. For each compression-saving ratio, we examine different thresholds by running trial-and-error and choose those that produce the highest top-1 accuracy. Tables A1–A3 in Appendix B summarize the tuples of threshold values. Table 2 summarizes the maximum compression-saving ratio for each group of specialized tasks and each CNN model that produces the same accuracy as the original uncompressed model.

The experimental results indicate that VELCRO produces a compression-saving ratio of 20.00–27.73% in ResNet-18 and 25.46–30.00% in GoogLeNet. The higher compression-saving ratio in GoogLeNet is attributed to the fact that GoogLeNet uses significantly a greater number of parameters and thereby has higher potential to leverage value locality. This explains why GoogLeNet better leverages value locality when the network is employed for special tasks. Conversely, MobileNet V2 produces a smaller compression-saving ratio, 13.50–19.76%, for the specialized tasks examined. These results comply with our previous measurements of the distribution of the variance tensor elements, which imply that the potential of leveraging value locality in MobileNet V2 is smaller than that of the other CNNs examined. This is explained by the fact that MobileNet V2 is much more compact than the other CNNs examined and thereby has a lower potential to leverage value locality.

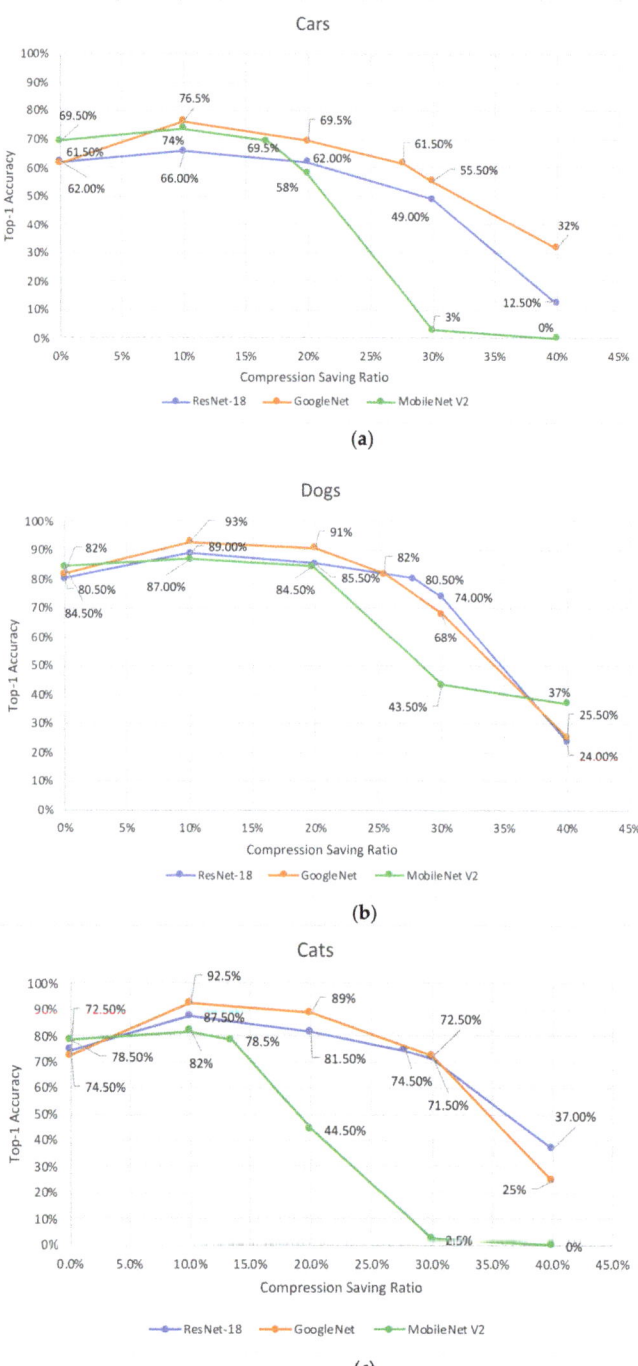

Figure 9. Accuracy for ResNet-18, GoogLeNet, and MobileNet V2 versus compression-saving ratio for specialized tasks: (**a**) Cars, (**b**) Dogs, and (**c**) Cats.

Table 2. Maximum compression-saving ratio achieved while maintaining the accuracy of the original uncompressed CNN model.

Specialized Task	ResNet-18	GoogLeNet	MobileNet V2
Cats	27.73%	30.00%	13.50%
Dogs	27.70%	25.46%	19.76%
Cars	20.00%	27.70%	16.80%

Note that VELCRO does not aim to compress the network memory footprint but rather to reduce the computational requirements. Therefore, any comparison of VELCRO to pruning approaches should consider computation aspects rather than the number of parameters in the network. Table 3 compares the VELCRO algorithm with other pruning approaches for both specialized CNNs and general-purpose ones. Although VELCRO achieves smaller computation savings, it requires significantly fewer computational resources than common pruning techniques [61] due to the avoidance of back propagation training.

Table 3. Comparison summary of VELCRO with respect to punning techniques. We also examine the output of the activation functions compressed by VELCRO.

Compression Method	Network	Specialized Task	Training Required	Computation Acceleration	Accuracy Loss
Taylor criterion [42]	AlexNet	Yes	Yes	1.9X	0.3%
CURL [43]	MobileNet V2	Yes	Yes	3X	Up to 4%
	ResNet-50	Yes	Yes	4X	Up to 2%
Deep compression [22]	Various CNN models	No	Yes	3X	None
Weights and connection learning [21]	AlexNet	No	Yes	3X	None
KSE [59]	ResNet-50	No	Yes	3.8–4.7X	0.84–0.64%
VELCRO	ResNet-18	Yes	No	1.25–1.38X	None
	GoogLeNet	Yes	No	1.38–1.42X	None
	MobileNet V2	Yes	No	1.15–1.24X	None

Table 4 presents the percent compression of activation elements with zero value out of all the compressed activation elements. The results in Table 3 correspond to the compression-saving ratios in Table 2 (i.e., when the network achieves maximum compression without losing accuracy). With ResNet-18 and GoogLeNet, the fraction of compressed zero values is in the range 0.08–0.31% and 0.56–0.64%, respectively. In contrast, MobileNet V2 produces a significantly larger fraction of compressed zero values: 10.48–14.91%, which is attributed to the fact that MobileNet V2 is a much more compact model than the other CNNs. These results indicate that VELCRO offers an extended level of compression with respect to pruning, which aims to remove weak connections of zero values.

Table 4. Compressed activation elements with zero value as a percent of all compressed activation elements.

Specialized Task	ResNet-18	GoogLeNet	MobileNet V2
Cats	0.08%	0.56%	14.91%
Dogs	0.20%	0.63%	10.48%
Cars	0.31%	0.64%	12.00%

Another important result gained from Figure 9a–c is that, when VELCRO is used with a relatively moderate compression ratio, it produces a significant increase in accuracy. The results are presented in Table 5, which summarizes the maximum top-1 accuracy achieved by VELCRO. These results are attributed to the fact that a relatively moderate level of compression helps the network leverage value locality to strengthen connections, thereby increasing the probability of favoring the prediction of classes that are in the scope of the specialized tasks.

Table 5. The maximum top-1 accuracy increase produced by VELCRO with respect to the uncompressed model when used for specialized tasks.

Specialized Task	ResNet-18	GoogLeNet	MobileNet V2
Cats	13.00%	20.00%	3.50%
Dogs	8.50%	11.00%	2.50%
Cars	4.00%	15.00%	4.50%

4.4. Hardware Implementation

In the last part of our experimental analysis, we demonstrate the computational optimization and energy savings of VELCRO through hardware implementation on the Xilinx® Alveo™ U280 Data Center accelerator card [63]. Our hardware implementation, which is illustrated in Figure 10, consists of 16 instance modules where each is comprised of a two-dimensional convolution layer with a 64 × 64 input feature map (IFMAP), 3 × 3 filter, and a ReLU activation function. In addition, each module also includes a compression control logic which skips the compressed computations and replaces them with their corresponding arithmetic averages. Our hardware implementation was designed in Verilog and implemented using the Xilinx® Vivado™ [64] design suite.

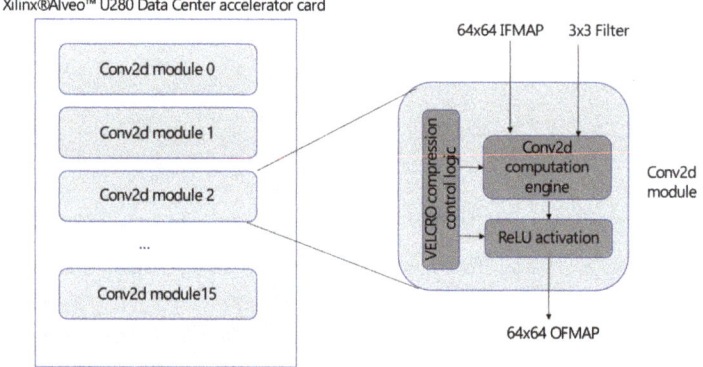

Figure 10. VELCRO compression implementation on Xilinx® Alveo™ U280 Accelerator Card.

Figure 11 presents the (normalized) throughput and energy consumption of a single module instance, denoted as conv2d, which consists of the hardware implementation of a two-dimensional convolution layer and ReLU activation. As expected, the computational throughput of the conv2d layer, which is measured as the number of conv2d operations per second, exhibits a growth rate proportional to $\frac{1}{1-C}$, where C is the compression saving ratio). In addition, it can be observed that the energy consumption related to the computation of a single conv2d layer decays linearly with the compression saving ratio. Thereby, for the compression saving results presented in Table 2, VELCRO can achieve 13.5–30% energy consumption savings while maintaining the same accuracy of the uncompressed model.

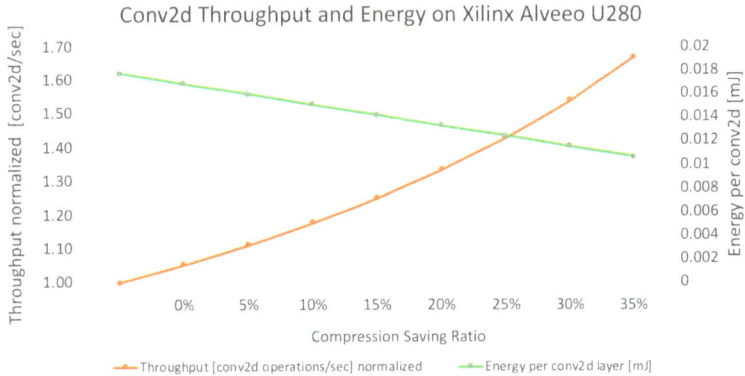

Figure 11. VELCRO throughput and energy consumption Xilinx® Alveo™ U280 Accelerator Card.

5. Conclusions

We present herein value-locality-based compression algorithm (VELCRO), wherein a compression approach is introduced for general-purpose deep neural networks deployed for a small subset of specialized tasks. We introduce the notion of value locality in the context of neural networks for specialized tasks and show that CNNs that are used for specialized tasks produce a high degree of value locality. An analysis of the experimental results indicates that VELCRO leverages value locality to compress the network and thereby saves up to 30% of the computations in ResNet-18 and GoogLeNet and up to 20% in MobileNet V2. The analysis also indicates that, for specialized tasks, VELCRO significantly improves the accuracy by 2–20% when given a relatively small compression-saving target. Finally, a major advantage of VELCRO is that it offers a fast compression process that is based on inference rather than backpropagation training, thereby liberating VELCRO from a significant computational load. We demonstrate the feasibility of VELCRO by designing the algorithm in hardware on the Xilinx® Alveo™ U280 Data Center accelerator card. Our hardware implementation indicates that VELCRO translates the computation compression into an energy consumption savings of 13.5–30%, corresponding to the compression-saving ratio.

Author Contributions: Conceptualization, F.G.; methodology, F.G. and G.S.; software, G.S. and F.G.; validation, F.G. and G.S.; formal analysis, F.G. and G.S.; investigation, F.G. and G.S.; resources, F.G. and G.S.; data curation, F.G. and G.S.; writing—original draft preparation, F.G. and G.S.; writing—review and editing, F.G. and G.S.; visualization, F.G. and G.S. All authors have read and agreed to the published version of the manuscript.

Funding: This research received no external funding.

Data Availability Statement: The ImageNet data sets used in our experiments are publicly available at https://image-net.org (accessed on 11 March 2021).

Conflicts of Interest: The authors declare no conflict of interest.

Appendix A

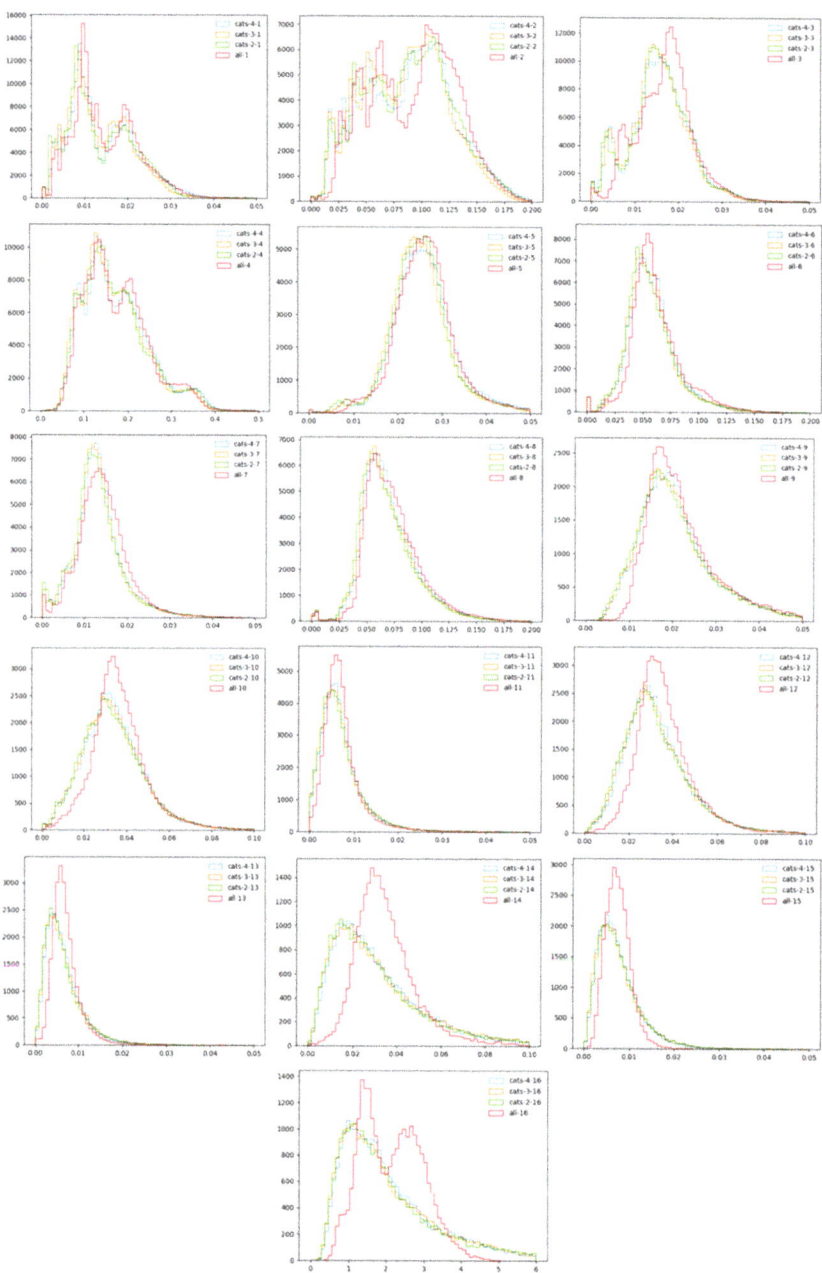

Figure A1. Distribution of ResNet-18 variance tensor elements for specialized tasks: all ImageNet classes, Cats-2, Cats-3, and Cats-4.

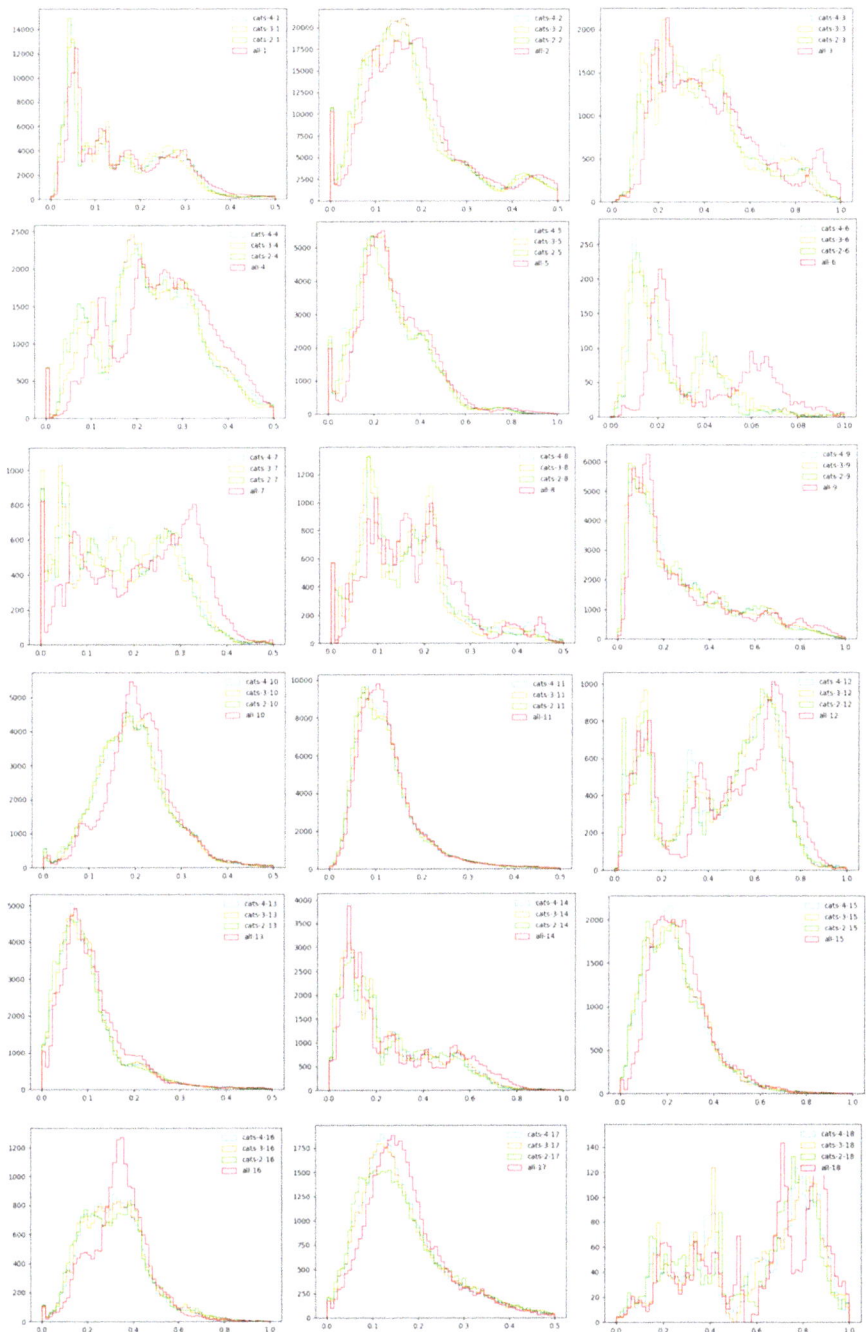

Figure A2. *Cont.*

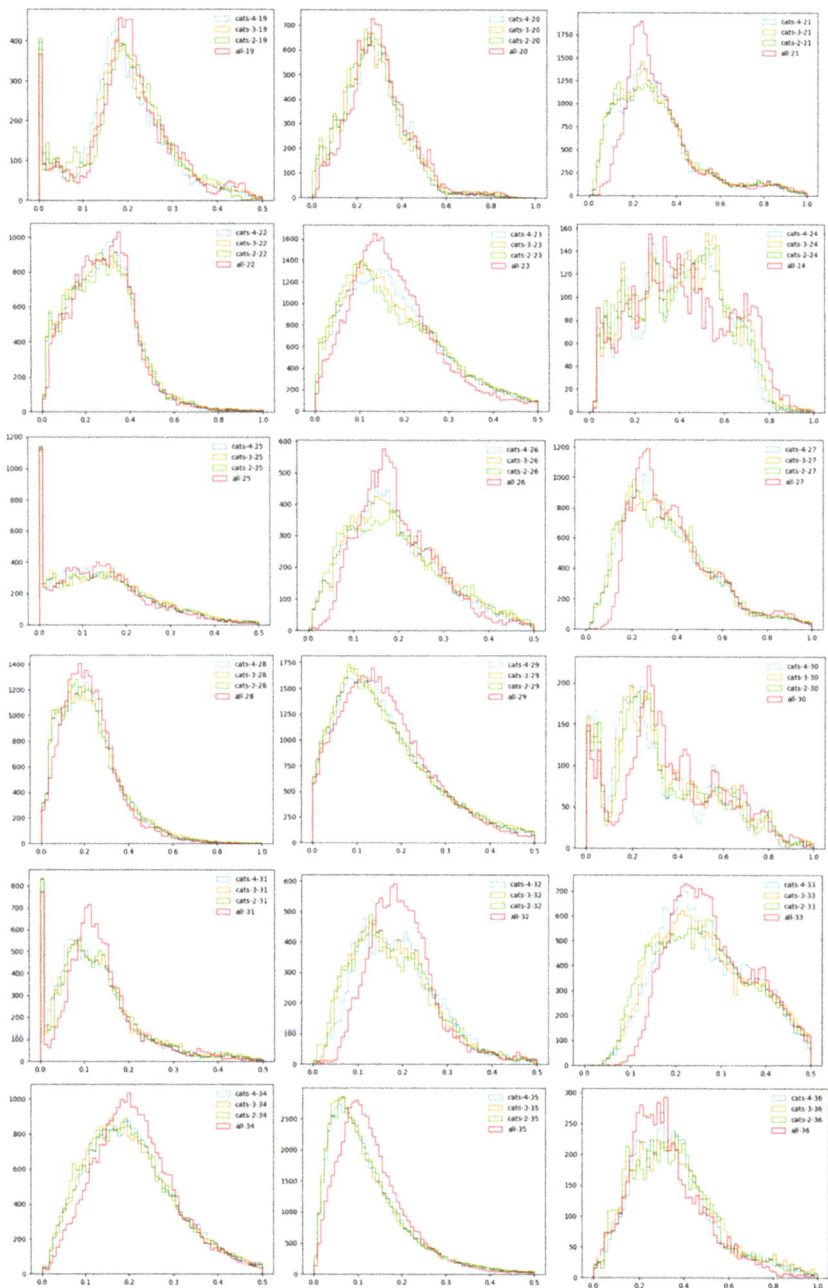

Figure A2. *Cont.*

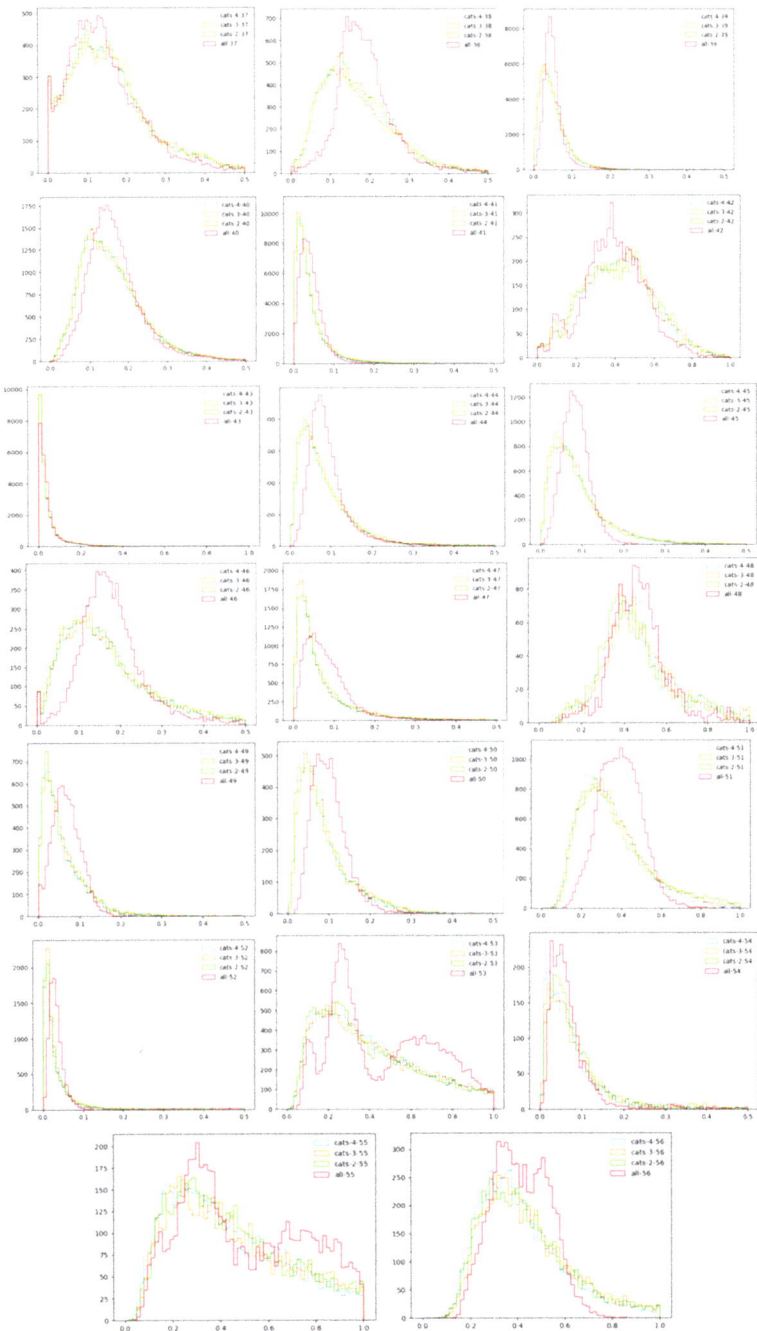

Figure A2. Distribution of GoogLeNet variance tensor elements for specialized tasks: all ImageNet classes, Cats-2, Cats-3, and Cats-4.

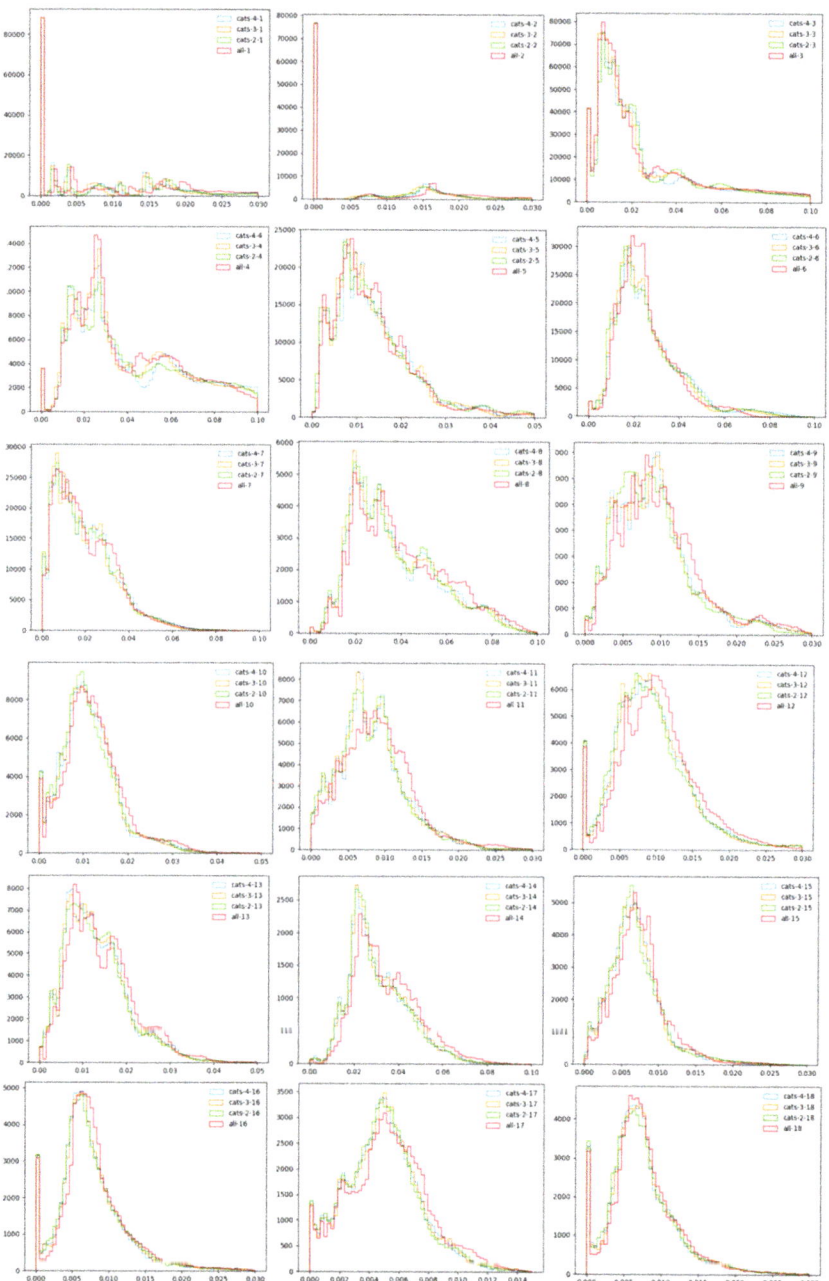

Figure A3. *Cont.*

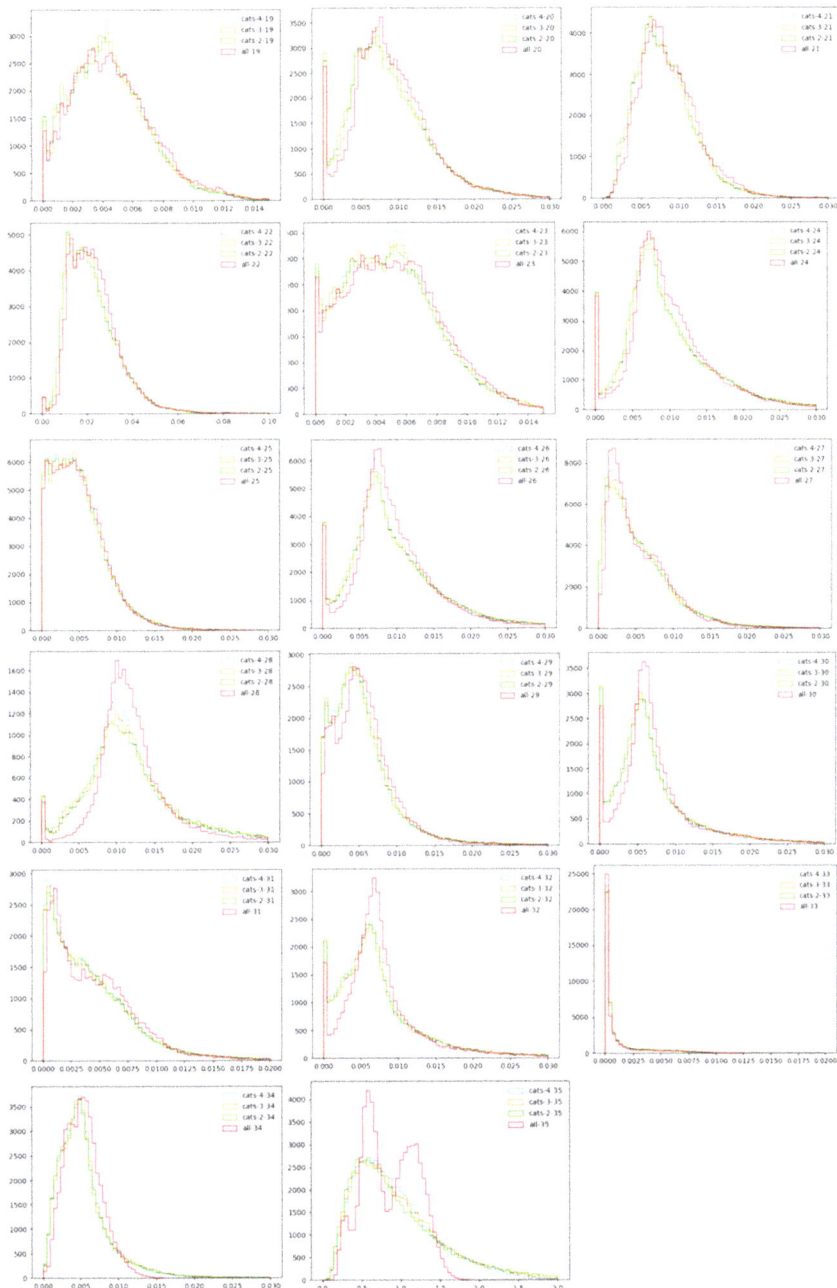

Figure A3. Distribution of MobileNet V2 variance tensor elements for specialized tasks: all ImageNet classes, Cats-2, Cats-3, and Cats-4.

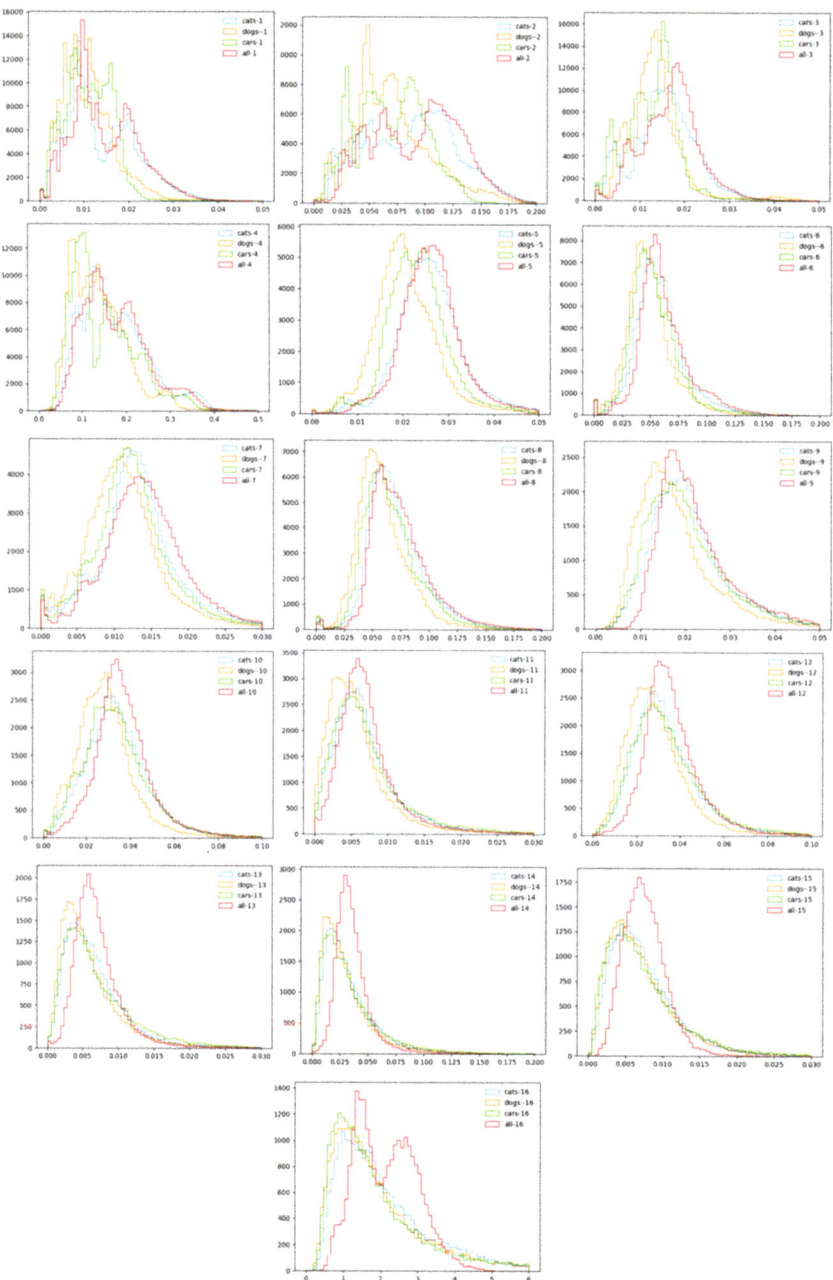

Figure A4. Distribution ResNet-18 variance tensor elements for specialized tasks: Cats, Dogs, Cars, and all ImageNet classes.

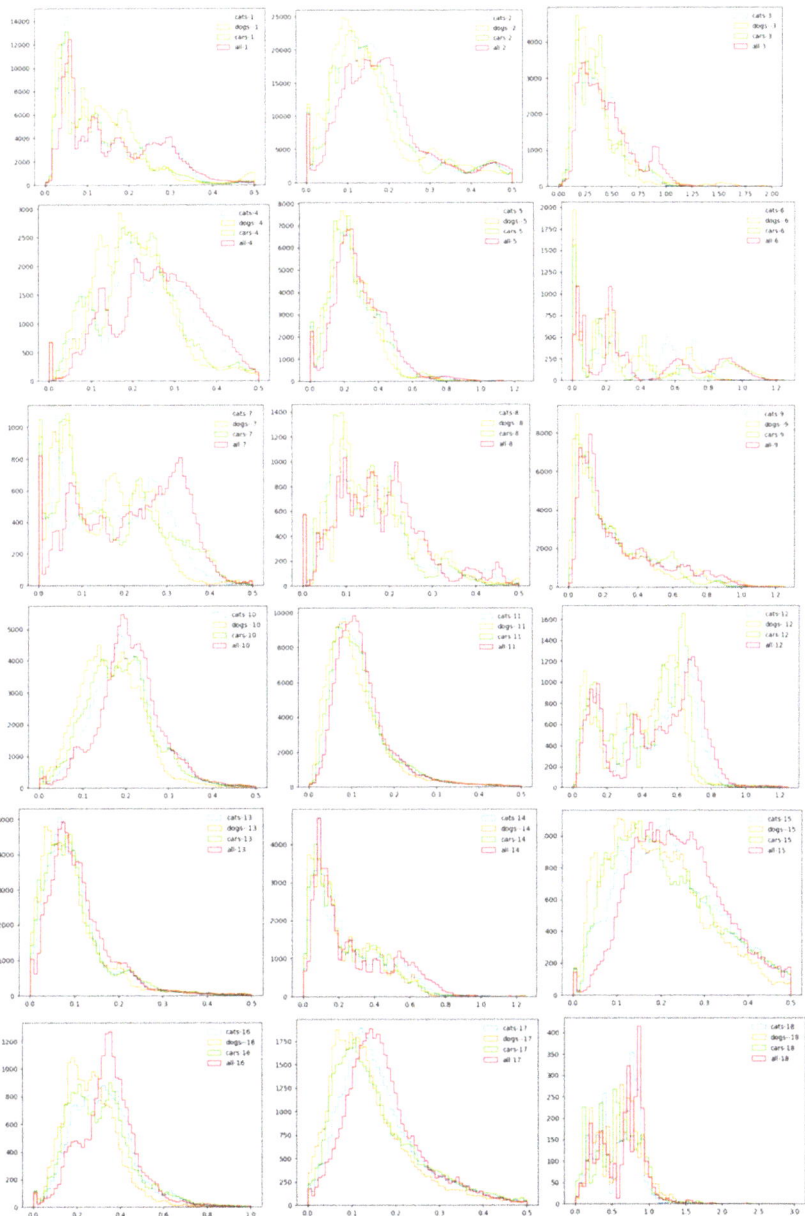

Figure A5. *Cont.*

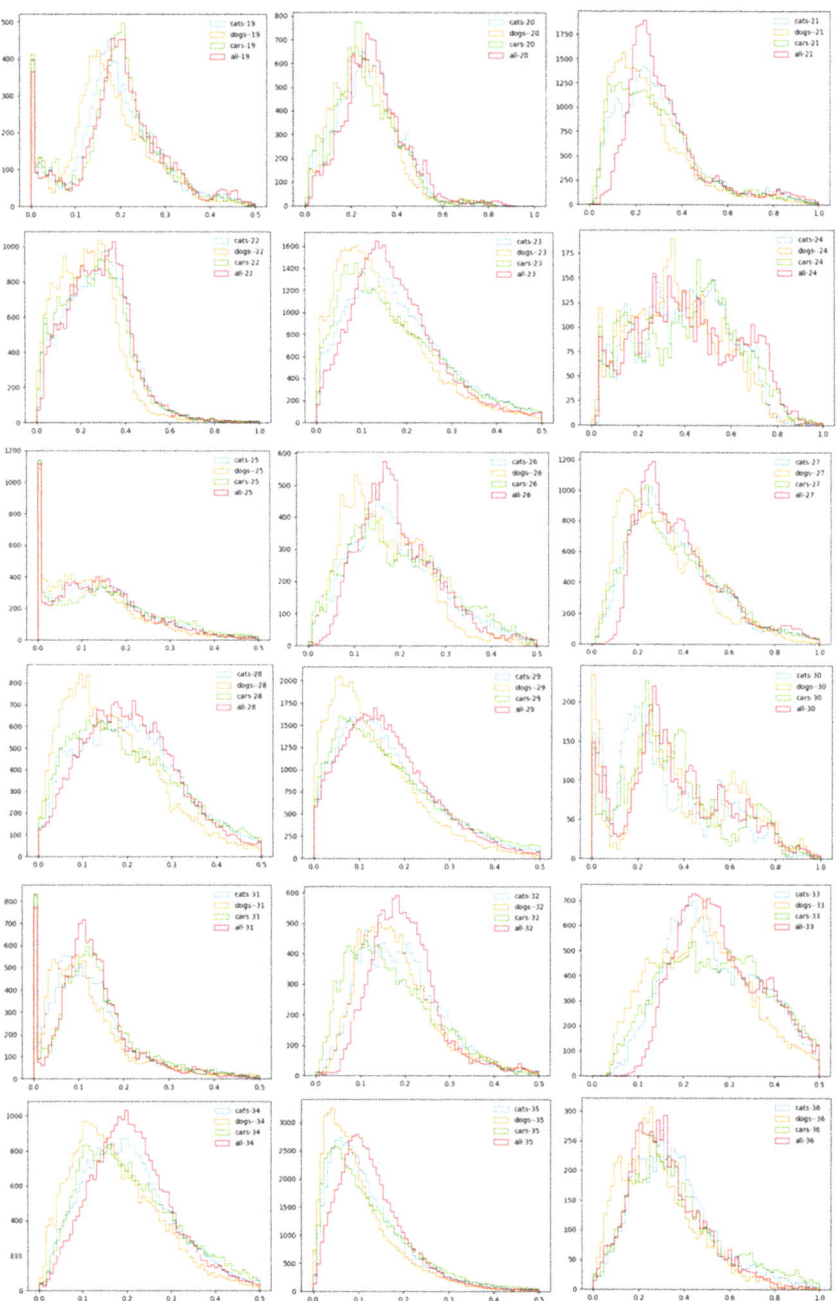

Figure A5. *Cont.*

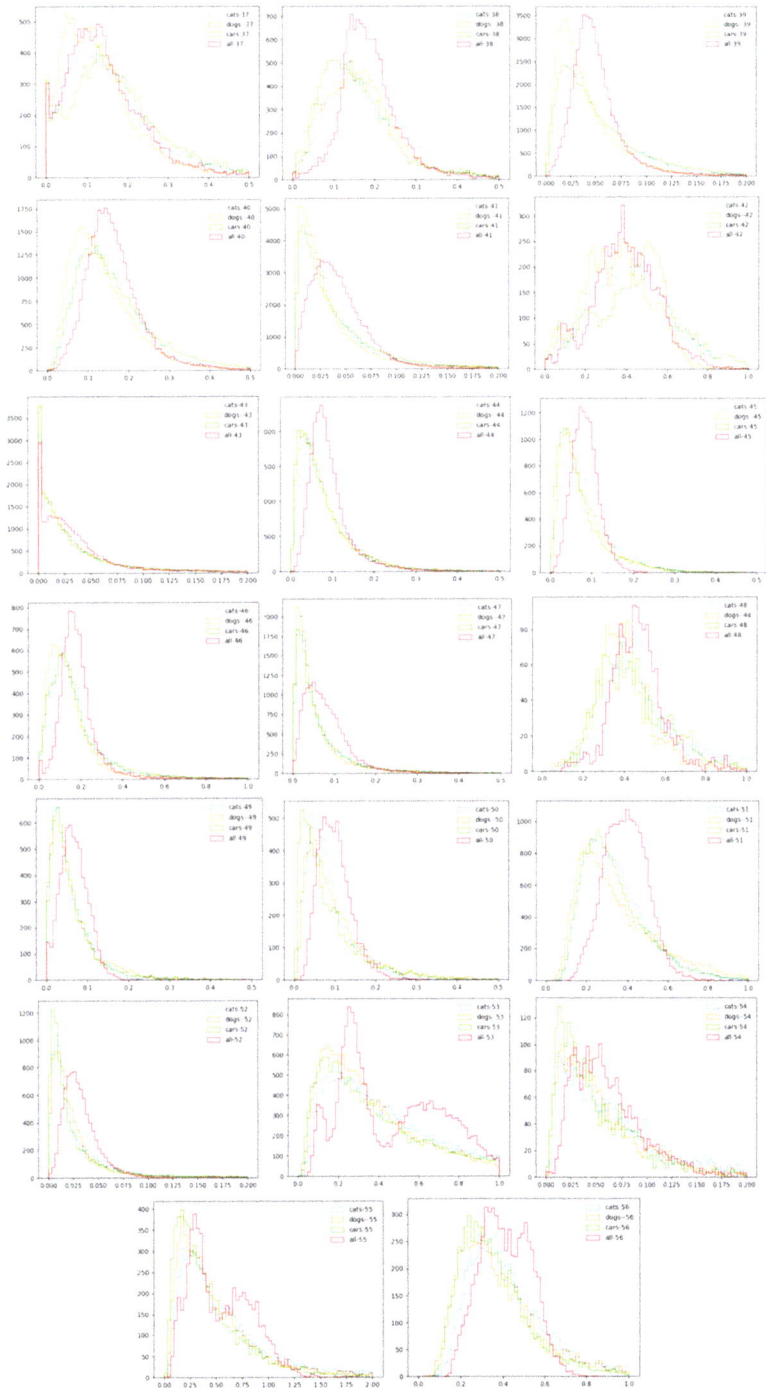

Figure A5. Distribution of GoogLeNet variance tensor elements for specialized tasks: Cats, Dogs, Cars, and all ImageNet classes.

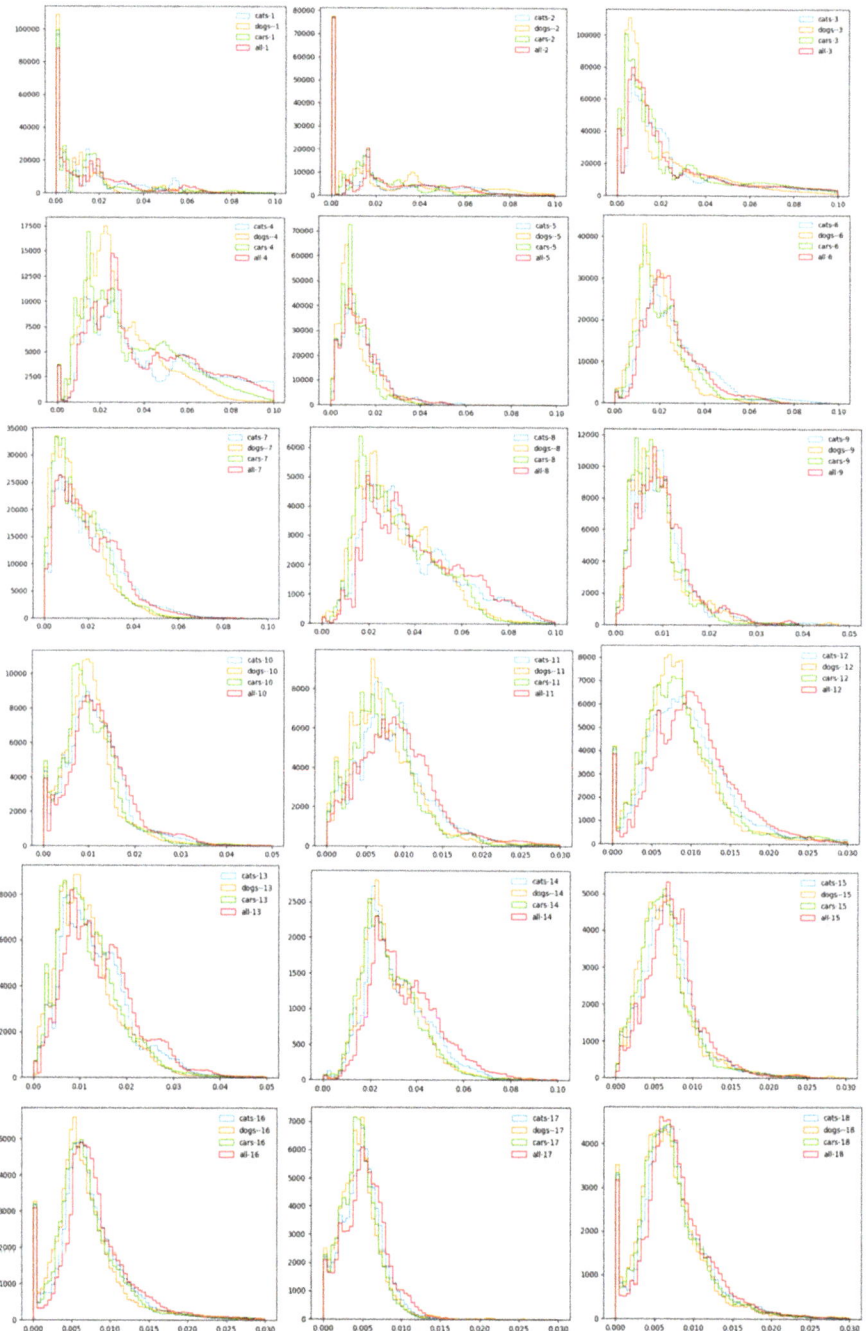

Figure A6. *Cont.*

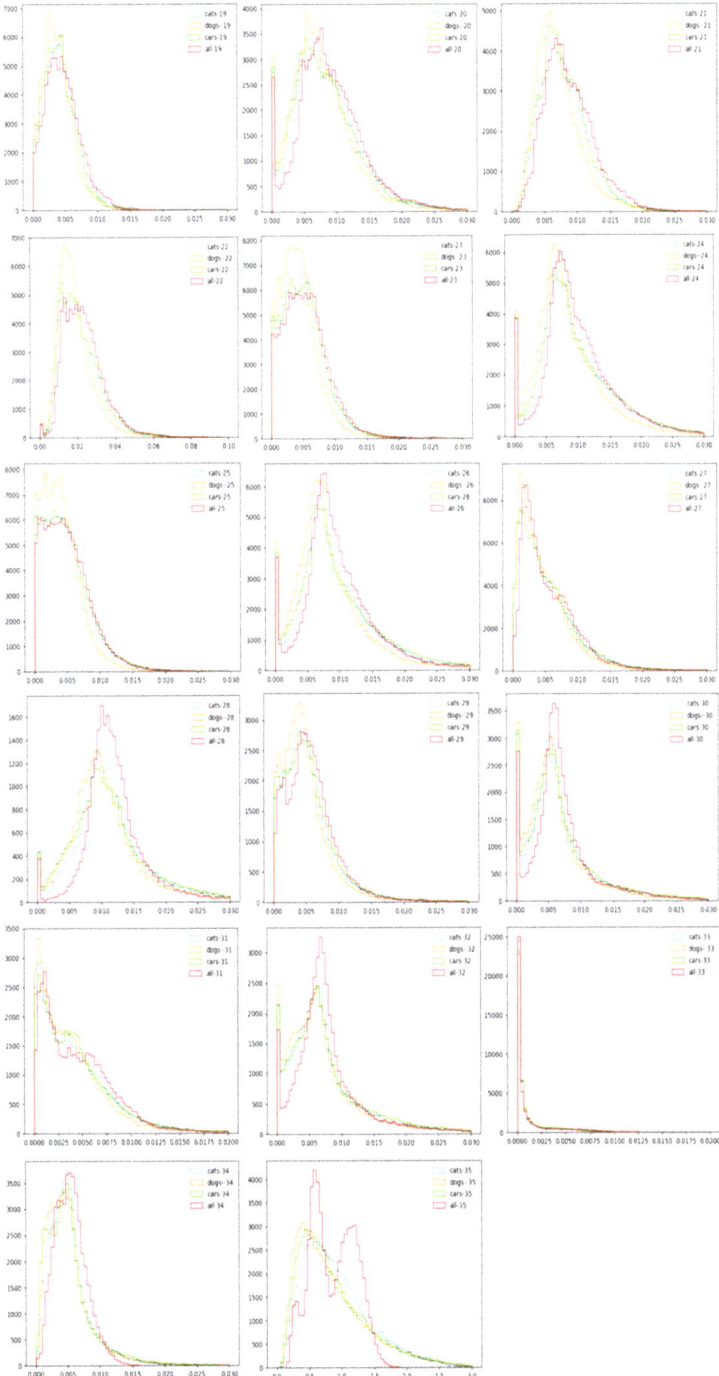

Figure A6. Distribution of MobileNet V2 variance tensor elements for specialized tasks: Cats, Dogs, Cars, and all ImageNet classes.

Appendix B

Table A1. Threshold tuple for Cats.

Compression Saving Ratio	ResNet-18 Threshold Tuple	GoogLeNet Threshold Tuple	MobileNet V2 Threshold Tuple
10%	(3, 3, 4, 4, 10, 10, 10, 10, 10, 10, 10, 10, 70, 10, 80, 90)	(8, 7, 7, 7, 7, 7, 8, 7, 7, 7, 7, 0, 0, 0, 0, 0, 0, 0, 0, 5, 6, 5, 5, 5, 5, 5, 5, 14, 14, 5, 5, 5, 8, 5, 5, 5,15, 15, 15, 36, 37, 18, 20, 34, 34, 34, 25, 40, 34, 34, 90, 90, 96, 90, 92, 92)	(22, 21, 10, 11, 11, 11, 5, 0, 0, 0, 0, 0, 0, 0, 0, 0, 0, 0, 0, 0, 0, 0, 0, 0, 0, 0, 40, 40, 40, 40, 40, 40, 90)
13.5%	n/a	n/a	(40, 40, 7, 2, 5, 2, 1, 11, 2, 10, 10, 15, 10, 15, 20, 10, 10, 15, 2, 5, 5, 10, 10, 10, 10, 15, 7, 5, 40, 42, 40, 40, 40, 40, 90)
20%	(17, 16, 17, 10, 10, 20, 20, 20, 20, 20, 30, 27, 70, 10, 80, 90)	(17, 15, 15, 15, 15, 15, 17, 15, 15, 15, 15, 19, 16, 16, 16, 16, 16, 16, 16, 16, 16, 22, 16, 16, 16, 16, 16, 16, 16, 16, 16, 16, 16, 17, 37, 37, 37, 36, 37, 37, 40, 34, 34, 34, 34, 40, 34, 34, 90, 90, 96, 90, 92, 92)	(35, 35, 18, 17, 17, 20, 8, 8, 16, 16, 16, 16, 16, 16, 16, 16, 16, 16, 4, 16, 0, 16, 3, 16, 16, 16, 23, 16, 42, 40, 40, 42, 40, 45, 90)
27.3%	(34, 36, 30, 15, 10, 25, 20, 20, 20, 20, 30, 27, 70, 10, 80, 90)	n/a	n/a
30%	(40, 40, 36, 15, 12, 25, 20, 20, 20, 20, 30, 27, 70, 10, 80, 90)	(34, 24, 24, 24, 24, 24, 27, 24, 24, 24, 24, 29, 25, 24, 25, 24, 24, 25, 26, 24, 25, 24, 33, 25, 24, 24, 24, 24, 24, 24, 24, 24, 24, 25, 24, 27, 56, 56, 54, 52, 56, 54, 60, 52, 52, 52, 52, 60, 52, 52, 90, 90, 96, 90, 92, 92)	(38, 38, 33, 32, 32, 31, 20, 20, 25, 31, 31, 26, 26, 26, 26, 26, 26, 26, 4, 28, 0, 27, 6, 26, 26, 25, 24, 17, 42, 40, 40, 42, 40, 45, 90)
40%	(70, 61, 60, 15, 12, 25, 20, 20, 20, 20, 30, 27, 70, 10, 80, 90)	(46, 32, 32, 32, 32, 32, 36, 32, 32, 32, 32, 36, 32, 32, 32, 32, 32, 32, 32, 32, 32, 32, 44, 32, 32, 32, 32, 32, 32, 32, 32, 32, 32, 32, 32, 36, 75, 75, 75, 75, 75, 75, 80, 75, 75, 75, 75, 80, 75, 75, 92, 92, 96, 92, 94, 94)	(66, 66, 35, 17, 32, 20, 17, 38, 13, 52, 52, 64, 44, 58, 58, 42, 42, 54, 13, 33, 33, 36, 36, 36, 36, 54, 26, 21, 80, 80, 80, 80, 80, 80, 90)

Table A2. Threshold tuple for Dogs.

Compression Saving Ratio	ResNet-18 Threshold Tuple	GoogLeNet Threshold Tuple	MobileNet V2 Threshold Tuple
10%	(8, 8, 8, 3, 8, 9, 9, 9, 9, 9, 9, 17, 17, 6, 60, 80)	(7, 7, 7, 7, 7, 7, 7, 7, 7, 7, 7, 7, 7, 8, 8, 8, 8, 8, 8, 8, 8, 8, 8, 8, 8, 8, 8, 8, 8, 8, 8, 8, 10, 20, 20, 20, 20, 20, 20, 20, 20, 20, 20, 20, 27, 16, 72, 72, 90, 92, 90, 91)	(20, 20, 10, 10, 10, 10, 5, 0, 10, 0, 10, 0, 0, 0, 0, 0, 0, 0, 0, 0, 0, 0, 0, 0, 0, 0, 0, 40, 40, 40, 40, 40, 40, 90)
19.76%	n/a	n/a	(29, 29, 16, 11, 15, 18, 13, 13, 36, 35, 43, 21, 13, 5, 16, 30, 23, 8, 5, 6, 0, 1, 7, 17, 10, 10, 5, 13, 40, 47, 41, 40, 42, 42, 90)
20%	(21, 21, 20, 6, 10, 21, 20, 20, 20, 18, 33, 24, 30, 6, 71, 80)	(15, 16, 16, 16, 16, 16, 16, 16, 16, 16, 16, 16, 16, 16, 16, 16, 16, 16, 16, 16, 16, 16, 16, 16, 16, 16, 16, 16, 16, 20, 40, 40, 40, 40, 40, 40, 40, 40, 40, 40, 40, 55, 32, 72, 72, 90, 92, 90, 91)	n/a

Table A2. Cont.

Compression Saving Ratio	ResNet-18 Threshold Tuple	GoogLeNet Threshold Tuple	MobileNet V2 Threshold Tuple
25.46%	n/a	(20, 20, 20, 20, 20, 20, 20, 20, 20, 20, 20, 20, 22, 22, 20, 20, 20, 22, 20, 20, 20, 20, 20, 20, 20, 20, 20, 22, 20, 20, 20, 20, 20, 20, 20, 25, 50, 52, 50, 50, 50, 52, 50, 52, 50, 50, 55, 55, 68, 42, 92, 90, 90, 92, 90, 91)	n/a
27.7%	(44, 28, 38, 12, 12, 21, 20, 20, 20, 20, 33, 24, 32, 10, 71, 87)	n/a	n/a
30%	(49, 33, 44, 13, 12, 21, 20, 20, 20, 20, 33, 24, 32, 12, 72, 90)	(25, 24, 24, 24, 24, 24, 24, 24, 24, 24, 24, 24, 24, 26, 24, 24, 24, 26, 24, 24, 24, 24, 24, 24, 24, 24, 24, 26, 24, 24, 24, 24, 24, 24, 24, 30, 60, 60, 60, 60, 60, 60, 60, 60, 60, 60, 65, 65, 68, 50, 92, 90, 90, 92, 90, 91)	(45, 45, 25, 27, 23, 28, 19, 19, 54, 53, 64, 31, 19, 7, 24, 45, 34, 12, 7, 9, 0, 1, 10, 25, 15, 15, 7, 20, 60, 70, 61, 60, 62, 62, 90)
40%	(50, 50, 50, 50, 40, 21, 20, 20, 20, 20, 33, 24, 32, 10, 70, 90)	(36, 36, 32, 32, 32, 32, 32, 32, 32, 32, 32, 32, 32, 38, 36, 36, 36, 38, 36, 36, 36, 36, 36, 36, 36, 36, 36, 36, 36, 36, 36, 36, 36, 36, 36, 45, 70, 70, 70, 70, 70, 70, 70, 70, 70, 70, 65, 65, 68, 50, 92, 90, 90, 92, 90, 91)	(67, 67, 34, 37, 31, 38, 28, 28, 73, 71, 64, 42, 27, 9, 32, 60, 45, 16, 9, 12, 0, 1, 13, 33, 20, 20, 9, 26, 65, 75, 65, 65, 65, 70, 90)

Table A3. Threshold tuple for Cars.

Compression Saving Ratio	ResNet-18 Threshold Tuple	GoogLeNet Threshold Tuple	MobileNet V2 Threshold Tuple
10%	(3, 5, 5, 10, 10, 10, 10, 10, 10, 10, 10, 12, 30, 13, 50, 80)	(6, 9, 9, 9, 9, 9, 15, 7, 7, 7, 7, 10, 7, 8, 7, 10, 7, 7, 7, 8, 8, 8, 10, 8, 8, 8, 10, 8, 10, 12, 9, 7, 7, 2, 8, 10, 7, 7, 5, 12, 10, 30, 1, 27, 27, 30, 1, 32, 27, 23, 40, 90, 90, 95, 90)	(20, 20, 20, 10, 10, 10, 5, 0, 0, 0, 0, 0, 0, 0, 0, 0, 0, 0, 0, 0, 0, 0, 0, 0, 0, 0, 0, 0, 0, 0, 20, 20, 30, 75)
16.8%	n/a	n/a	(30, 31, 20, 10, 15, 15, 11, 10, 10, 21, 16, 11, 1, 0, 26, 17, 21, 20, 2, 10, 0, 2, 8, 5, 5, 20, 4, 5, 2, 22, 20, 23, 20, 30, 75)
20%	(21, 21, 20, 20, 20, 20, 10, 20, 20, 10, 10, 12, 30, 13, 50, 80)	(15, 20, 20, 20, 20, 20, 30, 14, 14, 14, 14, 14, 20, 14, 16, 14, 20, 14, 14, 14, 17, 17, 17, 20, 17, 17, 17, 20, 17, 20, 24, 18, 14, 14, 4, 17, 20, 14, 14, 10, 24, 20, 60, 3, 55, 55, 60, 2, 65, 55, 50, 90, 90, 90, 95, 90)	(35, 35, 25, 12, 19, 19, 14, 13, 13, 26, 24, 13, 1, 0, 30, 17, 21, 20, 2, 12, 0, 2, 9, 6, 6, 23, 5, 6, 2, 22, 20, 23, 20, 30, 75)
27.70%	n/a	(20, 30, 30, 30, 30, 30, 45, 20, 20, 20, 20, 20, 30, 20, 25, 20, 30, 20, 20, 20, 25, 25, 25, 30, 25, 25, 25, 30, 25, 30, 35, 25, 20, 20, 7, 25, 30, 20, 30, 15, 35, 20, 60, 3, 55, 55, 60, 2, 65, 55, 50, 90, 90, 90, 95, 90)	
30%	(35, 30, 30, 30, 30, 30, 30, 30, 10, 12, 12, 30, 30, 50, 85)	(24, 32, 32, 32, 32, 32, 49, 22, 22, 22, 22, 22, 32, 22, 27, 22, 33, 22, 22, 22, 28, 28, 27, 33, 28, 28, 27, 33, 27, 33, 38, 27, 22, 22, 7, 27, 32, 22, 33, 16, 38, 20, 66, 3, 60, 60, 66, 2, 71, 60, 55, 90, 90, 90, 94, 89)	(52, 52, 42, 20, 29, 29, 21, 19, 19, 39, 36, 19, 1, 0, 45, 25, 31, 30, 3, 18, 0, 2, 13, 9, 9, 34, 7, 8, 3, 22, 20, 23, 20, 30, 75)

173

Table A3. *Cont.*

Compression Saving Ratio	ResNet-18 Threshold Tuple	GoogLeNet Threshold Tuple	MobileNet V2 Threshold Tuple
40%	(50, 40, 40, 40, 40, 40, 40, 40, 40, 10, 20, 20, 35, 35, 80, 90)	(36, 44, 44, 44, 44, 44, 55, 32, 32, 32, 30, 30, 43, 30, 37, 30, 44, 30, 30, 30, 40, 40, 38, 44, 36, 36, 36, 44, 30, 44, 52, 37, 30, 30, 12, 40, 44, 40, 45, 18, 55, 30, 67, 10, 65, 65, 66, 60, 71, 60, 55, 90, 90, 90, 94, 89)	(67, 67, 34, 37, 31, 38, 28, 28, 73, 71, 64, 42, 27, 9, 32, 60, 45, 16, 9, 12, 0, 1, 13, 33, 20, 20, 9, 26, 65, 75, 65, 65, 65, 70, 90)

References

1. Krizhevsky, A.; Sutskever, I.; Hinton, G.E. ImageNet Classification with Deep Convolutional Neural Networks. *Commun. ACM* **2017**, *60*, 84–90. [CrossRef]
2. He, K.; Zhang, X.; Ren, S.; Sun, J. Deep Residual Learning for Image Recognition. In Proceedings of the 2016 IEEE Conference on Computer Vision and Pattern Recognition (CVPR), Las Vegas, NV, USA, 27–30 June 2016.
3. Bianco, S.; Cadene, R.; Celona, L.; Napoletano, P. Benchmark Analysis of Representative Deep Neural Network Architectures. *IEEE Access* **2018**, *6*, 64270–64277. [CrossRef]
4. Russakovsky, O.; Deng, J.; Su, H.; Krause, J.; Satheesh, S.; Ma, S.; Huang, Z.; Karpathy, A.; Khosla, A.; Bernstein, M.; et al. ImageNet Large Scale Visual Recognition Challenge. *Int. J. Comput. Vis.* **2015**, *115*, 211–252. [CrossRef]
5. Reed, R. Pruning algorithms—A survey. *IEEE Trans. Neural Netw.* **1993**, *4*, 740–747. [CrossRef]
6. LeCun, Y.; Denker, J.S.; Solla, S.; Howard, R.E.; Jackel, L.D. Optimal brain damage. In *Advances in Neural Information Processing Systems (NIPS 1989)*; Touretzky, D., Ed.; Morgan Kaufmann: Denver, CO, USA, 1990; Volume 2.
7. Hassibi, B.; Stork, D.G.; Wolff, G.J. Optimal Brain Surgeon and general network pruning. In Proceedings of the IEEE International Conference on Neural Networks, San Francisco, CA, USA, 28 March–1 April 1993; Volume 1, pp. 293–299.
8. Zhang, C.; Bengio, S.; Hardt, M.; Recht, B.; Vinyals, O. Understanding Deep Learning Requires Rethinking Generalization. *arXiv* **2016**, arXiv:1611.03530.
9. Vanhoucke, V.; Senior, A.; Mao, M.Z. Improving the speed of neural networks on CPUs. In *Deep Learning and Unsupervised Feature Learning Workshop*; NIPS: Granada, Spain, 2011.
10. Gong, Y.; Liu, L.; Yang, M.; Bourdev, L. Compressing deep convolutional networks using vector quantization. *arXiv* **2014**, arXiv:1412.6115.
11. Courbariaux, M.; Bengio, Y.; David, J.-P. BinaryConnect: Training Deep Neural Networks with binary weights during propagations. In Proceedings of the 28th International Conference on Neural Information Processing Systems, Bali, Indonesia, 8–12 December 2015.
12. Lin, Z.; Courbariaux, M.; Memisevic, R.; Bengio, Y. Neural networks with few multiplications. *arXiv* **2015**, arXiv:1510.03009.
13. Shen, H.; Han, S.; Philipose, M.; Krishnamurthy, A. Fast Video Classification via Adaptive Cascading of Deep Models. In Proceedings of the 2017 IEEE Conference on Computer Vision and Pattern Recognition (CVPR), Honolulu, HI, USA, 21–26 July 2017.
14. Kang, D.; Emmons, J.; Abuzaid, F.; Bailis, P.; Zaharia, M. NoScope: Optimizing Neural Network Queries over Video at Scale. *Proc. VLDB Endow.* **2017**, *10*, 1586–1597. [CrossRef]
15. Kosaian, J.; Phanishayee, A.; Philipose, M.; Dey, D.; Vinayek, R. Boosting the Throughput and Accelerator Utilization of Specialized CNN Inference beyond Increasing Batch Size. In Proceedings of the Proceedings of the 38th International Conference on Machine Learning, PMLR 139, Long Beach, CA, USA, 18–24 July 2021.
16. Violaand, P.; Jones, M. Rapid object detection using a boosted cascade of simple features. In Proceedings of the 2001 IEEE Computer Society Conference on Computer Vision and Pattern Recognition. CVPR 2001, Kauai, HI, USA, 8–14 December 2001; Volume 1, pp. I-511–I-518.
17. Shazeer, N.; Mirhoseini, A.; Maziarz, K.; Davis, A.; Le, Q.V.; Hinton, G.E.; Dean, J. Outrageously large neural networks: The sparsely-gated mixture-of-experts layer. *arXiv* **2017**, arXiv:1701.06538.
18. Howard, A.G.; Zhu, M.; Chen, B.; Kalenichenko, D.; Wang, W.; Weyand, T.; Andreetto, M.; Adam, H. MobileNets: Efficient Convolutional Neural Networks for Mobile Vision Applications. *arXiv* **2017**, arXiv:1704.04861.
19. Szegedy, C.; Liu, W.; Jia, Y.; Sermanet, P.; Reed, S.; Anguelov, D.; Erhan, D.; Vanhoucke, V.; Rabinovich, A. Going Deeper with Convolutions. In Proceedings of the 2015 IEEE Conference on Computer Vision and Pattern Recognition (CVPR), Boston, MA, USA, 7–12 June 2015.
20. Deng, J.; Dong, W.; Socher, R.; Li, L.-J.; Li, K.; Fei-Fei, L. ImageNet: A Large-Scale Hierarchical Image Database. In Proceedings of the 2009 IEEE Conference on Computer Vision and Pattern Recognition, Miami, FL, USA, 22–24 June 2009.
21. Han, S.; Pool, J.; Tran, J.; Dally, W.J. Learning both weights and connections for efficient neural networks. *arXiv* **2015**, arXiv:1506.02626.
22. Han, S.; Mao, H.; Dally, W.J. Deep compression: Compressing deep neural networks with pruning, trained quantization and Huffman coding. *arXiv* **2015**, arXiv:1510.00149.

23. Castellano, G.; Fanelli, A.M.; Pelillo, M. An iterative pruning algorithm for feedforward neural networks. *IEEE Trans. Neural Netw.* **1997**, *8*, 519–531. [CrossRef] [PubMed]
24. Collins, M.D.; Kohli, P. Memory bounded deep convolutional networks. *arXiv* **2014**, arXiv:1412.1442.
25. Stepniewski, S.W.; Keane, A.J. Pruning backpropagation neural networks using modern stochastic optimisation techniques. *Neural. Comput. Appl.* **1997**, *5*, 76–98. [CrossRef]
26. Liu, Z.; Sun, M.; Zhou, T.; Huang, G.; Darrell, T. Rethinking the Value of Network Pruning. *arXiv* **2018**, arXiv:1810.05270.
27. Anwar, S.; Hwang, K.; Sung, W. Structured pruning of deep convolutional neural networks. *ACM J. Emerg. Technol. Comput. Syst.* **2017**, *13*, 1–18. [CrossRef]
28. Lebedev, V.; Lempitsky, V. Fast ConvNets using group-wise brain damage. In Proceedings of the 2016 IEEE Conference on Computer Vision and Pattern Recognition (CVPR), Las Vegas, NV, USA, 26 June–1 July 2016.
29. Zhou, H.; Alvarez, J.M.; Porikli, F. Less is more: Towards compact CNNs. In *Computer Vision—ECCV 2016*; Springer International Publishing: Cham, Switzerland, 2016; pp. 662–677. ISBN 9783319464923.
30. Wen, W.; Wu, C.; Wang, Y.; Chen, Y.; Li, H. Learning structured sparsity in Deep Neural Networks. *Adv. Neural Inf. Process. Syst.* **2016**, *29*, 2074–2082.
31. Li, H.; Kadav, A.; Durdanovic, I.; Samet, H.; Graf, H.P. Pruning Filters for Efficient ConvNets. *arXiv* **2016**, arXiv:1608.08710.
32. Srinivas, S.; Babu, R.V. Data-Free Parameter Pruning for Deep Neural Networks. In Proceedings of the British Machine Vision Conference 2015, Swansea, UK, 7–10 September 2015; British Machine Vision Association: Guildford, UK, 2015.
33. Rao, Y.; Lu, J.; Lin, J.; Zhou, J. Runtime Neural Pruning. In Proceedings of the Advances in Neural Information Processing Systems, Long Beach, CA, USA, 4–9 December 2017; pp. 2181–2191.
34. Shomron, G.; Weiser, U. Spatial Correlation and Value Prediction in Convolutional Neural Networks. *IEEE Comput. Arch. Lett.* **2019**, *18*, 10–13. [CrossRef]
35. Shomron, G.; Banner, R.; Shkolnik, M.; Weiser, U. Thanks for Nothing: Predicting Zero-Valued Activations with Lightweight Convolutional Neural Networks. In *Computer Vision—ECCV 2020*; Springer International Publishing: Cham, Switzerland, 2020; pp. 234–250.
36. See, A.; Luong, M.-T.; Manning, C.D. Compression of Neural Machine Translation Models via Pruning. *arXiv* **2016**, arXiv:1606.09274.
37. Narang, S.; Elsen, E.; Diamos, G.; Sengupta, S. Exploring Sparsity in Recurrent Neural Networks. *arXiv* **2017**, arXiv:1704.05119.
38. Zhu, M.; Gupta, S. To prune, or not to prune: Exploring the efficacy of pruning for model compression. *arXiv* **2017**, arXiv:1710.01878.
39. Yu, R.; Li, A.; Chen, C.-F.; Lai, J.-H.; Morariu, V.I.; Han, X.; Gao, M.; Lin, C.-Y.; Davis, L.S. NISP: Pruning Networks Using Neuron Importance Score Propagation. In Proceedings of the 2018 IEEE/CVF Conference on Computer Vision and Pattern Recognition, Salt Lake City, UT, USA, 18–22 June 2018.
40. Cireşan, D.C.; Meier, U.; Masci, J.; Gambardella, L.M.; Schmidhuber, J. High-Performance Neural Networks for Visual Object Classification. *arXiv* **2011**, arXiv:1102.0183.
41. Chen, W.; Wilson, J.T.; Tyree, S.; Weinberger, K.Q.; Chen, Y. Compressing Neural Networks with the Hashing Trick. In Proceedings of the 32nd International Conference on Machine Learning, Lille, France, 6–11 July 2015.
42. Molchanov, P.; Tyree, S.; Karras, T.; Aila, T.; Kautz, J. Pruning Convolutional Neural Networks for Resource Efficient Inference. *arXiv* **2016**, arXiv:1611.06440.
43. Luo, J.-H.; Wu, J. Neural Network Pruning with Residual-Connections and Limited-Data. In Proceedings of the 2020 IEEE/CVF Conference on Computer Vision and Pattern Recognition (CVPR), Seattle, WA, USA, 14–19 June 2020.
44. Choi, J.; Wang, Z.; Venkataramani, S.; Chuang, P.I.-J.; Srinivasan, V.; Gopalakrishnan, K. PACT: Parameterized Clipping acTivation for quantized neural networks. *arXiv* **2018**, arXiv:1805.06085.
45. Park, E.; Yoo, S.; Vajda, P. Value-aware quantization for training and inference of neural networks. In *Computer Vision—ECCV 2018*; Springer International Publishing: Cham, Switzerland, 2018; pp. 608–624; ISBN 9783030012243.
46. Zhou, S.; Wu, Y.; Ni, Z.; Zhou, X.; Wen, H.; Zou, Y. DoReFa-Net: Training low bitwidth convolutional neural networks with low bitwidth gradients. *arXiv* **2016**, arXiv:1606.06160.
47. Banner, R.; Nahshan, Y.; Hoffer, E.; Soudry, D. Post-training 4-bit quantization of convolution networks for rapid-deployment. *arXiv* **2018**, arXiv:1810.05723.
48. Choukroun, Y.; Kravchik, E.; Yang, F.; Kisilev, P. Low-bit quantization of neural networks for efficient inference. In Proceedings of the 2019 IEEE/CVF International Conference on Computer Vision Workshop (ICCVW), Seoul, Korea, 27–28 October 2019.
49. Fang, J.; Shafiee, A.; Abdel-Aziz, H.; Thorsley, D.; Georgiadis, G.; Hassoun, J.H. Post-training piecewise linear quantization for deep neural networks. In *Computer Vision—ECCV 2020*; Springer International Publishing: Cham, Switzerland, 2020; pp. 69–86; ISBN 9783030585358.
50. Shomron, G.; Gabbay, F.; Kurzum, S.; Weiser, U. Post-Training Sparsity-Aware Quantization. *arXiv* **2021**, arXiv:2105.11010.
51. Buciluă, C.; Caruana, R.; Niculescu-Mizil, A. Model Compression. In Proceedings of the 12th ACM SIGKDD International Conference on Knowledge Discovery and Data Mining—KDD'06, Philadelphia, PA, USA, 20–23 August 2006; ACM Press: New York, NY, USA.
52. Hinton, G.; Vinyals, O.; Dean, J. Distilling the Knowledge in a Neural Network. *arXiv* **2015**, arXiv:1503.02531.
53. Gou, J.; Yu, B.; Maybank, S.J.; Tao, D. Knowledge Distillation: A Survey. *Int. J. Comput. Vis.* **2021**, *129*, 1789–1819. [CrossRef]

54. Cai, H.; Gan, C.; Wang, T.; Zhang, Z.; Han, S. Once-for-All: Train One Network and Specialize It for Efficient Deployment. *arXiv* **2019**, arXiv:1908.09791.
55. Zeiler, M.D.; Fergus, R. Visualizing and Understanding Convolutional Networks. In *Computer Vision—ECCV 2014*; Springer International Publishing: Cham, Switzerland, 2014; pp. 818–833.
56. Zhou, B.; Khosla, A.; Lapedriza, A.; Oliva, A.; Torralba, A. Object Detectors Emerge in Deep Scene CNNs. *arXiv* **2014**, arXiv:1412.6856.
57. Morcos, A.S.; Barrett, D.G.T.; Rabinowitz, N.C.; Botvinick, M. On the Importance of Single Directions for Generalization. *arXiv* **2018**, arXiv:1803.06959.
58. Zhou, B.; Sun, Y.; Bau, D.; Torralba, A. Revisiting the Importance of Individual Units in CNNs via Ablation. *arXiv* **2018**, arXiv:1806.02891.
59. Boone-Sifuentes, T.; Robles-Kelly, A.; Nazari, A. Max-Variance Convolutional Neural Network Model Compression. In Proceedings of the 2020 Digital Image Computing: Techniques and Applications (DICTA), Melbourne, Australia, 29 November–2 December 2020; pp. 1–6.
60. Li, Y.; Lin, S.; Zhang, B.; Liu, J.; Doermann, D.; Wu, Y.; Huang, F.; Ji, R. Exploiting kernel sparsity and entropy for interpretable CNN compression. In Proceedings of the IEEE/CVF Conference on Computer Vision and Pattern Recognition, California, CA, USA, 16–20 June 2019; pp. 2800–2809.
61. Wang, Y.; Zhang, X.; Xie, L.; Zhou, J.; Su, H.; Zhang, B.; Hu, X. Pruning from Scratch. In Proceedings of the AAAI Conference on Artificial Intelligence, New York, NY, USA, 7–12 February 2020; Volume 34, pp. 12273–12280.
62. Paszke, A.; Gross, S.; Massa, F.; Lerer, A.; Bradbury, J.; Chanan, G.; Killeen, T.; Lin, Z.; Gimelshein, N.; Antiga, L.; et al. PyTorch: An Imperative Style, High-Performance Deep Learning Library. *Adv. Neural Inf. Process. Syst.* **2019**, *32*, 8026–8037.
63. Xilinx. Breathe New Life into Your Data Center with Alveo Adaptable Accelerator Cards. Xilinx White Paper, WP499 (v1.0). Available online: https://www.xilinx.com/support/documentation/white_papers/wp499-alveo-intro.pdf (accessed on 19 November 2018).
64. Xilinx. Vivado Design Suite. Xilinx White Paper, WP416 (v1.1). Available online: https://www.xilinx.com/support/documentation/white_papers/wp416-Vivado-Design-Suite.pdf (accessed on 22 June 2012).

Article

Early Prediction of DNN Activation Using Hierarchical Computations

Bharathwaj Suresh [1,*], Kamlesh Pillai [1,*], Gurpreet Singh Kalsi [1], Avishaii Abuhatzera [2] and Sreenivas Subramoney [1]

[1] Processor Architecture Research (PAR) Lab, Intel Labs, Bangalore 560048, India; gurpreet.s.kalsi@intel.com (G.S.K.); sreenivas.subramoney@intel.com (S.S.)
[2] Corporate Strategy Office, Intel, Haifa 3508409, Israel; avishaii.abuhatzera@intel.com
* Correspondence: bharathwaj@ucla.edu (B.S.); kamlesh.r.pillai@intel.com (K.P.)

Abstract: Deep Neural Networks (DNNs) have set state-of-the-art performance numbers in diverse fields of electronics (computer vision, voice recognition), biology, bioinformatics, etc. However, the process of learning (training) from the data and application of the learnt information (inference) process requires huge computational resources. Approximate computing is a common method to reduce computation cost, but it introduces loss in task accuracy, which limits their application. Using an inherent property of Rectified Linear Unit (ReLU), a popular activation function, we propose a mathematical model to perform MAC operation using reduced precision for predicting negative values early. We also propose a method to perform hierarchical computation to achieve the same results as IEEE754 full precision compute. Applying this method on ResNet50 and VGG16 shows that up to 80% of ReLU zeros (which is 50% of all ReLU outputs) can be predicted and detected early by using just 3 out of 23 mantissa bits. This method is equally applicable to other floating-point representations.

Keywords: DNN; ReLU; floating-point numbers; hardware acceleration

1. Introduction

Ever since its inception, deep learning has evolved into one of the most widely used technique to solve problems in the area of speech recognition [1], pattern recognition [2], and natural language processing [1]. The effectiveness of Deep Neural Networks (DNNs) is pronounced when there is a huge amount of data with minimal features which are not easily apparent to humans [2]. This makes DNNs valuable tools to meet future data processing needs. However, producing accurate results using a large dataset comes at a cost. DNN inference requires a huge amount of computing power, and, as a result, consumes a large amount of energy. In a study by Strubell et al., it was estimated that training a single deep learning model can emit the same amount of CO_2 as five cars do throughout their lifetime [3]. Due to this fact, optimizing DNN implementations has become an urgent requirement, and has been receiving widespread attention from the research community [4–6].

In their basic form, DNNs consist of simple mathematical operations like addition and multiplication, which are combined together to form the multiply and accumulate (MAC) operation. In fact, up to 95% of the computational workload of a DNN is due to MAC operations [7]. In a typical DNN, about a billion MAC operations are required to process each input sample [8]. This fact suggests that improving the efficiency of the MAC operations would contribute significantly towards reducing the computational requirement of DNNs. One way to do this is to reduce the number of bits used to perform the MAC operations, an idea that has been widely explored in the field of approximate computing [9]. Some studies have shown that using approximate computing techniques for DNN implementation can reduce power consumption by as much as 88% [10]. However, the majority of the approximate computing techniques result in a decrease of accuracy, which may not be acceptable for some applications. In particular, the training of DNNs, which could take many days even using GPUs, require high-precision floating-point values to achieve best results [11]. Hence, it is important to come up with methods that can make computation of DNNs more efficient without reducing the accuracy of the output.

A typical DNN consists of many convolution and fully connected layers. Each of these layers perform MAC operation on the input using weights that are trained to generate a unique feature representation as the output [1]. Many such layers placed in succession can be used to approximate a target function. While the convolution and fully-connected layers alone are sufficient to represent linear functions, they cannot be used directly for applications that need nonlinear representations. To introduce nonlinearity into the model, the outputs of the convolution and fully-connected layers are passed through a nonlinear operator called an activation function [12]. As every output value is required to pass through an activation function, choosing the right activation function is an important factor for the effectiveness of DNNs [13].

One of the most widely used activation function is the Rectified Linear Unit (ReLU) [14]. The simple, piece-wise linear nature of ReLU can enable faster learning, and maintain stable values when using gradient-descent methods [12]. The output of a ReLU function is the same as the input when the input is positive, and is zero for negative inputs. This means that the precision of output is important only when the input is a positive value. Input to a ReLU function is usually the output from a fully-connected or convolution layer of the DNN which consist of a large number of MAC operations [8]. Studies have found that between 50% to 95% of ReLU outputs in DNNs are zero [15]. Hence, a lot of high precision compute in DNNs is wasted where output elements are reduced to zero after ReLU function. Early detection of these negative values can result in reducing the energy spent on high precision MAC operations, which would ultimately result in an efficient DNN implementation.

To this end, our work proposes a method for early detection of negative input values to the ReLU function, accounting for the maximum possible error while performing MAC with reduced precision. Using these values, we develop a mathematical model that provides a threshold below which a negative output value is guaranteed, irrespective of the remaining bits to be computed. It is shown that a proposed model can detect up to 80% negative values for popular CNN models using just three mantissa bits of floating-point number. This mathematical model can be used as the basis to implement low-precision MAC operations for DNNs adopting ReLU functions, which would result in efficient DNN implementation without a loss in accuracy. In summary, our contributions are threefold:

- Study of the fraction of ReLU zeros in two popular CNN models—VGG16 [16] and ResNet50 [17]
- A mathematical model that can accurately detect negative values based on the number of mantissa bits used a low precision MAC operation.
- Implementation of the developed model to detect ReLU zeros early in the VGG16 [16] and ResNet50 [17] inference stage.

2. Literature Review

The training and inference of DNNs is a compute intensive task, and has resulted in the need for various hardware accelerators [18–23]. Memory performance can be optimized through data locality by maximizing the reuse of data at buffers close to the compute block, as shown by Chen et al. [20]. A bit serial approach was considered by Judd et al. to reduce overall computations required by reducing activation precision [21]. Unnecessary multiplication with zero values was eliminated in Cnvlutin, which resulted in improved performance [19]. TETRIS used a high bandwidth 3D memory, which lead to reduced internal buffer size, for dealing with the memory bottleneck issue and overcoming the memory bottleneck [23]. Pruning techniques have also been studied to maximize compute saving by exploiting the sparsity in DNNs [24]. However, these methods are typically used for very specific applications and are expensive to generalize.

Approximate computing has emerged as one of the most effective solutions for generic DNNs, and it can exploit the inherent resilience of the CNN model (i.e., its ability to handle variations in data and still be able to figure out the pattern) and reduce the computation costs [25]. The level of approximation could be varied for different DNN models and datasets, hence approximate computing gained popularity [9]. This has led many researchers to investigate methods to perform low-precision computations in DNNs [26–33].

One of the most commonly applied technique is quantization, which is the process of replacing floating point numbers by numbers with reduced bit width. A study by Gupta et al. [26] demonstrated DNN training using 16-bit wide floating-point number with a very small reduction in accuracy as compared to a 32-bit floating-point number. Another study explored the effect of variable precision across different CNN layers, and demonstrated accuracy close to the benchmarks [27]. Venkatesh et al. studied the possibility of using 2-bit weights and space computing methods to produce state-of-the-art results. The study employed few iterations of full-precision training, followed by reduced precision training and inference [30]. The study on compute complexity is reduction using a 1D kernel factorized network is presented in the work [34].

Another approach in approximate computing is the use of multipliers and adders that compute results in a simplified manner. The work by Sarwar et al. [29] highlighted the use of simplified add and shift operations for power savings in DNNs. Another study explored the use of alternate full-adder implementation for efficient CNN hardware [28]. Stochastic computing based circuits have also been studied as potential candidates to implement a low-power DNN hardware accelerator [32].

Approximate computing has also been pursued at the software level, by simplifying the DNN architectures to reduce compute. Pruning the synaptic weights, reducing bit width of the synapse, and minimizing the number of hidden layers or neurons within these layers were demonstrated as effective methods to develop energy efficient DNNs [29]. Wei et al. came up with a more structured approach with pattern-based weight pruning for real-time DNN execution [33].

While all these studies have highlighted the relevance and requirement of approximate computing, they also mentioned that it comes at the cost of reduced accuracy. However, DNNs often require high precision floating point values during training to achieve high accuracy and reduced training time [35–37]. Such a reduction in accuracy may be unacceptable in real-life applications like self-driving cars [38] or medical diagnosis [39,40], where errors could be life threatening. Hence, most commercial DNNs still use floating point precision in their computations [41,42]. Hence, it is important to come up with a method to perform low precision computations in DNNs without reducing the accuracy of the model.

Shormann et al. proposed a method to reduce convolution operations in CNNs by dynamically predicting zero-valued outputs [43]. SnaPEA performs a reordering of weights and keeps track of the partial sum to predict zero outputs early [44]. A similar method was employed by Asadikouhanjani et al. to propose an efficient DNN accelerator [45]. By considering the spatial surroundings of an output feature map, Mosaic-CNN performs reduced precision compute to predict zero values early [46]. Other studies have explored methods to predict the zero values in an output feature map using the sign values [47–49]. Our study attempts further research in this direction by proposing a novel method to predict ReLU zeros with reduced precision compute.

3. Background
3.1. Convolutional Neural Networks

Among the different types of DNNs, Convolutional Neural Networks (CNNs) are extensively used in image processing, computer vision, and speech processing applications, often resulting in superior performance [50]. The convolution layer, which converts the input image into a form that is easier to process by the next layer, is at the heart of a CNN.

Convolution is the application of a filter to an input to produce an output feature map to indicate a detected feature in the input data. Both the input values and the filter values are represented as matrices, with the filter dimensions typically being much smaller that the input. The values in the filter matrix are multiplied with the corresponding values in the input matrix, and the values are added to produce a single output value. This MAC operation is repeated by shifting the filter by a fixed amount each time, resulting in an output feature map. The number of element shifts by the weight matrix on the input matrix is called the stride. This convolution process is demonstrated in Figure 1. As shown in Figure 1, each term from the input layer is multiplied with every term in the filter matrix, and these values are added together (accumulate) to generate one value in the output feature map. This process is repeated by moving the filter matrix across the input matrix, until it has been traversed completely. Once the output feature map is generated, it is passed through an activation function (like ReLU) to introduce nonlinearity.

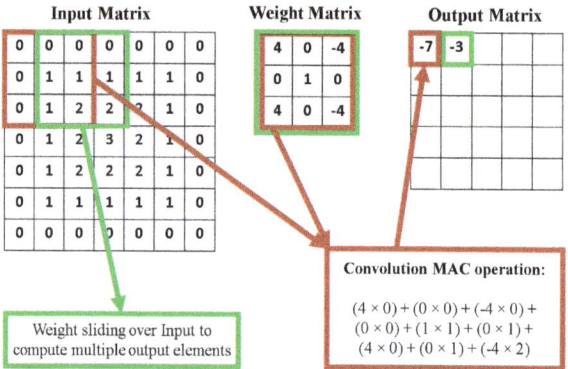

Figure 1. Example of the convolution operation. In this example, the stride is assumed to be 1. A 5 × 5 output is produced from the 7 × 7 input when a 3 × 3 weight matrix is considered.

3.2. ReLU Activation Function

The ReLU activation function is one of the most popular activation functions used in DNNs today [14]. The function returns zero for all negative inputs, and returns the input if it is a non-negative value. It can be written as:

$$f(x) = max(0, x) \qquad (1)$$

where max returns larger of the two inputs. The graphical representation is shown in Figure 2. The success of ReLU can be attributed to its simple implementation, which in turn reduces the computation time of the DNN model [51]. In addition, a majority of the ReLU outputs are zero [15], which makes the output matrix sparse and results in better prediction and reduced chances of overfitting [52]. Both the ReLU function and its derivative are monotonic, which ensures that the vanishing gradient problem is avoided when the gradient-descent training process is employed [53]. These factors have contributed to the widespread use of ReLU activation function in DNNs. Hence, the study of the ReLU activation function is important to implement DNNs more efficiently.

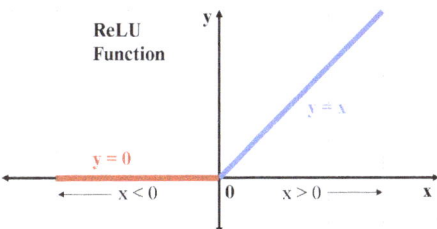

Figure 2. Graphical representation of the ReLU function. If x is the input and y is the output, then $y = 0$ for $x < 0$, and $y = x$ for $x \geq 0$.

3.3. Floating Point Number Representation

In any typical DNN, the input, output, and intermediate values are stored in the floating-point format. The standard format used in a majority of applications is the IEEE-754 floating point number format [54]. In this format, the Most Significant Bit (MSB) is the Sign bit (S) which is 0 for positive numbers, and 1 for negative numbers. This is followed by a fixed number of bits assigned to store the Exponent E, and the remaining bits are allotted to the Mantissa M. The fractional part is stored in the normalized form—i.e., the actual values in binary is 1 plus the fractional value represented by M. In order to accommodate negative exponent values, 127 is added (called excess-127). Hence, the actual exponent is E–127. Based on these rules, the floating point value represented using the S, E and M values in the IEEE-754 format is:

$$F = (-1)^S \times 2^{(E-127)} \times (1 + M) \qquad (2)$$

The two commonly used forms of the IEEE-754 format are the single and double precision format. In the single precision representation, there are 8 exponent bits and 23 mantissa bits to make a total of 32 bits. The double precision is a 64-bit representation with 11 exponents and 52 mantissa bits [54]. Figure 3 graphically depicts both the single and double precision representations.

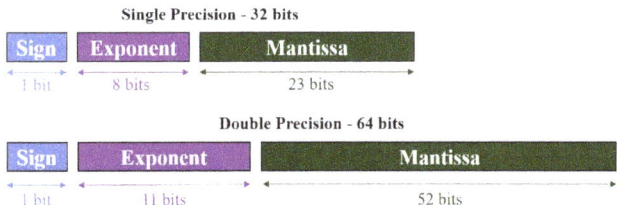

Figure 3. IEEE 754 floating point representation [54]. The total bits are divided into sign, exponent, and mantissa. The single precision format has 1 sign, 8 exponents, and 23 mantissa bits, while the double precision has 1 sign, 11 exponents, and 52 mantissa bits.

4. Methodology

4.1. Dataset and Framework

As image recognition is one of the most widely used and researched applications of CNNs, we focus our analysis on models within this domain. VGG-16 is one of the pioneer CNN models for large scale image recognition tasks [16]. It takes a 224 × 224 RGB image as input and passes it through different convolution, max-pooling, and fully connected layers. The final classification is implemented using a softmax layer. Figure 4 describes the VGG-16 architecture. As evident from the figure, there are 13 convolution layers, and each convolution layer is followed by a ReLU activation layer. A set of convolution layers are followed by pooling layers to reduce the dimensions of the input before sending it to the next convolution sets. Finally, a set of fully-connected layers are added to produce the output classification probability.

As DNNs like VGG16 became difficult to train, Residual Networks (ResNets) emerged as improved alternatives. In ResNets, shortcut (or identity) connections were introduced between different layers to perform quick identity mapping with no additional model parameters [17]. One such ResNet model is the ResNet-50, which has 50 different convolution and fully-connected layers along the path from input to output. Like VGG-16, ResNet-50 also takes 224 × 224 RGB images as its input. The ResNet50 architecture is shown in Figure 5. A convolution operation that is applied on the input and the layer size is reduced before it is sent to the residual layers. Each of the residual layers is comprised of three sets, each with a convolution layer followed by a ReLU activation layer. Before the last ReLU operation, an identity connection is added to train identity mappings in some of the layers. The Res 2–1, Res 3–1, and Res 4–1 groups shown in Figure 5 have a convolution layer in the identity path. These residual layers are followed by a pooling and fully-connected layer, which give the classification probabilities as the output.

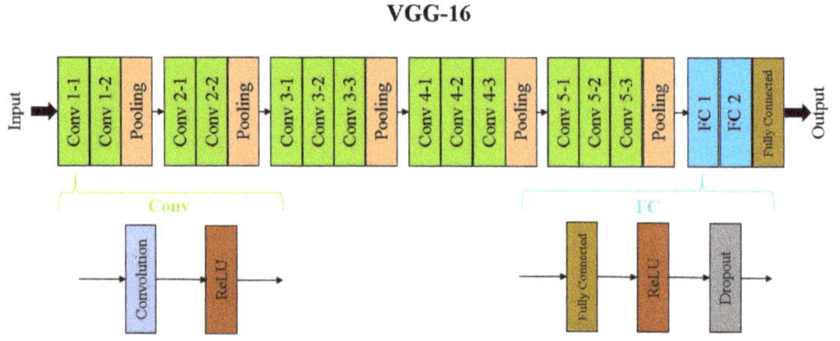

VGG-16

Figure 4. VGG-16 CNN architecture. There are 16 computation layers (13 convolution −3 × 3 kernel and three fully connected layers without dropout). Pooling layers are present in the intermediate stages to reduce the layer size as the network gets deeper. Regularization, normalization, and other layers may be present but have not been shown in this figure for simplicity.

ResNet-50

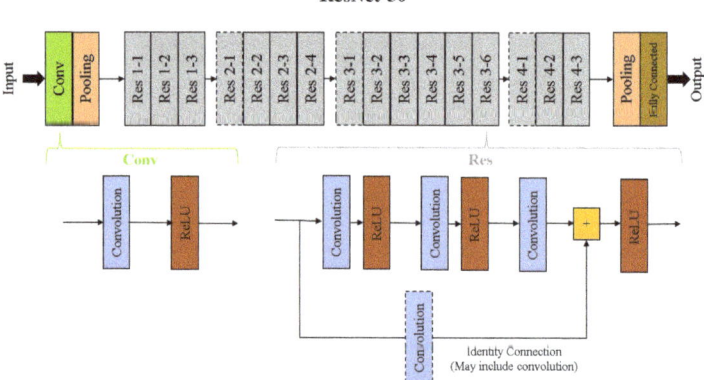

Figure 5. ResNet-50 Architecture. There are 50 computations layers (excluding convolution layers in the identity path) between the input and output. This includes 49 convolution layers and the fully-connected layer at the end. Res 2–1 (conv with 1 × 1, 64; 3 × 3, 64; 1 × 1, 256), Res 3–1 (1 × 1, 128; 3 × 3, 128; 1 × 1, 512) and Res 4–1 (1 × 1, 256; 3 × 3, 256; 1 × 1, 1024) are shown with a dotted boundary to indicate that they include a convolution layer along their identity path (also shown with a dotted boundary in the elaboration below without dropout). Regularization, normalization, and other layers may be present but have not been shown in this figure for simplicity.

These models were tested using the ImageNet Large Scale Visual Recognition Challenge 2012 (ILSVRC2012) inference dataset, which includes 50,000 images belonging to 1000 different classes [55]. These images were converted to the 224 × 224 RGB format, and the pixel values were normalized. To ensure that the training methods were standard, the pretrained models of ResNet-50 and VGG-16 were used from the Keras library [56] running on top of the TensorFlow [57] backend.

4.2. Proposed Hierarchical Computation

It is evident that each convolution layer involves MAC operations between the input values and a filter with the trained weight values. The result of this MAC operation is passed through the ReLU activation function and negative values are made zero. Our implementation includes an intermediate step that predicts negative values early using a reduced number of mantissa bits. The MAC operation is performed using reduced mantissa bits, and the output is obtained. Then, based on the number of mantissa bits used for the computation, the proposed model predicts whether a value estimated is definitely negative or not. If the value is determined to be negative, the output is made to be zero. For the other values, we perform MAC using the full precision and obtain the output like in typical implementations. Hence, we reduce the total number of cases for which the expensive full precision compute must be performed, while simultaneously ensuring no loss in accuracy.

The steps are described using a flowchart in Figure 6. In the case presented, for every workload, we first perform computation without any mantissa bits. Since only exponent values are present, this can be achieved directly by adding the exponent bits. If output can be predicted to be negative at this step, then set the output as zero and move on to the next set of element of the input workload. If inconclusive, the first 8 MSB mantissa bits (bits 23 to 16) are considered for further computation. Once again, predict the accumulated negative value, and set those outputs to zero. For cases where the accumulated element sign is still ambiguous, the remaining mantissa bits (bits 15 to 0) are also used and the full precision compute is performed. The remaining outputs are obtained after this step, and this whole cycle is repeated for the other input workloads. This way, the total compute can be split into multiple levels by adding additional mantissa bits. At each level, some negative values can be detected with reduced precision compute. At the same time, full precision compute can be performed for all positive outputs, ensuring no loss in accuracy. The selection of levels of compute and bits selection for each level can be determined based on model, workload, and underlying compute hardware availability. The impact of selected mantissa bits on correctly predicted negative values is described later in the Results section.

Figure 7 intuitively describes the proposed hierarchical computation approach to estimate ReLU output with reduced precision compute. Here, the "Ideal" is the value that is computed with full precision, while "Reduced" is the output with only a few MSB mantissa bits considered. If "Reduced" is a large negative value, the output can be estimated to be negative irrespective of the mantissa bits. Our model detects such values until it reaches a threshold, where "Reduced" is negative but close to zero. To estimate these values correctly, more mantissa bits (next set of MSB bits) need to be considered. Similarly, when "Reduced" is a large positive value, it can be estimated to be positive without using the mantissa bits. However, as the value approaches zero, more mantissa bits are required to correctly estimate the sign of the element. A threshold is estimated along the positive axis too, beyond which values are always positive. Our model determines both the positive and negative thresholds, which gives rise to the region of interest where full precision compute needs to be performed, as shown in Figure 7. These thresholds are obtained by considering the maximum error contribution from each mantissa bit of a floating point number. As shown in Figure 7, the error is inversely proportional to the number of mantissa bits "n", which means that the region of interest gets smaller as the value of "n" increases.

The next section derives a mathematical model that can perform the ReLU checks shown in Figure 6, based on the intuitive model proposed in Figure 7. We use error calculations to prove that the model can determine ReLU zeros with no loss in accuracy.

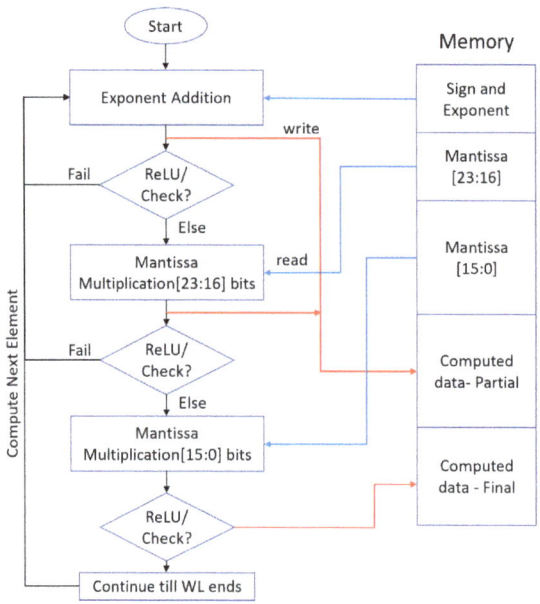

Figure 6. Flow chart depicting the steps to perform hierarchical compute (three steps) and detect ReLU zeros with reduced precision. The first step is to perform MAC using exponent and predict ReLU output; if undetermined, compute most significant 8-bits of mantissa and check ReLU again if still not conclusive perform compute using remaining mantissa bits (every next step uses previously computed values). Here, the red arrows depict writing to the memory, and blue arrows indicate read from memory. Black arrows indicate that the computation has been completed for the given input.

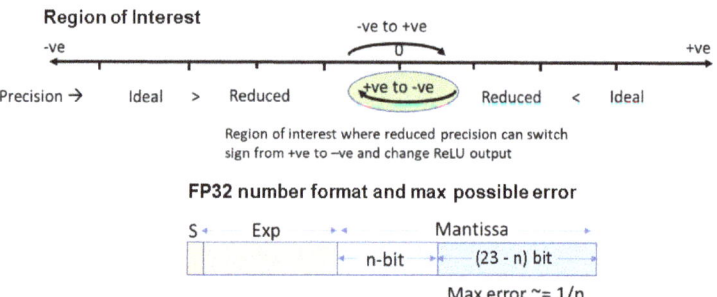

Figure 7. Intuition behind estimating ReLU zeros based on reduced precision compute. In the hierarchical compute method, the value of "n" (number of MSB mantissa bits) is increased at each step, resulting in a decrease in the region of interest, until only positive values are remaining.

4.3. Mathematical Model

In this section, the mathematical model of the proposed solution is presented. There are three theorems, which cover all scenarios of the proposed solution. Theorem 1 presents an important scenario where errors due to addition of positive values are the main contributors for a sign change of resultant from positive to negative, which impacts the threshold calculation. Theorem 2 describes the max error that is needed to detect negative values out of MAC operations, and Theorem 3 talks about the major condition that needs to be satisfied for predicting the ReLu output.

Theorem 1. *Let*

$$X_S(a) = \sum_{k=-n}^{n} (ifm(k) * wt(a-k)) \quad (3)$$

where a = number of terms involved in the convolution, if m(k) and wt(k) are input feature map and weight kernel in single precision floating point representation (FP32) with a reduced number of mantissa bits (number of mantissa bits after reduction = m).

$$X_{SPOS}(n) = \sum_{k=0}^{n} (ifm(k) * wt(n-k)) \quad (4)$$

is responsible to convert a positive $X_S(n)$ with m = 23 (FP32) to negative $X_S(n)$ with m < 23.

Proof of Theorem 1. Let $X_S(b)$ where $b < a$, with $m < 23$. Let $X_{Reduced}$ and X_{Ideal} be the values of the next term to be added to the convolution sum, with $m = 23$ and $m < 23$, respectively. When this term is added to the existing sum, two different sums are obtained depending on the presence or absence of all mantissa bits. Let these be called X_S^{Ideal} and $X_S^{Reduced}$, respectively. That is,

$$X_S^{Ideal} = X_S + X_{Ideal} \quad (5)$$

$$X_S^{Reduced} = X_S + X_{Reduced} \quad (6)$$

It is evident that reducing the number of mantissa bits in a floating point number results in a number having lower magnitude. However, the sign remains unaffected as the sign bit is unchanged. Hence, if

$$X_{Ideal} < 0$$
$$\implies X_{Reduced} > X_{Ideal}$$
$$\implies X_S + X_{Reduced} > X_S + X_{Ideal}$$

From (5) and (6), we have

$$X_S^{Reduced} > X_S^{Ideal} \quad (7)$$

From (7), it is evident that, if $X_S^{Reduced} < 0$, it can be concluded that $X_S^{Ideal} < 0$. In other words, error due to addition of a negative value cannot alter the sign of the sum from positive to negative. On the contrary, if

$$X_{Ideal} < 0$$
$$\implies X_{Reduced} < X_{Ideal}$$
$$\implies X_S + X_{Reduced} < X_S + X_{Ideal}$$

From (5) and (6), we have

$$X_S^{Reduced} < X_S^{Ideal} \quad (8)$$

In the case of (8), $X_S^{Reduced} < 0$ does not guarantee that $X_S^{Ideal} < 0$. Hence, errors due to the addition of positive values contribute towards sign change from positive to negative, and are important in determining the threshold to conclude that the convolution sum is negative when reduced-mantissa is considered. □

Theorem 2. *If a positive term in the convolution sum is given by $C_{Mul} = 2^{E_{Mul}} \times M_{Mul}$, where E_{Mul} and M_{Mul} are the unbiased exponent and mantissa value of the term, the maximum error that is possible when the number of mantissa bits is reduced to n is given by $C_{ErrMax} = 2^{E_{Mul}-n+1} \times M_{Mul}$.*

Proof of Theorem 2. For any floating point number given by

$$N = (-1)^S \times 2^E \times M$$

where S, E, M represent the sign, unbiased exponent, and mantissa value, the maximum possible error when only n mantissa bits are included is given by

$$E_{Max} = -2^{(E-n)} \times (-1)^S \tag{9}$$

Consider an activation input (I) and weight (W) of a convolution layer. They are represented as

$$I = (-1)^{S_I} \times 2^{E_I} \times M_I \tag{10}$$

$$W = (-1)^{S_W} \times 2^{E_W} \times M_W \tag{11}$$

From (9), the most erroneous values that could result from reducing the number of mantissa bits to n in I (10) and W (11) are given by

$$I_{Reduced} = (-1)^{S_I} \times 2^{E_I} \times M_I - 2^{(E_I-n)} \times (-1)^{S}_{I} \tag{12}$$

$$W_{Reduced} = (-1)^{S_W} \times 2^{E_W} \times M_W - 2^{(E_W-n)} \times (-1)^{S}_{W} \tag{13}$$

The convolution term when I (10) and W (10) are multiplied is given by

$$C_{Ideal} = (-1)^{S_I+S_W} \times 2^{E_I+E_W} \times (M_I \times M_W) \tag{14}$$

With reduced mantissa in the convolution step, (12) and (13) give

$$\begin{aligned} C_{Reduced} &= I_{Reduced} \times W_{Reduced} \\ &= (-1)^{S_I+S_W} \times 2^{E_I+E_W} \times (M_I \times M_W) \\ &\quad - (-1)^{S_I+S_W} \times 2^{E_I+E_W-n} \times (M_I + M_W) \\ &\quad + 2^{E_I+E_W-2n} \end{aligned}$$

Hence,

$$C_{Reduced} = 2^{E_I+E_W} \times (M_I \times M_W - (2^{-n} \times (M_I + M_W - 2^{-n}))) \tag{15}$$

The error in convolution terms due to reduced mantissa can be obtained from (14) and (15)

$$\begin{aligned} C_{Error} &= C_{Ideal} - C_{Reduced} \\ &- 2^{E_I+E_W-n} \times (M_I + M_W + 2^{-n}) \end{aligned}$$

As 2^{-n} is always positive,

$$C_{Error} \leq 2^{E_I+E_W-n} \times (M_I + M_W). \tag{16}$$

Since M_I and M_W represent the mantissa values,

$$1 \leq M_I, M_W \leq 2$$
$$\longrightarrow M_I + M_W \leq 2 \times M_I \times M_W$$

Hence, (16) can be rewritten as

$$\begin{aligned} C_{Error} &\leq 2^{E_I+E_W-n} \times (2 \times M_I \times M_W) \\ &= 2^{E_I+E_W-n+1} \times (M_I \times M_W) \end{aligned}$$

From (14), we get

$$C_{Error} \leq 2^{-n+1} \times C_{Ideal} \tag{17}$$

It is evident from Theorem 1 that only positive terms will contribute to errors that can contribute to incorrectly identifying a negative value. Hence, $S_I + S_W = 0$ (Either both I and W are positive or both are negative). Including this in (14), we can rewrite C_{Ideal} as

$$C_{Ideal} = 2^{E_{Mul}} \times M_{Mul} \tag{18}$$

where $E_{Mul} = E_I + E_W$ and $M_{Mul} = M_I \times M_W$. Hence, the maximum error in a positive term in the convolution sum is

$$C_{ErrMax} = 2^{E_{Mul}-n+1} \times M_{Mul} \tag{19}$$

Hence, we obtain the maximum error, which is needed to detect negative values from a MAC operation. □

Theorem 3. *If the convolution sum before the ReLU activation layer is given by $C_{Tot} = (-1)^{S_{Tot}} \times 2^{E_{Tot}} \times M_{Tot}$, and the sum of positive terms in the summation (including the bias value) is given by $C_{Pos} = 2^{E_{Pos}} \times M_{Pos}$, then the value of C_{Tot} can be concluded to be negative if $S_{Tot} = 1$ and $E_{Tot} > E_{Pos} - n$, where n is the number of mantissa bits used in the computation.*

Proof of Theorem 3. Let the sum of all product terms in the convolution be given by

$$C_{Tot} = \sum_i (-1)^{S_i} \times 2^{E_i} \times M_i = (-1)^{S_{Tot}} \times 2^{E_{Tot}} \times M_{Tot} \tag{20}$$

From (19) in Theorem 2, the maximum error due positive terms in the convolution is given by $C^i_{ErrMax} = 2^{E_i-n+1} \times M_i$. Hence, when these errors are accumulated for all positive terms (including bias), we get

$$C_{ErrTot} = \sum_{i:S_i=0} C^i_{ErrMax} = \sum_{i:S_i=0} 2^{E_i-n+1} \times M_i \tag{21}$$

Note that, unlike other terms in the convolution sum, the bias does not involve multiplication of reduced mantissa numbers. Hence, the maximum error for bias values will be lower. However, the same error has been considered (as an upper bound) to simplify calculations.

We can represent the sum of positive terms (including bias) in the convolution sum as

$$C_{Pos} = \sum_{i:S_i=0} 2^{E_i} \times M_i = 2^{E_{Pos}} \times M_{Pos} \tag{22}$$

Using (22), the total error in (21) can be rewritten as

$$C_{ErrTot} = 2^{-n} \times C_{Pos} \tag{23}$$

To conclude that a convolution sum is zero/negative, the following two conditions should hold:

$$|C_{Tot}| \geq |C_{Pos}| \tag{24}$$

$$S_{Tot} = 1 \tag{25}$$

(24) can be expanded using (20) and (22) to give

$$2^{E_{Tot}} \times M_{Tot} \geq 2^{E_{Pos}-n+1} \times M_{Pos} \tag{26}$$

Note that, if $E_{Tot} = E_{Pos} - n + 1$, then the condition $M_{Tot} \geq M_{Pos}$ must hold (as the total convolution sum (C_{Tot}) must be greater than or equal to the sum of positive convolution terms and bias (C_{Pos})) As a consequence, (26) now becomes

$$E_{Tot} \geq E_{Pos} - n + 1 \tag{27}$$

$$\Rightarrow E_{Tot} > E_{Pos} - n \qquad (28)$$

Hence, from (25) and (28), we can conclusively say that a convolution sum computed using reduced-mantissa bits is negative (In addition, its ReLU output is zero) if $S_{Tot} = 1$ and $E_{Tot} > E_{Pos} - n$. □

4.4. Early Negative Value Prediction

The theorems derived above can be used to implement the proposed model for hierarchical computation. The steps to find out if a reduced precision value is a ReLU zero can be represented as an algorithm, as shown here:

1. Consider inputs and weights of convolution with reduced "n" mantissa bits
2. Compute C_{Pos}, the sum of positive convolution terms, as per (22)
3. Obtain E_{Pos}, the exponent value of C_{Pos}, as per (22)
4. Compute $C_{Tot} = C_{Pos} + C_{Neg}$, where C_{Neg} is the sum of negative convolution terms
5. Obtain E_{Tot}, the exponent value of C_{tot}, as per (20)
6. If $E_{Tot} > (E_{Pos} - n)$, then assign the ReLU output as zero. The computation is complete.
7. If ReLU zero is not assigned, repeat steps 1–6 for higher values of "n".

5. Results

In order to motivate the use of the hierarchical compute method to detect ReLU zeros early, we first identify the number of ReLU zeros that are present when a typical image is processed using the ResNet-50 and VGG-16 CNN models; findings are shown in Figure 8. It is evident from the figure that, in a majority of the layers, more than 50% of the ReLU outputs are zero, with many of the deeper layers having up to 90% ReLU zeros. Considering all the layers, we found that, on an average, 61.77% of the ReLU outputs were zeros in VGG-16, while 61.24% ReLU zeros were seen for ResNet-50. These results indicate that a large portion of compute is wasted on computing ReLU zeros, which can be avoided using the proposed method.

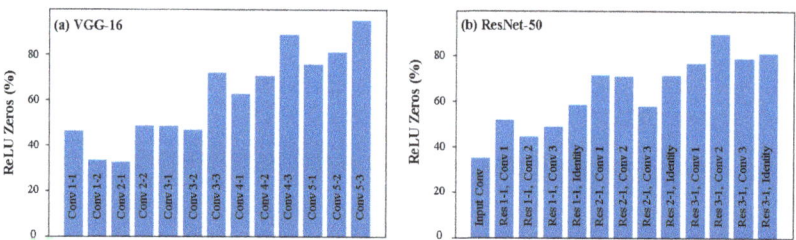

Figure 8. Percentage of ReLU zeros present in (**a**) VGG16; (**b**) ResNet50 when a typical image is processed through the models. Only a few layers of ResNet-50 are shown for clarity—a similar trend is observed in all the layers.

In addition to the percentage of ReLU zeros, it is also important to understand the distribution of values seen by the ReLU layer. The results from ResNet-50 layers are shown in Figure 9. A total of 10 bins were chosen—values below -8, -8 to -4, -4 to -2, -2 to -1, -1 to 0, 0 to 1, 1 to 2, 2 to 4, 4 to 8 and values above 8. We notice that, in all layers, about 50% of the values fall between -1 and 1, and more than 80% between -2 and 2. This implies that the majority of the values are close to zero. As a result, it is not practical to use a fixed threshold value along with reduced precision compute. A large negative threshold (say -2) can ensure that a value computed with reduced precision will have the correct sign. However, we can see from the distribution that only very few values (under 20%) can be detected with such a fixed threshold. If the threshold is pushed closer to zero, the chances of incorrectly detecting ReLU zeros increase. This study demonstrates the importance of the variable threshold derived using our model.

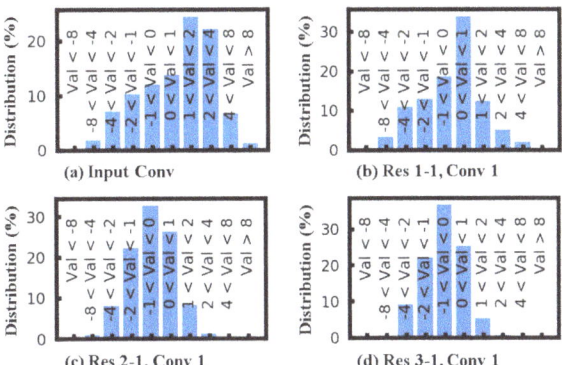

Figure 9. Distribution of ReLU inputs in different layers of ResNet-50. Here, Val is the input to the ReLU function. A total of 10 bins have been considered, and the range in each bin in mentioned in the figure. The different layers shown in the figure are: (**a**) first convolution layer from the input image (**b**) first convolution layer in the Res 1-1 block; (**c**) first convolution layer in the Res 2-1 block; (**d**) first convolution layer in the Res 3-1 block.

The proposed model was tested by evaluating the ReLU output values at different layers of both the VGG-16 and ResNet-50 CNN implementation. This was done by comparing the outputs from the convolution layer using (25) and (28). The total number of negative values that were detected using our model were noted and compared with the total number of output values to provide the percentage of negative values that are detected early. This was repeated for different layers, with different numbers of mantissa bits. Figure 10 shows the percentage of ReLU zeros detected by our model across different layers of ResNet-50 with different numbers of mantissa bits. It is evident that, as the number of mantissa bits considered increases, our model is able to detect the majority of ReLU zeros in all layers.

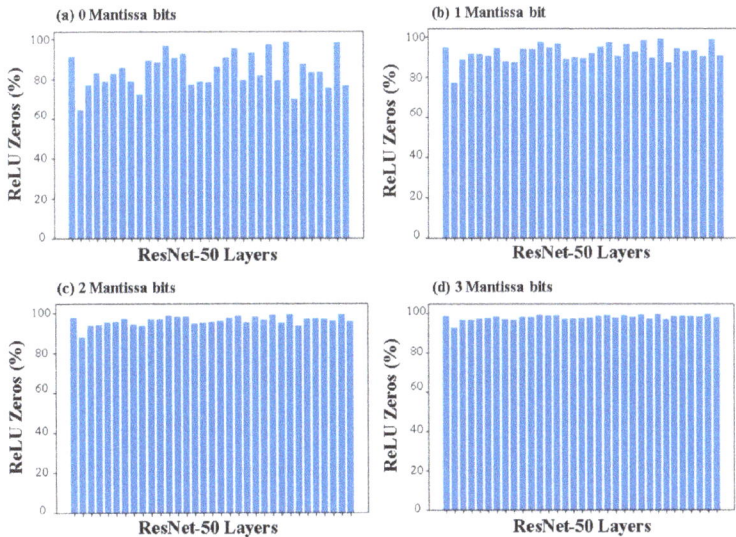

Figure 10. Percentage of ReLU values detected using our model across different ResNet-50 layers. The first 33 convolution layers are shown in the figure. The number of MSB mantissa bits used were (**a**) 0; (**b**) 1; (**c**) 2; and (**d**) 3.

To get a closer look at the impact of increasing the number of mantissa bits, we plotted the percentage of ReLU zeros detected with 0, 1, 2, and 3 mantissa bits for randomly chosen layers in VGG-16 and ResNet-50. This is shown in Figure 11. As expected, the fraction of negative values detected increases as the number of mantissa bits used for computation is increased. Close to 80% of negative values can be detected early using just three mantissa bits, which can result in a significant increase in the efficiency of the network. Due to the nature of weights, range of values, and so on, we observe that the results across different layers vary. However, as seen in Figure 10, the amount of variation decreases as we use more mantissa bits. Additionally, we note similar effectiveness of our model for both VGG-16 and ResNet-50, which shows that the model does not depend on the type of CNN implementation—it works based on the fundamental characteristics of MAC operations and floating-point numbers, which makes it a generalized solution for any CNN layer with a ReLU activation function.

From the results presented, we see that about 60% of the outputs of the ReLU activation function are zero values in CNNs like VGG-16 and ResNet-50. If three mantissa bits are used for computation and our model is deployed, 80% of these ReLU zeros can be detected. Hence, we can expect about 50% of the all ReLU outputs to be estimated early. This way, almost half of the total computations can be carried out in low precision and the other half can be computed in full precision, while ensuring no loss in accuracy.

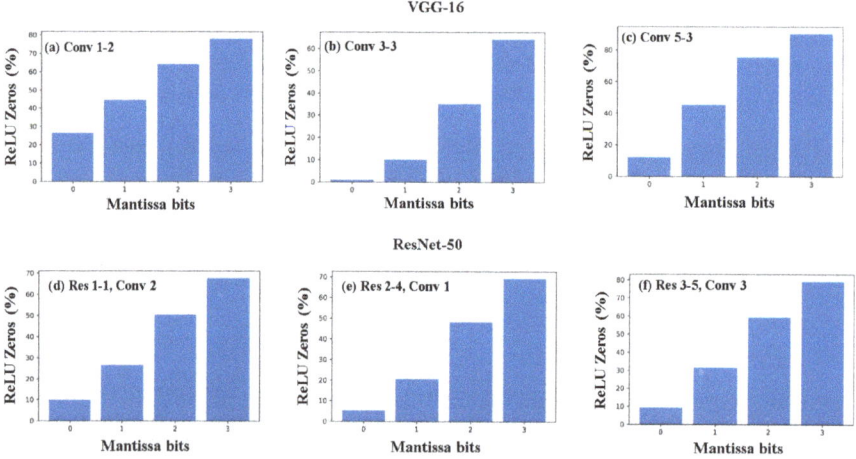

Figure 11. Percentage of ReLU zeros identified by our model when different mantissa bits were considered. The figure shows the results in (**a**) Conv 1-2; (**b**) Conv 3-3; and (**c**) Conv 5-3 layers of VGG-16, and (**d**) second convolution layer of Res 1-1 block; (**e**) first convolution layer of Res 2-4 block; and (**f**) third convolution layer of Res 3-5 block. Similar results were observed in other layers of both ResNet-50 and VGG-16.

6. Discussion

6.1. Generalization to Other DNNs

The results presented in this work utilize CNNs as the end application due to their ubiquitous nature and applicability to various fields. However, the model we have proposed is built on fundamental properties of floating-point numbers, MAC operations, and the ReLU activation function. Hence, the model can be extended to other applications too. When there are no negative values in the whole process, the algorithm will not predict any outcome and bypass all MAC output as valid output. However, these computations that are bypassed by algorithms as valid will be reused as a partial product for computing the actual output with the remaining mantissa bits. Since the compute used for prediction is re-purposed as a partial product and also for the cases which are slightly uncertain, the algorithm tends to predict them as positive value (no approximation) and bypass them out of the algorithm as valid outputs such that it will always go through the full precision compute. Hence, no accuracy drop is expected with the use of a proposed solution.

The proposed solution will be applicable across various networks with activation functions which displays a nonlinear behavior for either positive or negative numbers (not both) and the other one must be a zero or any constant value. To support activation functions like sigmoid, we might need to redevelop mathematical constraints to predict values between [−ve, +ve] range, while all other values outside this range can be set to a constant.

6.2. Implementation on GPU/Other Accelerators

This method is implementable on any compute engine that supports DNN workload. Since the proposed solution supports the reusing of partially computed elements that were used for early prediction, this will not impose a heavy tax on the existing hardware. On GPU and other accelerators, the proposed solution will need fine-tuning of data flow, data storage pattern and control logic, etc.

6.3. Extension to Training

The results presented in this work demonstrate the effectiveness of our model during the inference stage in a DNN. However, the process of training also involves the same set of steps, along with the additional step of adjusting the parameters. Hence, our model can be used in every layer with a ReLU activation layer. Since DNN training is a time-consuming and compute intensive process, this method can provide a significant improvement. It is also noteworthy to mention that, unlike inference, training must be done with high precision values. As a result, many of the approximate computing methods that have been studied cannot be extended to training. However, since our method ensures that there is no loss in accuracy, it can be applied to training as well.

7. Conclusions

In this work, we proposed a mathematical model that can detect zero outputs of the ReLU activation function using low-precision MAC operations. Our model takes into account the error resulting from the reduction of the number of bits in a floating-point representation, and identifies values that would be negative even when full-precision compute is performed. Our model can adapt based on the number of mantissa bits considered in the computation, ensuring its suitability for different number formats used in DNNs. We show that around 80% of ReLU zeros can be detected using just three mantissa bits, which corresponds to a total of 50% of all ReLU outputs in VGG16 and ResNet50 CNN implementations. As the model is developed with no assumption about the nature of the network or the application, we claim that the model can be extended to all DNNs that use the ReLU activation function. In addition, as the MAC operation and the activation layer in DNN training is identical to inference, this model can be adopted to make the compute-hungry training process more efficient. We also propose a system level model to implement this method and perform hardware acceleration of DNNs. The widespread use of DNNs with the ReLU activation function means that our model can be used as an error-free way to reduce computations in numerous applications.

Author Contributions: Conceptualization, B.S., K.P. and G.S.K.; writing—original draft preparation, B.S.; writing—review and editing, K.P. and G.S.; supervision, A.A. and S.S. All authors have read and agreed to the published version of the manuscript.

Funding: This research received no external funding.

Institutional Review Board Statement: Not applicable.

Informed Consent Statement: Not applicable.

Data Availability Statement: The dataset used for this study is publicly available and can be downloaded at https://www.image-net.org (accessed on 1 May 2021).

Conflicts of Interest: The authors declare no conflict of interest.

Abbreviations

The following abbreviations are used in this manuscript:

DNN	Deep Neural Network
ReLU	Rectified Linear Unit
VGG	Visual Geometry Group
ResNet	Residual Network
MAC	Multiply And Accumulate
GPU	Graphics Processing Unit
CNN	Convolutional Neural Network
IEEE	Institute of Electrical and Electronics Engineers
ILSVRC	ImageNet Large Scale Visual Recognition Challenge

References

1. Liu, W.; Wang, Z.; Liu, X.; Zeng, N.; Liu, Y.; Alsaadi, F.E. A survey of deep neural network architectures and their applications. *Neurocomputing* **2017**, *234*, 11–26. [CrossRef]
2. Zhang, L.; Zhang, Y. Big data analysis by infinite deep neural networks. *Jisuanji Yanjiu Yu Fazhan/Comput. Res. Dev.* **2016**, *53*, 68–79. [CrossRef]
3. Strubell, E.; Ganesh, A.; McCallum, A. Energy and Policy Considerations for Deep Learning in NLP. In Proceedings of the ACL 2019—57th Annual Meeting of the Association for Computational Linguistics, Florence, Italy, 28 July–2 August 2019; pp. 3645–3650.
4. Harlap, A.; Narayanan, D.; Phanishayee, A.; Seshadri, V.; Devanur, N.; Ganger, G.; Gibbons, P. PipeDream: Fast and Efficient Pipeline Parallel DNN Training. *arXiv* **2018**, arXiv:1806.03377.
5. Deng, C.; Liao, S.; Xie, Y.; Parhi, K.K.; Qian, X.; Yuan, B. PERMDNN: Efficient Compressed DNN Architecture with Permuted Diagonal Matrices. In Proceedings of the 2018 51st Annual IEEE/ACM International Symposium on Microarchitecture (MICRO), Fukuoka, Japan, 20–24 October 2018; pp. 189–202.
6. Duggal, J.K.; El-Sharkawy, M. Shallow squeezenext: An efficient shallow DNN. In Proceedings of the 2019 IEEE International Conference on Vehicular Electronics and Safety (ICVES), Cairo, Egypt, 4–6 September 2019. [CrossRef]
7. Wei, W.; Xu, L.; Jin, L.; Zhang, W.; Zhang, T. AI Matrix—Synthetic Benchmarks for DNN. *arXiv* **2018**, arXiv:1812.00886.
8. Hanif, M.A.; Javed, M.U.; Hafiz, R.; Rehman, S.; Shafique, M. Hardware–Software Approximations for Deep Neural Networks. In *Approximate Circuits*; Springer International Publishing: Cham, Switzerland, 2019; pp. 269–288. [CrossRef]
9. Agrawal, A.; Choi, J.; Gopalakrishnan, K.; Gupta, S.; Nair, R.; Oh, J.; Prener, D.A.; Shukla, S.; Srinivasan, V.; Sura, Z. Approximate computing: Challenges and opportunities. In Proceedings of the 2016 IEEE International Conference on Rebooting Computing (ICRC), San Diego, CA, USA, 17–19 October 2016. [CrossRef]
10. Liu, B.; Wang, Z.; Guo, S.; Yu, H.; Gong, Y.; Yang, J.; Shi, L. An energy-efficient voice activity detector using deep neural networks and approximate computing. *Microelectron. J.* **2019**, *87*, 12–21. [CrossRef]
11. Zhu, H.; Akrout, M.; Zheng, B.; Pelegris, A.; Jayarajan, A.; Phanishayee, A.; Schroeder, B.; Pekhimenko, G. Benchmarking and Analyzing Deep Neural Network Training. In Proceedings of the 2018 IEEE International Symposium on Workload Characterization (IISWC), Raleigh, NC, USA, 30 September–2 October 2018; pp. 88–100. [CrossRef]
12. Nwankpa, C.; Ijomah, W.; Gachagan, A.; Marshall, S. Activation Functions: Comparison of trends in Practice and Research for Deep Learning. *arXiv* **2018**, arXiv:1811.03378.
13. Wang, Y.; Li, Y.; Song, Y.; Rong, X. The Influence of the Activation Function in a Convolution Neural Network Model of Facial Expression Recognition. *Appl. Sci.* **2020**, *10*, 1897. [CrossRef]
14. Alom, M.Z.; Taha, T.M.; Yakopcic, C.; Westberg, S.; Sidike, P.; Nasrin, M.S.; Van Esesn, B.C.; Awwal, A.A.S.; Asari, V.K. The history began from alexnet: A comprehensive survey on deep learning approaches. *arXiv* **2018**, arXiv:1803.01164.
15. Shi, S.; Chu, X. Speeding up convolutional neural networks by exploiting the sparsity of rectifier units. *arXiv* **2017**, arXiv:1704.07724.
16. Simonyan, K.; Zisserman, A. Very deep convolutional networks for large-scale image recognition. In Proceedings of the International Conference on Learning Representations (ICLR), San Diego, CA, USA, 7–9 May 2015.
17. He, K.; Zhang, X.; Ren, S.; Sun, J. Deep residual learning for image recognition. In Proceedings of the IEEE Conference on Computer Vision and Pattern Recognition, Las Vegas, NV, USA, 27–30 June 2016; pp. 770–778. [CrossRef]
18. Albericio, J.; Delmás, A.; Judd, P.; Sharify, S.; O'Leary, G.; Genov, R.; Moshovos, A. Bit-pragmatic deep neural network computing. In Proceedings of the 50th Annual IEEE/ACM International Symposium on Microarchitecture, Cambridge, MA, USA, 14–18 October 2017; pp. 382–394.

19. Albericio, J.; Judd, P.; Hetherington, T.; Aamodt, T.; Jerger, N.E.; Moshovos, A. Cnvlutin: Ineffectual-neuron-free deep neural network computing. *ACM SIGARCH Comput. Archit. News* **2016**, *44*, 1–13. [CrossRef]
20. Chen, T.; Du, Z.; Sun, N.; Wang, J.; Wu, C.; Chen, Y.; Temam, O. Diannao: A small-footprint high-throughput accelerator for ubiquitous machine-learning. *ACM SIGARCH Comput. Archit. News* **2014**, *42*, 269–284. [CrossRef]
21. Judd, P.; Albericio, J.; Hetherington, T.; Aamodt, T.M.; Moshovos, A. Stripes: Bit-serial deep neural network computing. In Proceedings of the 2016 49th Annual IEEE/ACM International Symposium on Microarchitecture (MICRO), Taipei, Taiwan, 15–19 October 2016; pp. 1–12.
22. Chen, Y.H.; Krishna, T.; Emer, J.S.; Sze, V. Eyeriss: An energy-efficient reconfigurable accelerator for deep convolutional neural networks. *IEEE J. Solid-State Circuits* **2016**, *52*, 127–138. [CrossRef]
23. Gao, M.; Pu, J.; Yang, X.; Horowitz, M.; Kozyrakis, C. Tetris: Scalable and efficient neural network acceleration with 3d memory. In Proceedings of the Twenty-Second International Conference on Architectural Support for Programming Languages and Operating Systems, Xi'an, China, 8–12 April 2017; pp. 751–764.
24. Hua, W.; Zhou, Y.; De Sa, C.; Zhang, Z.; Suh, G.E. Boosting the performance of cnn accelerators with dynamic fine-grained channel gating. In Proceedings of the 52nd Annual IEEE/ACM International Symposium on Microarchitecture, Columbus, OH, USA, 12–16 October 2019; pp. 139–150.
25. Chen, C.Y.; Choi, J.; Gopalakrishnan, K.; Srinivasan, V.; Venkataramani, S. Exploiting approximate computing for deep learning acceleration. In Proceedings of the 2018 Design, Automation & Test in Europe Conference & Exhibition (DATE), Dresden, Germany, 19–23 March 2018; pp. 821–826.
26. Gupta, S.; Agrawal, A.; Gopalakrishnan, K.; Narayanan, P. Deep Learning with Limited Numerical Precision. In Proceedings of the 32nd International Conference on Machine Learning (ICML 2015), Lille, France, 6–11 July 2015; Volume 3, pp. 1737–1746.
27. Judd, P.; Albericio, J.; Hetherington, T.; Aamodt, T.; Jerger, N.E.; Urtasun, R.; Moshovos, A. Reduced-Precision Strategies for Bounded Memory in Deep Neural Nets. *arXiv* **2015**, arXiv:1511.05236.
28. Shafique, M.; Hafiz, R.; Javed, M.U.; Abbas, S.; Sekanina, L.; Vasicek, Z.; Mrazek, V. Adaptive and Energy-Efficient Architectures for Machine Learning: Challenges, Opportunities, and Research Roadmap. In Proceedings of the 2017 IEEE Computer Society Annual Symposium on VLSI (ISVLSI), Bochum, Germany, 3–5 July 2017; pp. 627–632. [CrossRef]
29. Sarwar, S.S.; Srinivasan, G.; Han, B.; Wijesinghe, P.; Jaiswal, A.; Panda, P.; Raghunathan, A.; Roy, K. Energy efficient neural computing: A study of cross-layer approximations. *IEEE J. Emerg. Sel. Top. Circuits Syst.* **2018**, *8*, 796–809. [CrossRef]
30. Venkatesh, G.; Nurvitadhi, E.; Marr, D. Accelerating Deep Convolutional Networks using low-precision and sparsity. In Proceedings of the 2017 IEEE International Conference on Acoustics, Speech and Signal Processing (ICASSP), New Orleans, LA, USA, 5–9 March 2017; pp. 2861–2865.
31. Liu, B.; Guo, S.; Qin, H.; Gong, Y.; Yang, J.; Ge, W.; Yang, J. An energy-efficient reconfigurable hybrid DNN architecture for speech recognition with approximate computing. In Proceedings of the 2018 IEEE 23rd International Conference on Digital Signal Processing (DSP), Shanghai, China, 19–21 November 2018; pp. 1–5.
32. Ardakani, A.; Leduc-Primeau, F.; Onizawa, N.; Hanyu, T.; Gross, W.J. VLSI implementation of deep neural network using integral stochastic computing. *IEEE Trans. Very Large Scale Integr. (VLSI) Syst.* **2017**, *25*, 2688–2699. [CrossRef]
33. Niu, W.; Ma, X.; Lin, S.; Wang, S.; Qian, X.; Lin, X.; Wang, Y.; Ren, B. Patdnn: Achieving real-time dnn execution on mobile devices with pattern-based weight pruning. In Proceedings of the Twenty-Fifth International Conference on Architectural Support for Programming Languages and Operating Systems, Lausanne, Switzerland, 16–20 March 2020; pp. 907–922.
34. Sarker, M.M.K.; Rashwan, H.; Akram, F.; Singh, V.K.; Banu, S.F.; Chowdhury, F.; Choudhury, K.; Chambon, S.; Radeva, P.; Puig, D.; et al. SLSNet: Skin lesion segmentation using a lightweight generative adversarial network. *Expert Syst. Appl.* **2021**, *183*, 115433. [CrossRef]
35. Han, S.; Mao, H.; Dally, W.J. Deep compression: Compressing deep neural networks with pruning, trained quantization and huffman coding. *arXiv* **2015**, arXiv:1510.00149.
36. Courbariaux, M.; Bengio, Y.; David, J.P. Training deep neural networks with low precision multiplications. *arXiv* **2014**, arXiv:1412.7024.
37. Louizos, C.; Reisser, M.; Blankevoort, T.; Gavves, E.; Welling, M. Relaxed quantization for discretized neural networks. *arXiv* **2018**, arXiv:1810.01875.
38. Chernikova, A.; Oprea, A.; Nita-Rotaru, C.; Kim, B. Are self-driving cars secure? Evasion attacks against deep neural networks for steering angle prediction. In Proceedings of the 2019 IEEE Security and Privacy Workshops (SPW), San Francisco, CA, USA, 19–23 May 2019; pp. 132–137.
39. Pustokhina, I.V.; Pustokhin, D.A.; Gupta, D.; Khanna, A.; Shankar, K.; Nguyen, G.N. An effective training scheme for deep neural network in edge computing enabled Internet of medical things (IoMT) systems. *IEEE Access* **2020**, *8*, 107112–107123. [CrossRef]
40. Sarker, M.M.K.; Makhlouf, Y.; Craig, S.G.; Humphries, M.P.; Loughrey, M.; James, J.A.; Salto-Tellez, M.; O'Reilly, P.; Maxwell, P. A Means of Assessing Deep Learning-Based Detection of ICOS Protein Expression in Colon Cancer. *Cancers* **2021**, *13*, 3825. [CrossRef]
41. Li, Z.; Zhang, Y.; Wang, J.; Lai, J. A survey of FPGA design for AI era. *J. Semicond.* **2020**, *41*, 021402. [CrossRef]
42. Dean, J. Machine learning for systems and systems for machine learning. In Proceedings of the 2017 Conference on Neural Information Processing Systems, Long Beach, CA, USA, 4–9 December 2017.

43. Shomron, G.; Banner, R.; Shkolnik, M.; Weiser, U. Thanks for nothing: Predicting zero-valued activations with lightweight convolutional neural networks. In Proceedings of the European Conference on Computer Vision, Glasgow, UK, 23–28 August 2020; pp. 234–250.
44. Akhlaghi, V.; Yazdanbakhsh, A.; Samadi, K.; Gupta, R.K.; Esmaeilzadeh, H. Snapea: Predictive early activation for reducing computation in deep convolutional neural networks. In Proceedings of the 2018 ACM/IEEE 45th Annual International Symposium on Computer Architecture (ISCA), Los Angeles, CA, USA, 1–6 June 2018; pp. 662–673.
45. Asadikouhanjani, M.; Ko, S.B. A novel architecture for early detection of negative output features in deep neural network accelerators. *IEEE Trans. Circuits Syst. II Express Briefs* **2020**, *67*, 3332–3336. [CrossRef]
46. Kim, C.; Shin, D.; Kim, B.; Park, J. Mosaic-CNN: A combined two-step zero prediction approach to trade off accuracy and computation energy in convolutional neural networks. *IEEE J. Emerg. Sel. Top. Circuits Syst.* **2018**, *8*, 770–781. [CrossRef]
47. Chang, J.; Choi, Y.; Lee, T.; Cho, J. Reducing MAC operation in convolutional neural network with sign prediction. In Proceedings of the 2018 International Conference on Information and Communication Technology Convergence (ICTC), Jeju Island, Korea, 17–19 October 2018; pp. 177–182.
48. Lin, Y.; Sakr, C.; Kim, Y.; Shanbhag, N. PredictiveNet: An energy-efficient convolutional neural network via zero prediction. In Proceedings of the 2017 IEEE International Symposium on Circuits and Systems (ISCAS), Baltimore, MD, USA, 28–31 May 2017; pp. 1–4.
49. Song, M.; Zhao, J.; Hu, Y.; Zhang, J.; Li, T. Prediction based execution on deep neural networks. In Proceedings of the 2018 ACM/IEEE 45th Annual International Symposium on Computer Architecture (ISCA), Los Angeles, CA, USA, 1–6 June 2018; pp. 752–763.
50. Sze, V.; Chen, Y.H.; Yang, T.J.; Emer, J.S. Efficient Processing of Deep Neural Networks: A Tutorial and Survey. *Proc. IEEE* **2017**, *105*, 2295–2329. [CrossRef]
51. Agarap, A.F. Deep learning using rectified linear units (relu). *arXiv* **2018**, arXiv:1803.08375.
52. Kepner, J.; Gadepally, V.; Jananthan, H.; Milechin, L.; Samsi, S. Sparse deep neural network exact solutions. In Proceedings of the 2018 IEEE High Performance extreme Computing Conference (HPEC), Waltham, MA USA, 25–27 September 2018; pp. 1–8.
53. Talathi, S.S.; Vartak, A. Improving performance of recurrent neural network with relu nonlinearity. *arXiv* **2015**, arXiv:1511.03771.
54. Kahan, W. IEEE standard 754 for binary floating-point arithmetic. *Lect. Notes Status IEEE* **1996**, *754*, 11.
55. Russakovsky, O.; Deng, J.; Su, H.; Krause, J.; Satheesh, S.; Ma, S.; Huang, Z.; Karpathy, A.; Khosla, A.; Bernstein, M.; et al. ImageNet Large Scale Visual Recognition Challenge. *Int. J. Comput. Vis.* **2014**, *115*, 211–252. [CrossRef]
56. Chollet, F. and others.; Keras. 2015. Available online: https://keras.io (accessed on 1 May 2021).
57. Abadi, M.; Agarwal, A.; Barham, P.; Brevdo, E.; Chen, Z.; Citro, C.; Corrado, G.S.; Davis, A.; Dean, J.; Devin, M.; et al. TensorFlow: Large-Scale Machine Learning on Heterogeneous Systems. Software. 2015. Available online: tensorflow.org (accessed on 1 May 2021).

Article

Prediction of Hydraulic Jumps on a Triangular Bed Roughness Using Numerical Modeling and Soft Computing Methods

Mehdi Dasineh [1], Amir Ghaderi [2,*], Mohammad Bagherzadeh [3], Mohammad Ahmadi [4] and Alban Kuriqi [5,*]

1. Department of Civil Engineering, Faculty of Engineering, University of Maragheh, Maragheh 8311155181, Iran; mehdi.dasineh3180@gmail.com
2. Department of Civil Engineering, Faculty of Engineering, University of Zanjan, Zanjan 537138791, Iran
3. Department of Civil Engineering, Faculty of Engineering, Urmia University, Urmia 5756151818, Iran; bagherzadeh.mbz96@gmail.com
4. Department of Civil Engineering, Faculty of Engineering, Shabestar Branch, Islamic Azad University, Shabestar 1584743311, Iran; sthfar@gmail.com
5. CERIS, Instituto Superior Técnico, Universidade de Lisboa, 1049-001 Lisbon, Portugal
* Correspondence: amir_ghaderi@znu.ac.ir (A.G.); alban.kuriqi@tecnico.ulisboa.pt (A.K.); Tel.: +98-93845-03512 (A.G.)

Abstract: This study investigates the characteristics of free and submerged hydraulic jumps on the triangular bed roughness in various T/I ratios (i.e., height and distance of roughness) using CFD modeling techniques. The accuracy of numerical modeling outcomes was checked and compared using artificial intelligence methods, namely Support Vector Machines (SVM), Gene Expression Programming (GEP), and Random Forest (RF). The results of the FLOW-3D® model and experimental data showed that the overall mean value of relative error is 4.1%, which confirms the numerical model's ability to predict the characteristics of the free and submerged jumps. The SVM model with a minimum of Root Mean Square Error (RMSE) and a maximum of correlation coefficient (R^2), compared with GEP and RF models in the training and testing phases for predicting the sequent depth ratio (y_2/y_1), submerged depth ratio (y_3/y_1), tailwater depth ratio (y_4/y_1), length ratio of jumps (L_j/y_2^*) and energy dissipation ($\Delta E/E_1$), was recognized as the best model. Moreover, the best result for predicting the length ratio of free jumps (L_{jf}/y_2^*) in the optimal gamma is γ = 10 and the length ratio of submerged jumps (L_{js}/y_2^*) is γ = 0.60. Based on sensitivity analysis, the *Froude number* has the greatest effect on predicting the (y_3/y_1) compared with submergence factors (SF) and T/I. By omitting this parameter, the prediction accuracy is significantly reduced. Finally, the relationships with good correlation coefficients for the mentioned parameters in free and submerged jumps were presented based on numerical results.

Keywords: artificial intelligence; energy dissipation; FLOW-3D; hydraulic jumps; bed roughness; sensitivity analysis

Citation: Dasineh, M.; Ghaderi, A.; Bagherzadeh, M.; Ahmadi, M.; Kuriqi, A. Prediction of Hydraulic Jumps on a Triangular Bed Roughness Using Numerical Modeling and Soft Computing Methods. *Mathematics* 2021, 9, 3135. https://doi.org/10.3390/math9233135

Academic Editors: Freddy Gabbay and Florin Leon

Received: 13 September 2021
Accepted: 3 December 2021
Published: 5 December 2021

Publisher's Note: MDPI stays neutral with regard to jurisdictional claims in published maps and institutional affiliations.

Copyright: © 2021 by the authors. Licensee MDPI, Basel, Switzerland. This article is an open access article distributed under the terms and conditions of the Creative Commons Attribution (CC BY) license (https://creativecommons.org/licenses/by/4.0/).

1. Introduction

The hydraulic jump is a natural phenomenon in an open channel, sometimes regarded as an effective method of energy dissipation near structures such as gates, chutes, and spillways [1]. The hydraulic jump is specified by the expansion of large-scale turbulence, surface waves and spray, energy dissipation, and air entrainment [2]. If the tailwater depth equals the subcritical sequent depth, it is called a free hydraulic jump. Furthermore, if the tailwater depth is greater than the subcritical sequent depth, the jump is submerged (submerged hydraulic jump). A hydraulic jump has been widely studied, but only a few investigations have regarded the effect of bed roughness on the characteristics of hydraulic jumps. Enormous research studies dealing with the free and submerged hydraulic jumps such as McCorquodale and Khalifa [3], Smith [4], Graber et al. [5], Vallé and Pasternack [6], Dey and Sarkar [7], Tokyay et al. [8], and Samadi-Boroujeni et al. [9] were carried out. Ead and Rajaratnam [10] experimentally studied hydraulic jumps on corrugated beds.

The results showed that the length of jumps was about half of those on smooth beds. Carollo et al. [11] investigated the hydraulic jump properties on a bed roughened by gravel particles. The results indicate that the roughness reduces the sequent depth and the length of the jump. Pagliara et al. [12] studied the hydraulic jump on homogeneous and non-homogeneous rough beds. The results satisfactorily matched with the experimental data and presented new equations to estimate the length of jump and sequent depth. Abbaspour et al. [13] investigated the impact of a corrugated bed on hydraulic jumps. The results stated that the jump length and tailwater depth on corrugated beds are smaller than the smooth bed. Chanson [14] observed the flow resistance effects in decreasing the sequence depth ratio for a given Froude number. The results indicated that the Bélanger equation is not appropriate. In addition, the cross-sectional properties of irregular channels have an important influence on the flow characteristics. Ahmed et al. [15] investigated the effect of bed roughness on the submerged jump. Conclusions show that the length of a jump and tailwater depth on a bed roughness are smaller than on a smooth bed. Palermo and Pagliara [16] produced two general equations for evaluating relative energy dissipation across various hydraulic and geometrical conditions. Pourabdollah et al. [17] studied free and submerged jumps in different stilling basins. They showed that the sequent depth, the submerged depth, and the length of the jump decreased compared to the classical jump. Moreover, the average energy dissipation of the submerged jump on the bed roughness was more than those of the classical jump. Habibzadeh et al. [18] investigated characteristics of hydraulic jumps with and without blocks. The mean longitudinal velocity, turbulence intensity, Turbulent Kinetic Energy (TKE), and shear stress and water surface fluctuations were studied and compared for various flow regimes.

In addition to laboratory research, numerical works have been done on hydraulic jumps. Gharangik and Chaudhry [19] solved the 1D Boussinesq equations to simulate a hydraulic jump in a rectangular channel. The results showed that the equation terms have little influence in determining the location of the hydraulic jump. Ma et al. [20] investigated the turbulence characteristics of 2D submerged hydraulic jumps using the k–ε turbulence model. The results are compared with available experimental data and are acceptable. Mousavi et al. [21] investigated predictive modeling of the free hydraulic jumps pressure through advanced statistical methods. It was verified that maximum and minimum pressure fluctuation are located near the spillway toe and downstream of hydraulic jumps, respectively. Abbaspour et al. [22] numerically studied hydraulic jump on a corrugated bed using the standard k-ε and RNG turbulent models. Their results stated that the k-ε model was suitable for predicting the jump characteristics. Chern and Syamsuri [23] applied the Smoothed Particle Hydrodynamics (SPH) model to evaluate characteristics of the hydraulic jump in different corrugated beds and classified jump. Bayon et al. [24] investigated the performance of Open-FOAM and FLOW-3D® software in the numerical investigation of the hydraulic jump. Nikmehr and Aminpour [25] investigated the characteristics of a hydraulic jump over bed roughness with trapezoidal blocks using the CFD model. The results state that increasing the distance and the height of the roughness will decrease the velocity near the bed and increase the shear stress. Ghaderi et al. [26] numerically investigated the characteristics of the hydraulic jumps over various roughness shapes using the FLOW-3D® model. The results were compared with previous studies. Relationships with good correlation coefficients for the mentioned parameters in free and submerged jumps were presented based on numerical results. Ghaderi et al. [27] studied the effects of triangular microroughness on the characteristics of the submerged jump with the help of the FLOW-3D® model. To validate the present model, comparisons between numerical simulations and experimental results were performed for the smooth bed and triangular microroughness [27].

Recent advancements in data-driven models, i.e., Gene Expression Programming (GEP) and Artificial Neural Networks (ANN), and their application in hydraulics engineering have challenged the conventional techniques of the analysis. Several researchers the algorithm tends to predict them as positive value (no approximation) and bypass them out of the algorithm as valid outputs such that it will always go through the full precision compute. Hence, no accuracy drop is expected with the use of a proposed solution.

The proposed solution will be applicable across various networks with activation functions which displays a nonlinear behavior for either positive or negative numbers (not both) and the other one must be a zero or any constant value. To support activation functions like sigmoid, we might need to redevelop mathematical constraints to predict values between [−ve, +ve] range, while all other values outside this range can be set to a constant.

6.2. Implementation on GPU/Other Accelerators

This method is implementable on any compute engine that supports DNN workload. Since the proposed solution supports the reusing of partially computed elements that were used for early prediction, this will not impose a heavy tax on the existing hardware. On GPU and other accelerators, the proposed solution will need fine-tuning of data flow, data storage pattern and control logic, etc.

6.3. Extension to Training

The results presented in this work demonstrate the effectiveness of our model during the inference stage in a DNN. However, the process of training also involves the same set of steps, along with the additional step of adjusting the parameters. Hence, our model can be used in every layer with a ReLU activation layer. Since DNN training is a time-consuming and compute intensive process, this method can provide a significant improvement. It is also noteworthy to mention that, unlike inference, training must be done with high precision values. As a result, many of the approximate computing methods that have been studied cannot be extended to training. However, since our method ensures that there is no loss in accuracy, it can be applied to training as well.

7. Conclusions

In this work, we proposed a mathematical model that can detect zero outputs of the ReLU activation function using low-precision MAC operations. Our model takes into account the error resulting from the reduction of the number of bits in a floating-point representation, and identifies values that would be negative even when full-precision compute is performed. Our model can adapt based on the number of mantissa bits considered in the computation, ensuring its suitability for different number formats used in DNNs. We show that around 80% of ReLU zeros can be detected using just three mantissa bits, which corresponds to a total of 50% of all ReLU outputs in VGG16 and ResNet50 CNN implementations. As the model is developed with no assumption about the nature of the network or the application, we claim that the model can be extended to all DNNs that use the ReLU activation function. In addition, as the MAC operation and the activation layer in DNN training is identical to inference, this model can be adopted to make the compute-hungry training process more efficient. We also propose a system level model to implement this method and perform hardware acceleration of DNNs. The widespread use of DNNs with the ReLU activation function means that our model can be used as an error-free way to reduce computations in numerous applications.

Author Contributions: Conceptualization, B.S., K.P. and G.S.K.; writing—original draft preparation, B.S.; writing—review and editing, K.P. and G.S.; supervision, A.A. and S.S. All authors have read and agreed to the published version of the manuscript.

Funding: This research received no external funding.

Institutional Review Board Statement: Not applicable.

Informed Consent Statement: Not applicable.

Data Availability Statement: The dataset used for this study is publicly available and can be downloaded at https://www.image-net.org (accessed on 1 May 2021).

Conflicts of Interest: The authors declare no conflict of interest.

Abbreviations

The following abbreviations are used in this manuscript:

DNN	Deep Neural Network
ReLU	Rectified Linear Unit
VGG	Visual Geometry Group
ResNet	Residual Network
MAC	Multiply And Accumulate
GPU	Graphics Processing Unit
CNN	Convolutional Neural Network
IEEE	Institute of Electrical and Electronics Engineers
ILSVRC	ImageNet Large Scale Visual Recognition Challenge

References

1. Liu, W.; Wang, Z.; Liu, X.; Zeng, N.; Liu, Y.; Alsaadi, F.E. A survey of deep neural network architectures and their applications. *Neurocomputing* **2017**, *234*, 11–26. [CrossRef]
2. Zhang, L.; Zhang, Y. Big data analysis by infinite deep neural networks. *Jisuanji Yanjiu Yu Fazhan/Comput. Res. Dev.* **2016**, *53*, 68–79. [CrossRef]
3. Strubell, E.; Ganesh, A.; McCallum, A. Energy and Policy Considerations for Deep Learning in NLP. In Proceedings of the ACL 2019—57th Annual Meeting of the Association for Computational Linguistics, Florence, Italy, 28 July–2 August 2019; pp. 3645–3650.
4. Harlap, A.; Narayanan, D.; Phanishayee, A.; Seshadri, V.; Devanur, N.; Ganger, G.; Gibbons, P. PipeDream: Fast and Efficient Pipeline Parallel DNN Training. *arXiv* **2018**, arXiv:1806.03377.
5. Deng, C.; Liao, S.; Xie, Y.; Parhi, K.K.; Qian, X.; Yuan, B. PERMDNN: Efficient Compressed DNN Architecture with Permuted Diagonal Matrices. In Proceedings of the 2018 51st Annual IEEE/ACM International Symposium on Microarchitecture (MICRO), Fukuoka, Japan, 20–24 October 2018; pp. 189–202.
6. Duggal, J.K.; El-Sharkawy, M. Shallow squeezenext: An efficient shallow DNN. In Proceedings of the 2019 IEEE International Conference on Vehicular Electronics and Safety (ICVES), Cairo, Egypt, 4–6 September 2019. [CrossRef]
7. Wei, W.; Xu, L.; Jin, L.; Zhang, W.; Zhang, T. AI Matrix—Synthetic Benchmarks for DNN. *arXiv* **2018**, arXiv:1812.00886.
8. Hanif, M.A.; Javed, M.U.; Hafiz, R.; Rehman, S.; Shafique, M. Hardware–Software Approximations for Deep Neural Networks. In *Approximate Circuits*; Springer International Publishing: Cham, Switzerland, 2019; pp. 269–288. [CrossRef]
9. Agrawal, A.; Choi, J.; Gopalakrishnan, K.; Gupta, S.; Nair, R.; Oh, J.; Prener, D.A.; Shukla, S.; Srinivasan, V.; Sura, Z. Approximate computing: Challenges and opportunities. In Proceedings of the 2016 IEEE International Conference on Rebooting Computing (ICRC), San Diego, CA, USA, 17–19 October 2016. [CrossRef]
10. Liu, B.; Wang, Z.; Guo, S.; Yu, H.; Gong, Y.; Yang, J.; Shi, L. An energy-efficient voice activity detector using deep neural networks and approximate computing. *Microelectron. J.* **2019**, *87*, 12–21. [CrossRef]
11. Zhu, H.; Akrout, M.; Zheng, B.; Pelegris, A.; Jayarajan, A.; Phanishayee, A.; Schroeder, B.; Pekhimenko, G. Benchmarking and Analyzing Deep Neural Network Training. In Proceedings of the 2018 IEEE International Symposium on Workload Characterization (IISWC), Raleigh, NC, USA, 30 September–2 October 2018; pp. 88–100. [CrossRef]
12. Nwankpa, C.; Ijomah, W.; Gachagan, A.; Marshall, S. Activation Functions: Comparison of trends in Practice and Research for Deep Learning. *arXiv* **2018**, arXiv:1811.03378.
13. Wang, Y.; Li, Y.; Song, Y.; Rong, X. The Influence of the Activation Function in a Convolution Neural Network Model of Facial Expression Recognition. *Appl. Sci.* **2020**, *10*, 1897. [CrossRef]
14. Alom, M.Z.; Taha, T.M.; Yakopcic, C.; Westberg, S.; Sidike, P.; Nasrin, M.S.; Van Esesn, B.C.; Awwal, A.A.S.; Asari, V.K. The history began from alexnet: A comprehensive survey on deep learning approaches. *arXiv* **2018**, arXiv:1803.01164.
15. Shi, S.; Chu, X. Speeding up convolutional neural networks by exploiting the sparsity of rectifier units. *arXiv* **2017**, arXiv:1704.07724.
16. Simonyan, K.; Zisserman, A. Very deep convolutional networks for large-scale image recognition. In Proceedings of the International Conference on Learning Representations (ICLR), San Diego, CA, USA, 7–9 May 2015.
17. He, K.; Zhang, X.; Ren, S.; Sun, J. Deep residual learning for image recognition. In Proceedings of the IEEE Conference on Computer Vision and Pattern Recognition, Las Vegas, NV, USA, 27–30 June 2016; pp. 770–778. [CrossRef]
18. Albericio, J.; Delmás, A.; Judd, P.; Sharify, S.; O'Leary, G.; Genov, R.; Moshovos, A. Bit-pragmatic deep neural network computing. In Proceedings of the 50th Annual IEEE/ACM International Symposium on Microarchitecture, Cambridge, MA, USA, 14–18 October 2017; pp. 382–394.

have shown that soft computing techniques are more feasible and accurate than conventional techniques.

Karbasi and Azamathulla [28] studied free hydraulic jump characteristics in the bed roughness using Support Vector Regression (SVR), GEP, and ANN methods. The results showed the GEP model has better accuracy than other methods. Roushangar and Ghasempour [29] studied hydraulic jump characteristics in sudden expanding channels using GEP. The results showed the GEP model has better accuracy and was compared with existing empirical equations. Roushangar and Ghasempour [30] predicted the hydraulic jump energy dissipation using an SVM with channel geometry and roughness boundary conditions. The sensitivity analysis results stated that the Froude number had the most important impact on the modeling. Roushangar and Homayounfar [31] investigated the characteristics of hydraulic jumps on horizontal and sloping beds using the SVM method. Results verify that the upstream Froude number is the most critical and influential parameter for predicting the sequent depth in free and submerged jumps. At the same time, Naseri and Othman [32] predicted the length of jump on the smooth beds using ANN. Nasrabadi et al. [33] studied submerged hydraulic jump characteristics using machine learning methods. According to the evaluation, the Developed Group Method of Data Handling (DGMDH) model is more accurate than the Group Method Data Handling (GMDH) model and other previous research predicting the submergence depth and jump length relative energy dissipation.

Many studies have been carried out on hydraulic jumps over smooth beds. Nevertheless, few studies have investigated the effect of bed roughness and corresponding characteristics of free and submerged jumps numerically and predicted the outcomes from the numerical models using novel soft computing techniques. Hence, the main objectives of this study are the investigation of the effects of bed roughness parameters considering various roughness arrangements on characteristics of the free and submerged jumps, such as sequent depth and submerged depth, the length of jumps, and energy dissipation in triangular bed roughness through different hydraulic conditions with the CFD technique (CFD is a numerical methodology commonly used in engineering [34]), and verification of the prediction of this numerical model with the help of soft computing methods (SVM, GEP, and RF).

2. Materials and Methods

2.1. Dimensional Analysis

The hydraulic jumps characteristics on bed roughness are dependent on fluid properties, bed dimensions, and hydraulic state of flow. Therefore, subcritical of the free jump depth (y_2) and submerged of the submerged jump depth (y_3) will be a function of the following parameters:

$$y_2 = f_1(y_1, u_1, g, \mu, \rho, T, I) \quad (1)$$

$$y_3 = f_2(y_1, y_2, y_4, u_1, g, \mu, \rho, T, I) \quad (2)$$

Using the dimensional analysis, the following relationships are obtained:

$$\frac{y_2}{y_1} = f_3(Fr_1 = \frac{u_1}{\sqrt{gy_1}}, Re_1 = \frac{y_1 u_1}{\nu}, \frac{T}{y_1}, \frac{T}{I}) \quad (3)$$

$$\frac{y_3}{y_1} = f_4(Fr_1 = \frac{u_1}{\sqrt{gy_1}}, Re_1 = \frac{y_1 u_1}{\nu}, SF = \frac{y_4 - y_2}{y_2}, \frac{T}{y_1}, \frac{T}{I}) \quad (4)$$

where y_1 and y_4 are referred to as supercritical of the free jump depth and tailwater of the submerged jump depth; u_1 is inlet velocity; and g, ρ, μ, SF, and ν are the gravity acceleration, mass density of water, water dynamic viscosity, submergence factors, and water kinematic viscosity, respectively. T and I are height and distance of roughness, and Fr_1 and Re_1 are Froude and Reynolds numbers, respectively. The values of the Reynolds number (Re_1) were in the range of 39,884–59,825. For large values of the Reynolds number, viscous effects can be neglected [35–37]. Based on the Ead and Rajaratnam [10] and Abbaspour et al. [22] studies, T/y_1 does not significantly affect the hydraulic jumps' depth ratio y_2/y_1 and y_3/y_1. Then, relationships (3) and (4) become:

$$\frac{y_2}{y_1} = f_5(Fr_1, \frac{T}{I}) \tag{5}$$

$$\frac{y_3}{y_1} = f_6(Fr_1, SF = \frac{y_4 - y_2}{y_2}, \frac{T}{I}) \tag{6}$$

Using the Buckingham Π theorem, for the length of jump on the free and submerged jumps (L_{jf}/y_2 and L_{js}/y_2), the following relationships are obtained:

$$\frac{L_{jf}}{y_2} = f_7(Fr_1, \frac{T}{I}) \tag{7}$$

$$\frac{L_{js}}{y_2} = f_8(Fr_1, SF = \frac{y_4 - y_2}{y_2}, \frac{T}{I}) \tag{8}$$

Figure 1 shows a schematic view of free and submerged jumps on the triangular bed roughness, along with critical hydraulic parameters of the present study. In this figure, d is gate opening.

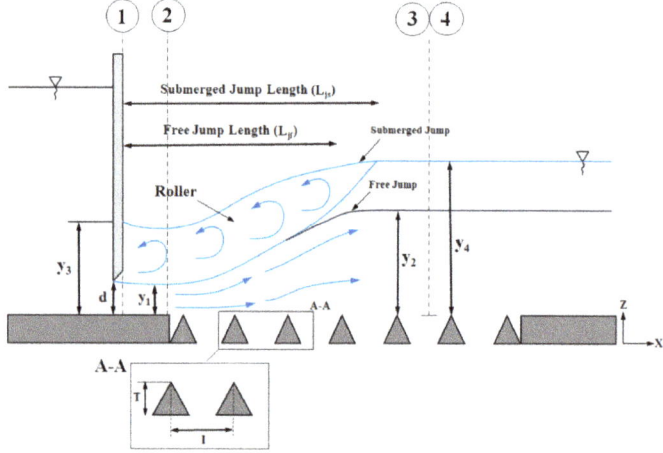

Figure 1. Definition sketch of the free and submerged hydraulic jumps on a triangular bed roughness after Ghaderi et al. [26].

2.2. The FLOW-3D® Model

Numerical simulations were carried out using FLOW-3D, a well-known and established computational fluid dynamics software. This software uses the finite volume method in a Cartesian, staggered grid to solve the RANS equations (Reynolds Average Navier–Stokes) that describe continuity and momentum and are expressed as:

$$\frac{\partial}{\partial x}(uA_x) + \frac{\partial}{\partial y}(vA_y) + \frac{\partial}{\partial z}(wA_z) = 0 \tag{9}$$

$$\frac{\partial u_i}{\partial t} + \frac{1}{V_F}\left(u_j A_j \frac{\partial u_i}{\partial x_j}\right) = -\frac{1}{\rho}\frac{\partial P}{\partial x_i} + G_i + f_i \tag{10}$$

where u, v, and w represent the components of velocity in the x, y, and z-direction; V_F is the volume fraction of fluid in each cell; A_x, A_y, and A_z are the fractional areas open to flow in the subscript's direction; ρ is the fluid density; P is the hydrostatic pressure; G_i is the gravitational acceleration in subscript direction; and f_i is the Reynolds stress. In FLOW-3D, free surfaces are modeled with the Volume of Fluid (VOF) technique and developed by Hirt and Nichols [37]. The VOF transport equation is expressed by the following equation:

$$\frac{\partial F}{\partial t} + \frac{1}{V_F}\left[\frac{\partial(FA_x u_1)}{\partial x} + \frac{\partial(FA_y u_2)}{\partial y} + \frac{\partial(FA_z u_3)}{\partial z}\right] = 0 \tag{11}$$

Here, F denotes the fraction function. In particular, as already stated, if a cell is empty, then $F = 0$, and if a cell is full, then $F = 1$ [38]. The free surface is determined at a position related to intermediate amounts of F (i.e., the user may usually determine $F = 0.5$, or another intermediate amount).

2.2.1. Turbulence Model

In this study, the RNG k-ε turbulence model is used to simulate the turbulence in the water flow. The RNG k-ε model improves the standard k-ε model (Equations (12) and (13)), reflecting small-scale effects by large-scale motion and modified viscosity terms, and can handle the flow with a large degree of curvature well [39]. This model showed satisfactory outcomes in previous studies on hydraulic engineering studies in complex geometry and flow fields [26,27,40–46].

$$\frac{\partial(\rho k)}{\partial t} + \frac{\partial(\rho k u_i)}{\partial x_i} = \frac{\partial}{\partial x_j}\left(\alpha_k \mu_{eff}\frac{\partial k}{\partial x_j}\right) + G_k + \rho\varepsilon \tag{12}$$

$$\frac{\partial(\rho\varepsilon)}{\partial t} + \frac{\partial(\rho\varepsilon u_i)}{\partial x_i} = \frac{\partial}{\partial x_j}\left(\alpha_\varepsilon \mu_{eff}\frac{\partial \varepsilon}{\partial x_j}\right) + \frac{C_{1\varepsilon}^*\varepsilon}{k}G_k - C_{2\varepsilon}\rho\frac{\varepsilon^2}{k} \tag{13}$$

Here, k is called turbulent kinetic energy (TKE); ε is the turbulence dissipation rate; G_k is the generation of turbulent kinetic energy caused by the average velocity gradient; G_b is the generation of turbulent kinetic energy caused by buoyancy. S_k and S_ε are source terms. α_k, α_ε and μ_{eff}, $C_{2\varepsilon}$, $C_{1\varepsilon}^*$ are model constants is effective viscosity.

2.2.2. Boundary Conditions

Corresponding to the physical conditions of the problem, four different boundary conditions were considered. Hence, the inlet and the exit boundary of the first mesh block needed to be set in the flow direction. The inlet boundary condition was set as discharge flow rate (Q) with flow depth at the channel's beginning. The boundary condition at the downstream end of the domain was described by a pressure boundary condition (P) corresponding to the tailwater depth in the flume. No-slip conditions were applied at the wall boundaries and the bottom, and they were treated as non-penetrative boundaries. Wall roughness has been neglected due to the slight roughness of the material of the experimental facility, which was used for validation. An atmospheric boundary condition is set to the upper boundary of the channel. This allows the flow to enter and leave the domain as null von Neumann conditions are imposed to all variables except for pressure, which is set to zero (i.e., atmospheric pressure). The symmetry (S) is used at the inner boundaries as well. Figure 2 shows the computational domain of the present study and the boundary conditions governing the simulation.

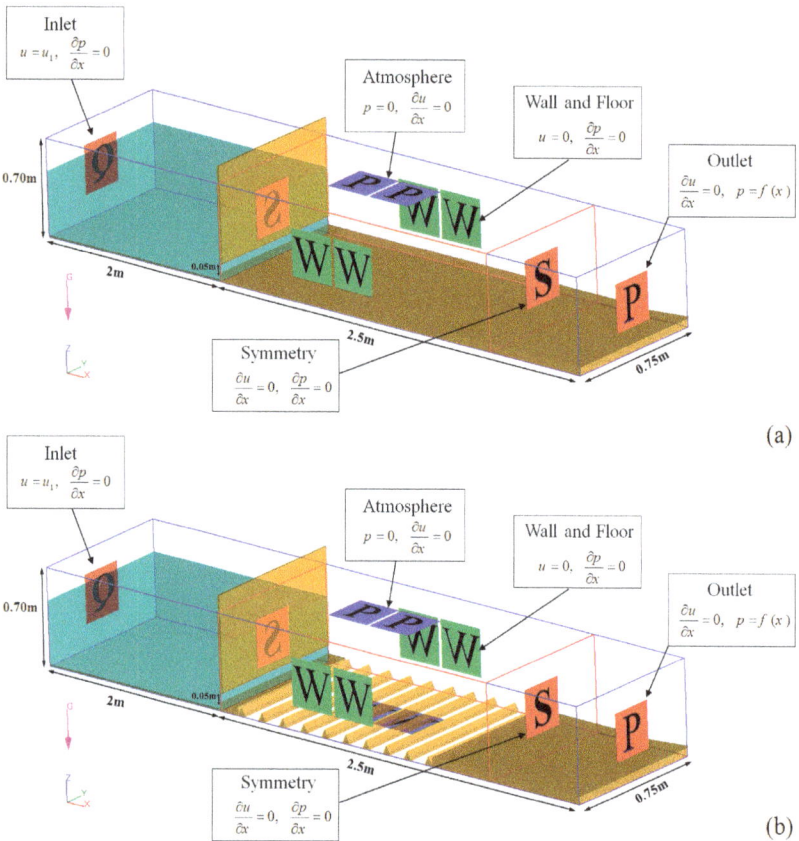

Figure 2. The boundary conditions governing the simulation, (**a**) smooth bed, (**b**) the triangular bed roughness.

2.2.3. Checking Stability and Convergence Criterion

To obtain the correct numerical or experimental model data values, it is necessary to reach a stable state. A stability criterion similar to the Courant number is used to calculate the allowed time-step size. The Courant Number tells how fast the fluid passes through a cell. If the Courant Number is greater than 1, the velocity of the fluid is so high that it passes through a cell in less than one time step. This leads to numerical instabilities: the stability criteria leading to time steps between 0.001 s and 0.0016 s. The evolution in time was used as a relaxation to the final steady state. During the simulations, the solutions' steady-state convergence was checked by monitoring the flow discharge variations at the inlet and outlet boundaries. Figure 3 shows that t = 16 s is appropriate to achieve a near steady-state condition for $Q = 0.03$ m^3/s and $Q = 0.045$ m^3/s. The computational time for the simulations was between 14–18 h using a personal computer with eight cores of a CPU (Intel Core i7-7700K @ 4.20 GHz and 16 GB RAM).

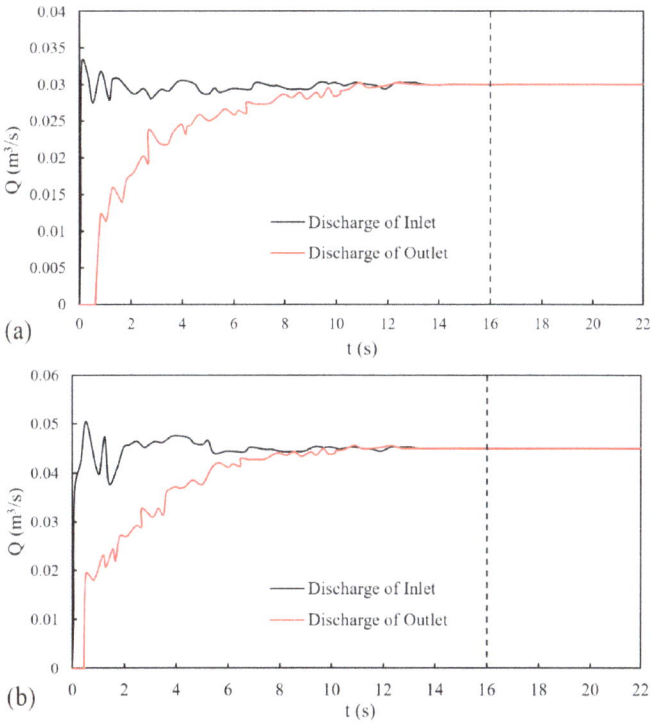

Figure 3. CFD flow discharge time variation in the inlet and outlet boundaries, (**a**) $Q = 0.03$ m^3/s, (**b**) $Q = 0.045$ m^3/s.

2.2.4. Numerical Domain

The research provided by Ahmed et al. [15] compares the numerical model and laboratory test results. Although the length of the experimental flume was 24.5 m, the present numerical study is set equal to 4.5 m to improve the performance in terms of computational effort and reduction in the number of overall cells [26] (for more details, see Ahmed et al. [15]). Table 1 shows the parameters of the numerical models.

Table 1. The parameters of the numerical models.

Bed Type	Q (m^3/s)	I (cm)	T (cm)	d (cm)	y_1 (cm)	y_4 (cm)	Fr_1	SF
Smooth	0.03, 0.045	-	-	5	1.62–3.83	9.64–32.10	1.7–9.3	0.26–0.50
Triangular roughness	0.03, 0.045	4–8–12–16–20	4	5	1.62–3.84	6.82–30.08	1.7–9.3	0.21–0.44

The geometry of the models is built represented through an STL (stereolithography) file. The numerical mesh is constructed to adopt two mesh blocks, a containing mesh block for the entire spatial domain and a nested block with refined cells for the area of interest. The hydraulic jump occurs (Figure 4). The best practice is to have fixed points aligning the mesh boxes and for the aspect ratios to be no greater than 2.

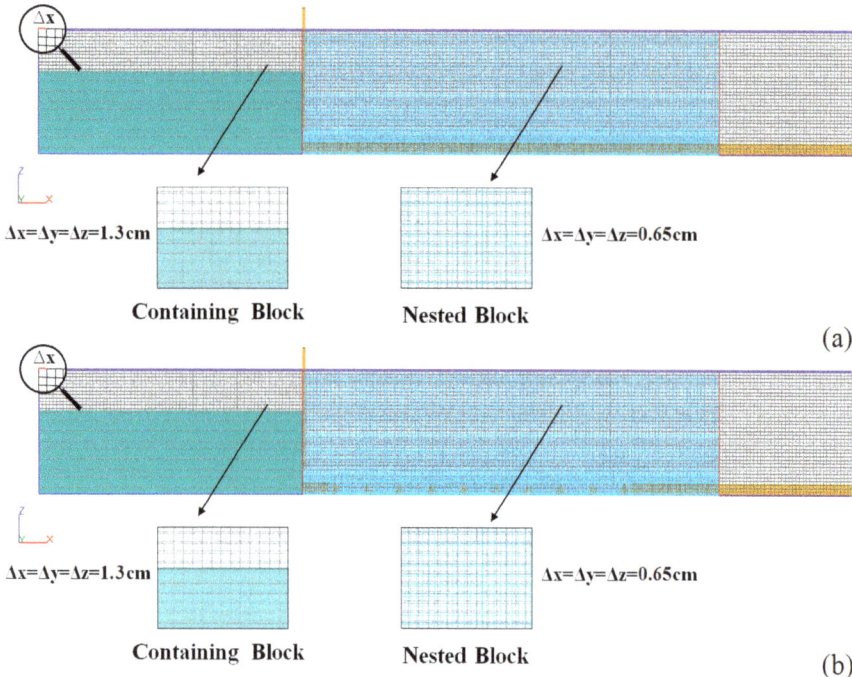

Figure 4. Structured rectangular hexahedral mesh with two different mesh blocks, (**a**) smooth bed, (**b**) the triangular bed roughness.

2.2.5. Mesh Size Sensitivity Analysis

According to the sensitivity mesh results and by comparing y_3/y_1 and y_2/y_1 ratios at $Fr_1 = 4.5$ for a submerged and free hydraulic jump, numerical solutions for five different mesh sizes at distances close to the computational grid were used. Table 2 provides a summary list of the results for three different mesh sizes. Figure 5 shows that the simulated y_3/y_1 and y_2/y_1 ratios exhibit better agreement with the measured y_3/y_1 and y_2/y_1 for the finer cell size of 0.60 cm. In addition, the variation of mean relative errors can be neglected by decreasing the cell size from 0.65 cm to 0.60 cm. As a result, the selected mesh consists of a containing block with 1.3 cm cells and a nested block with 0.65 cm cells. In the present research, the same mesh was utilized for all models to reduce the effect of computational mesh on simulation results. A distance of the first cell from the walls was selected to prevent computations in the viscous sub-layer.

Table 2. Mesh size sensitivity analysis for simulation.

Test No.	Coarser Cells Size (cm)	Finer Cells Size (cm)	Total Cells	$(y_3/y_1)_{Num}$	$(y_3/y_1)_{Exp}$	$(y_2/y_1)_{Num}$	$(y_2/y_1)_{Exp}$	MAPE [1]-y_3/y_1 (%)	MAPE-y_2/y_1 (%)
T1	2.00	0.95	910,358	8.55	6.88	7.43	5.88	26.36	24.27
T2	1.70	0.85	1,285,482	7.85	6.88	6.91	5.88	17.51	14.09
T3	1.50	0.75	1,871,649	7.38	6.88	6.44	5.88	9.52	7.26
T4	1.30	0.65	2,908,596	7.17	6.88	6.20	5.88	5.44	4.21
T5	1.15	0.60	3,812,035	7.10	6.88	6.08	5.88	3.41	3.34

[1] Mean Absolute Percentage Error = $100 \times \frac{1}{n} \sum_{1}^{n} \left| \frac{X_{Exp} - X_{Num}}{X_{Exp}} \right|$. X_{Exp}: the experimental value of X; X_{Num}: the numerical value of X; and n: the total amount of data.

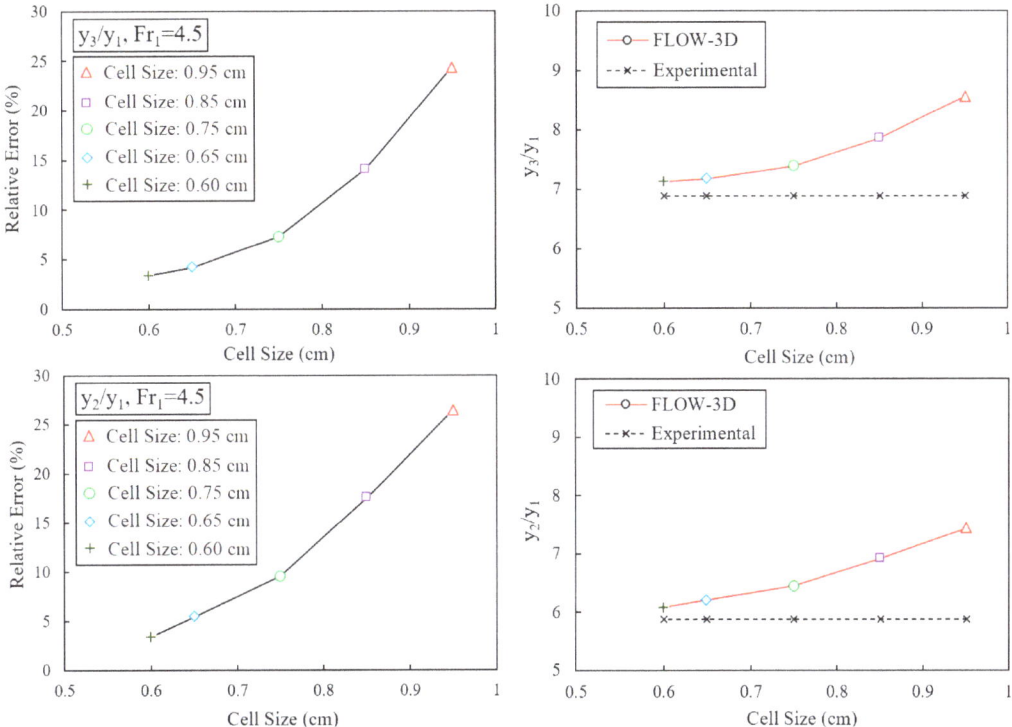

Figure 5. Variations of the relative error of y_3/y_1 and y_2/y_1 at Fr_1 versus cell size.

2.3. Artificial Intelligence Methods

2.3.1. Support Vector Machine (SVM)

SVM algorithm is a data mining algorithm that uses the regression method to solve classification and prediction problems. Like artificial neural networks, problem-solving steps are divided into two phases of training and testing (i.e., validation). First, the system is trained by a part of data, then the problem's solution is evaluated with test data. The SVM is based on linear data classification and tries to select a line with a high margin of confidence in the linear division of data. The training data closer to the separator page is called the support vector. The maximum distance between the two categories is known as the optimal separator page [47]. Based on the limited information of the samples, the SVM algorithm seeks the best option among the models with different complexities and the ability to train these models [48]. The SVM algorithm consists of four different kernels, which are presented in Table 3. The most widely used kernel functions in support vector machine problems are Gaussian (RBF) and ring kernel (ERBF) functions [49]. These functions are used when information on the data type and their nature is not available in problem-solving [50]. In the present study, the RBF function has been used to predict the parameters.

Here, X_i and X_j are two vectors in directions i and j, and a, c, and d are Kernel parameters. According to Figure 6, first, the input data is entered into the statistical software. Based on dimensional analysis, the dependent and independent parameters are defined in the software environment by selecting the function (RBF) and entering the main feature of the SVM model of this function (i.e., γ by trial-and-error method). Selecting the appropriate values of γ makes the results accurate and close to reality.

Table 3. Types of kernel functions [50].

Function	Expression
Linear Kernel	$K(x_i, x_j) = (x_i, x_j)$
Polynomial Kernel	$K(x_i, x_j) = ((x_i, x_j) + 1)^d$
Radial Basis Kernel	$K(x_i, x_j) = exp\left(-\frac{\|x_i - x_j\|^2}{2\sigma^2}\right)$
Sigmoid Kernel	$K(x_i, x_j) = tanh(-a(x_i, x_j) + c)$

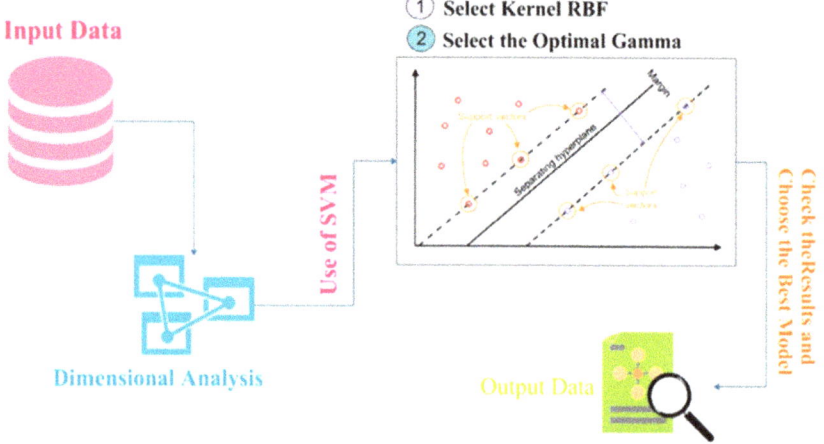

Figure 6. Schematic of the Support Vector Machine (SVM).

2.3.2. Gene Expression Programming (GEP)

The GEP method is a combined and developed Genetic Algorithm (GA) and Genetic Programming (GP) developed by Ferreira [51]. This method combines linear and simple chromosomes with constant length, similar to genetic algorithms, and branch structures of different sizes and shapes, similar to decomposition trees in genetic programming. The first step in the GEP is to form the initial population through solutions. Then, the chromosomes are shown as a tree (ETs). The fitness function determines the degree of compatibility of each member of the population of chromosomes. Next, the number of genes and chromosomes must be determined to run the GEP model. One of the strengths of the GEP is that the criterion for genetic diversity is very simple, so genetic operators operate at chromosomal levels. Another strength of this method is the unique nature of its multi-genes, which provides the basis for evaluating complex simulations [52]. The GEP algorithm consists of five steps: determining the fitness function, selecting the set of terminals and the set of functions to create the chromosomes, selecting the structure of the chromosomes, selecting the link function, and selecting the genetic operators and their rates [50,53]. In the present study, the GeneXproTools program was used to predict the parameters. The main steps of the GEP method are shown schematically in Figure 7.

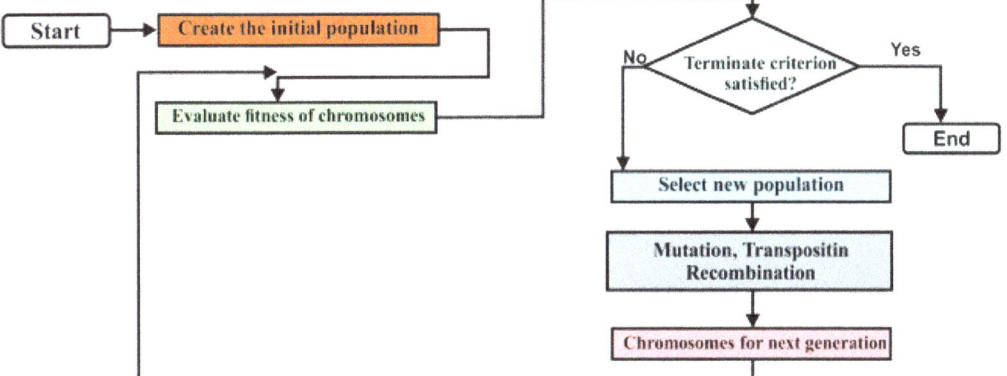

Figure 7. Schematic of the Gene Expression Programming (GEP).

2.3.3. Random Forest (RF)

RF algorithm is currently one of the learning algorithms. This is a cumulative learning algorithm for regression-based problems and grouping based on decision tree development [54]. An RF is a collection of unpruned trees in which each tree is generated by a recursive segmentation algorithm [51]. In other words, an RF is a combination of several decision trees in which several self-organizing samples of data participate. The self-organizing method is the sampling method with placement. None of the selected data are deleted from the input samples to generate the following subset. Therefore, some data may be used more than once in educational branches. Others that have little effect on modeling should never be used. For the selective self-organizing sample, a classification tree is grown using the recursive segmentation algorithm. The analysis operation is based on a random sample of the number of predictor variables in each node. The recursive decomposition algorithm continues until the tree reaches its maximum size without pruning it [54]. The performance of the RF algorithm is shown in Figure 8.

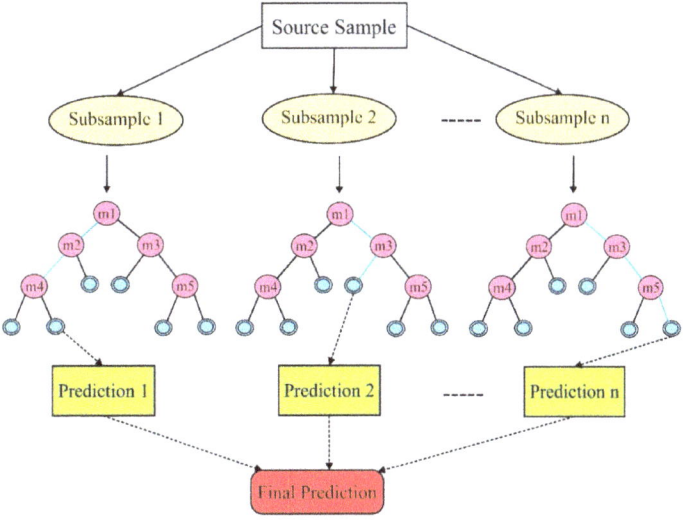

Figure 8. Performance of Random Forest (RF).

2.4. Evaluation Criteria

In the present study, the evaluation criteria of correlation coefficient (R^2), Root Mean Square Error (RMSE), Normalized Root Mean Square of Error (NRMSE), and Mean Absolute Percentage Error (MAPE) were used to compare the results of prediction models of hydraulic parameters of hydraulic jumps (Equations (14)–(17)).

$$R^2 = \left(\frac{n \sum X_{Num} X_{Pre} - (\sum X_{Num})(\sum X_{Pre})}{\sqrt{n(\sum X_{Num}^2) - (\sum X_{Num})^2} \sqrt{n(\sum X_{Pre}^2) - (\sum X_{Pre})^2}} \right)^2 \quad (14)$$

$$RMSE = \sqrt{\frac{1}{n} \sum_1^n (X_{Num} - X_{Pre})^2} \quad (15)$$

$$NRMSE(\%) = 100 \times \frac{\sqrt{\frac{1}{n} \sum_1^n (X_{Num} - X_{Pre})^2}}{\sum_1^n X_{Num}} \quad (16)$$

$$MAPE(\%) = 100 \times \frac{1}{n} \sum_1^n \left| \frac{X_{Num} - X_{Pre}}{X_{Num}} \right| \quad (17)$$

Here, the X_{Pre} and the X_{Num} are the predicted and the numerical values. It should be noted that the best model is the model in which RMSE is zero and R^2 is one, and also NRMSE and MAPE values are less than 10%.

3. Results

In the present study, the output results of the FLOW-3D® model were investigated using SVM, GEP, and RF methods. For this purpose, a total of 620 output data of numerical model were used to predict the parameters (y_2/y_1), (y_3/y_1), (y_4/y_1), (L_j/y_2^*), and ($\Delta E/E_1$) with artificial intelligence methods. To achieve accurate prediction and better results, the training process was repeated several times. Finally, a pattern of 25% data for testing and 75% data for training was used for all methods.

3.1. Validity of the FLOW-3D® Model Results

Although the CFD technique has been on the rise for more than half a century, computers have only allowed us to solve more complex 3D geometries in the recent decade. Because of that, it is very important to validate CFD results [55]. Hence, a comparison between numerical and experimental results on basic parameters including submerged ratio (y_3/y_1), tailwater ratio (y_4/y_1), and relative jump length (L_{js}/y_1) of a submerged hydraulic jump and the sequent depth ratio (y_2/y_1) of a free hydraulic jump on a smooth bed have been used to validate the numerical model and are plotted in Figure 9.

Moreover, the essential flow variables are summarized in Table 4.

From the graphs, a substantial agreement can be observed between numerical and experimental results by Ahmed et al. [15] as a function of Fr_1. The overall mean value of relative error is 4.1%, which confirms the ability of the numerical model to predict the specifications of free and submerged jumps. In general, the CFD model is in excellent agreement with the experimental data [56].

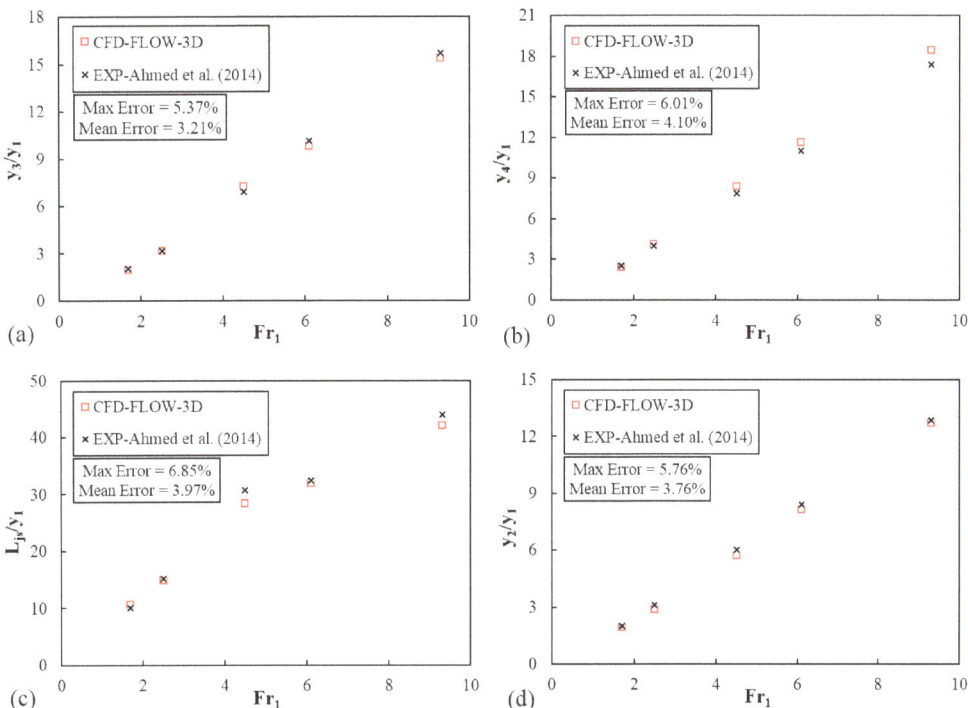

Figure 9. Numerical versus basic experimental parameters of submerged and free hydraulic jumps. (**a**) y_3/y_1, (**b**) y_4/y_1, (**c**) L_{js}/y_1, and (**d**) y_2/y_1.

Table 4. Basic flow variables for the numerical and physical models after Ahmed et al. [15].

Models	Bed	Q (m^3/s)	d (cm)	y_1 (cm)	u_1 (m/s)	Fr_1
Numerical and physical	Smooth	0.045	5	1.62–3.83	1.04–3.70	1.7–9.3

3.2. Sequent Depth Ratio in the Free Jump (y_2/y_1)

The y_2/y_1, which somehow represents the height of the jump, is directly related to the changes in the Fr_1 and the distance of the roughness element. By increasing these parameters, the value y_2/y_1 is increased. According to the results of the FLOW-3D® model, the most significant decrease y_2/y_1 with increasing Froude number compared to the smooth bed is at $T/I = 0.50$ with 17.83% as mean. The results showed that the y_2/y_1 for the jump on the bed roughness was smaller than that of the corresponding jumps on a smooth bed [26,27]. Table 5 summarizes the results of estimating the y_2/y_1. Comparing the results of three models, the SVM model with the lowest RMSE = 0.2075 and the highest R^2 = 0.9966 for the training phase and RMSE = 0.2990 and R^2 = 0.9960 for the testing phase in predicting the y_2/y_1 as a model the best was selected.

Figures 10 and 11 compare the results of the FLOW-3D® model and the SVM model to estimate the y_2/y_1 in the training and testing phase. It can be seen that the SVM model has a good performance in predicting this parameter, and the output results of the SVM model are in good agreement with the FLOW-3D® values and were recognized as the best model. It is also observed that during predicting y_2/y_1 in the testing phase, the SVM model estimates higher values at maximum points than the FLOW-3D® model.

Table 5. Prediction results for the sequent depth ratio (y_2/y_1).

	Training				Testing			
Model	R^2	RMSE	NRMSE (%)	MAPE (%)	R^2	RMSE	NRMSE (%)	MAPE (%)
GEP	0.9953	0.2356	4.03	5.35	0.9933	0.3335	5.56	8.83
RF	0.9682	0.5924	10.97	11.91	0.9275	1.0811	14.73	11.26
SVM	0.9966	0.2075	3.5481	5.07	0.9960	0.2990	4.98	8.46

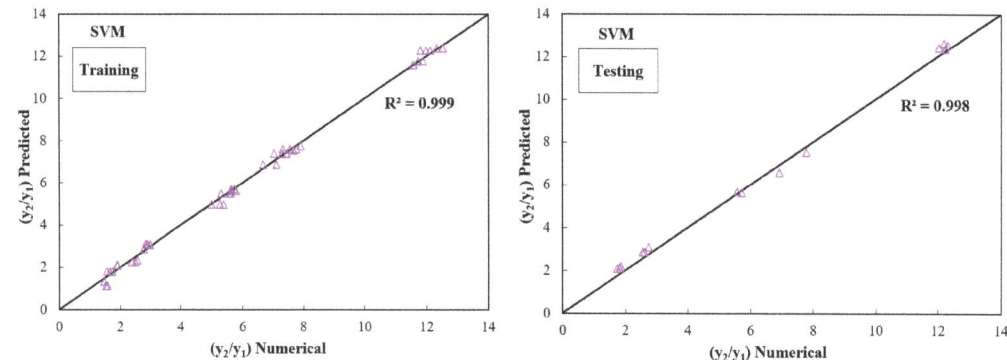

Figure 10. FLOW-3D® model versus SVM model predicted for the y_2/y_1.

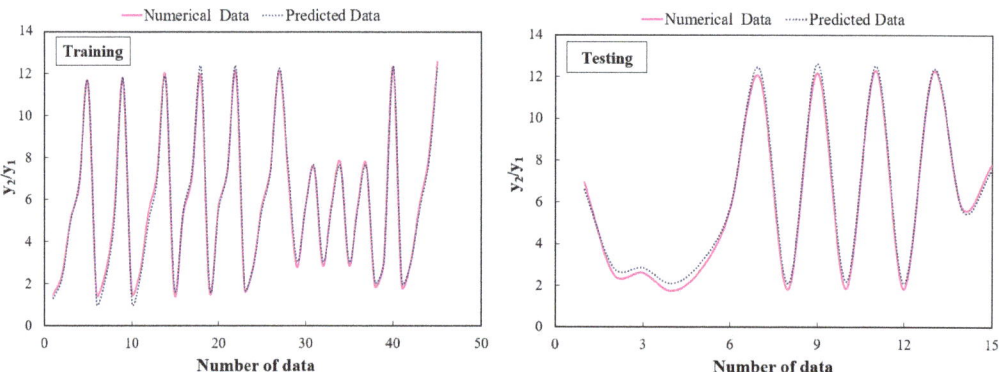

Figure 11. Comparison of FLOW-3D® model and SVM model for estimating the y_2/y_1.

In general, based on the numerical data of the present study, the equation provided for the y_2/y_1 in the free jump with a correlation coefficient equal to 0.997 is expressed as:

$$\frac{y_2}{y_1} = 1.338 Fr_1 - 2.458\left(\frac{T}{I}\right) + 0.0528 \tag{18}$$

3.3. Submerged Depth Ratio in Submerged Jump (y_3/y_1)

Based on dimensional analysis, the submerged depth ratio (y_3/y_1) and the tailwater ratio (y_4/y_1) depend on the Fr_1, T/I, and SF. According to the results of the FLOW-3D®, the most significant decrease y_3/y_1 and y_4/y_1 with increasing Froude number compared to the smooth bed are at $T/I = 0.50$ with 20.88% and 23.34% as mean, respectively [26,27]. Comparing the results of the three models presented in Table 6 shows that among the three models, for the y_3/y_1, the SVM model with values of RMSE = 0.3391 and R^2 = 0.9964 for the testing phase is close to the FLOW-3D® numerical model. The SVM model also performed better in predicting y_4/y_1 and had very little error. After the SVM model, the GEP model also provided acceptable results in estimating (y_3/y_1) and (y_4/y_1).

Table 6. Prediction results for the submerged depth ratio (y_3/y_1) and the tailwater depth ratio (y_4/y_1).

	Training				Testing			
y_3/y_1	R^2	RMSE	NRMSE (%)	MAPE (%)	R^2	RMSE	NRMSE (%)	MAPE (%)
GEP	0.9903	0.4016	5.96	8.63	0.9895	0.5379	7.61	9.92
RF	0.9815	0.5679	8.43	11.97	0.9750	0.7804	11.04	23.03
SVM	0.9978	0.2024	3.01	2.67	0.9964	0.3391	4.80	4.93
y_4/y_1	R^2	RMSE	NRMSE (%)	MAPE (%)	R^2	RMSE	NRMSE (%)	MAPE (%)
GEP	0.9972	0.2811	3.41	3.34	0.9963	0.3923	4.54	6.84
RF	0.9901	0.5157	6.26	9.68	0.9899	0.6462	7.48	10.04
SVM	0.9991	0.1639	1.99	1.63	0.9988	0.2806	3.25	5.08

Figures 12 and 13 present the results of comparing the FLOW-3D® model and predicting the SVM, GEP, and RF models in the testing phase (y_3/y_1) and (y_4/y_1). According to the graphs, it is clear that the SVM model has a better prediction than the other two models. At the maximum and minimum points, the (y_3/y_1) and (y_4/y_1), always accompanied by turbulence in the water surface, it can be seen that the SVM model has the highest efficiency and the lowest error over other models. The predicted values of these parameters by the SVM model have good adaptation. They overlap with the output values of the numerical model.

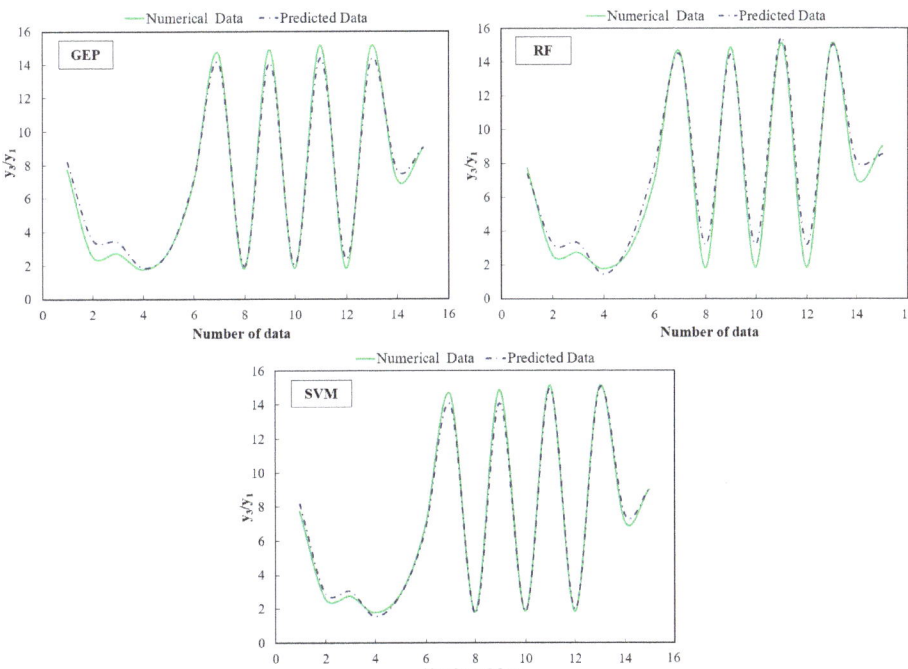

Figure 12. Comparison of the numerical results and the predicted models of (y_3/y_1) for the testing phase.

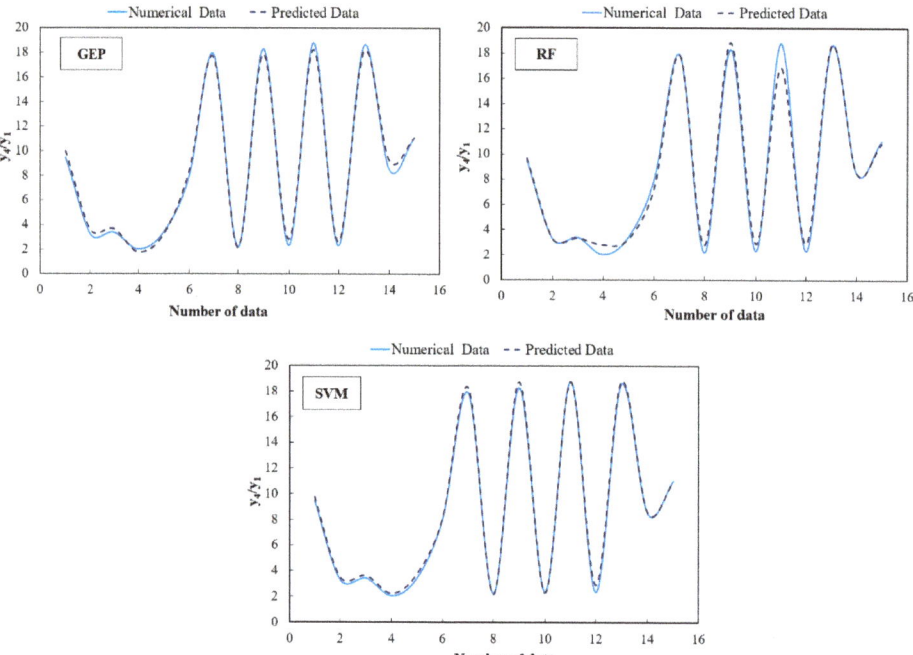

Figure 13. Comparison of the numerical results and the predicted models of (y_4/y_1) for the testing phase.

In general, based on the results drawn from this study, the following equation for the y_3/y_1 and y_4/y_1 in the submerged jump with a correlation coefficient equal to 0.993 and 0.989, respectively, on the triangular bed roughness was obtained:

$$\frac{y_3}{y_1} = 1.538 Fr_1 + 3.263 SF - 3.219\left(\frac{T}{I}\right) - 0.915 \quad (19)$$

$$\frac{y_4}{y_1} = 1.909 Fr_1 + 3.015 SF - 3.961\left(\frac{T}{I}\right) - 0.977 \quad (20)$$

3.4. The Length Ratio of Jumps (L_j/y_2^*)

In the present study, the subcritical depth of the classical hydraulic jump (y_2^*) can be obtained by the Bélanger equation, as explained by French [57]:

$$y_2^* = \frac{y_1}{2}\left[\sqrt{(1 + 8Fr_1^2)} - 1\right] \quad (21)$$

According to the results of the FLOW-3D® model, the (L_j/y_2^*) for the bed roughness is less than the smooth bed, and for the submerged jump it is larger than the free jump. For $T/I = 0.5$, the ratio length of free and submerged jumps decreases by about 25.52% and 21.65% as a mean, respectively [27]. Estimating the jump length reduces the volume of construction operations and ultimately reduces the project's overall cost. Therefore, an accurate estimation of the hydraulic jump length is essential to design the length of the stilling basin based on this parameter. The results of predicting (L_j/y_2^*) along with the evaluation criteria are presented in Table 7. According to the results, the SVM model has good statistical criteria among other models and has high accuracy in predicting the relative length of free and submerged hydraulic jumps.

Table 7. Prediction results for the length of the jumps (L_j/y_2^*).

	Training				Testing			
L_{jf}/y_2^*	R^2	RMSE	NRMSE (%)	MAPE (%)	R^2	RMSE	NRMSE (%)	MAPE (%)
GEP	0.829	0.234	4.39	3.66	0.766	0.249	4.54	4.54
RF	0.752	0.278	5.08	3.92	0.741	0.319	6.32	5.21
SVM	0.919	0.169	3.16	2.74	0.881	0.174	3.19	2.90
L_{js}/y_2^*	R^2	RMSE	NRMSE (%)	MAPE%	R^2	RMSE	NRMSE (%)	MAPE (%)
GEP	0.878	0.273	3.88	3.17	0.867	0.336	4.71	4.33
RF	0.787	0.316	4.37	3.51	0.764	0.425	6.50	5.46
SVM	0.961	0.154	2.19	1.93	0.940	0.212	2.97	2.37

Graphs of changes in R^2 and RMSE versus different gammas are presented for the best model of the L_{jf}/y_2^* and the L_{js}/y_2^* in the testing phase (Figure 14). In the support vector machine, selecting the appropriate gamma is one of the main parameters in determining the best model, which has been done by trial and error. Finally, the best result for predicting the L_{jf}/y_2^* in the optimal gamma is 10 ($\gamma = 10$), and for L_{js}/y_2^* in the optimal gamma it is 0.60 ($\gamma = 0.60$).

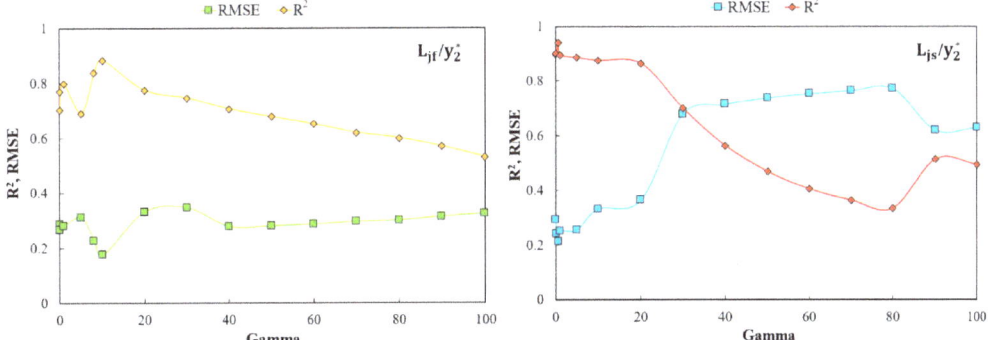

Figure 14. Variations R^2 and RMSE versus gamma for the best SVM model in jump length estimation.

Figures 15 and 16 show the results of the FLOW-3D® and the predicted models of L_{jf}/y_2^* and the L_{js}/y_2^* data for the best SVM model in the training and testing phases. According to Figure 15, it can be seen that when the values of the L_{jf}/y_2^* reach the maximum and minimum points, the prediction accuracy of the SVM model decreases. In other words, when the L_{jf}/y_2^* reaches the maximum and minimum jump values, the prediction error of the SVM model increases. Moreover, as shown in Figure 16 for the L_{js}/y_2^*, it can be seen that the SVM model always has values close to the FLOW-3D® model and has a better performance compared to the L_{jf}/y_2^*. On the other hand, most SVM model errors in both parameters occurred in the initial range of testing data. In the middle to the end of the data, the prediction error decreased.

The following equation shows the relationship between the L_j/y_2^* with a correlation coefficient equal to 0.724 and 0.944, respectively, for the free and submerged jumps:

$$\frac{L_{jf}}{y_2^*} = 0.065 Fr_1 - 3.757\left(\frac{T}{I}\right) + 6.103 \quad (22)$$

$$\frac{L_{js}}{y_2^*} = 0.037 Fr_1 + 5.568 SF - 2.556\left(\frac{T}{I}\right) + 5.579 \quad (23)$$

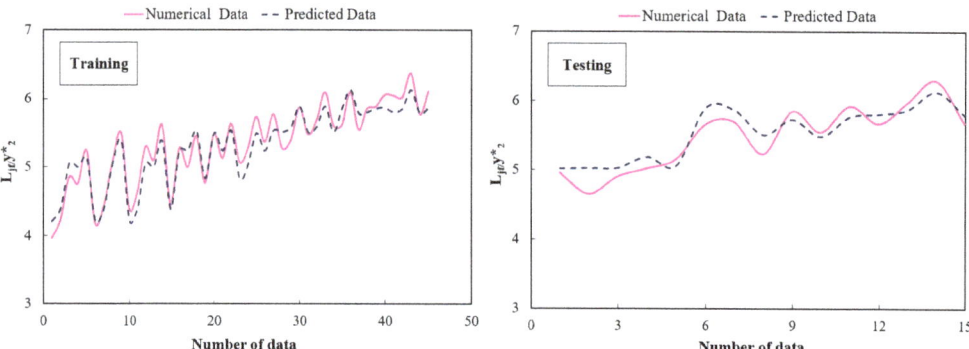

Figure 15. Comparison of FLOW-3D® and SVM model values to estimate the L_{jf}/y_2^*.

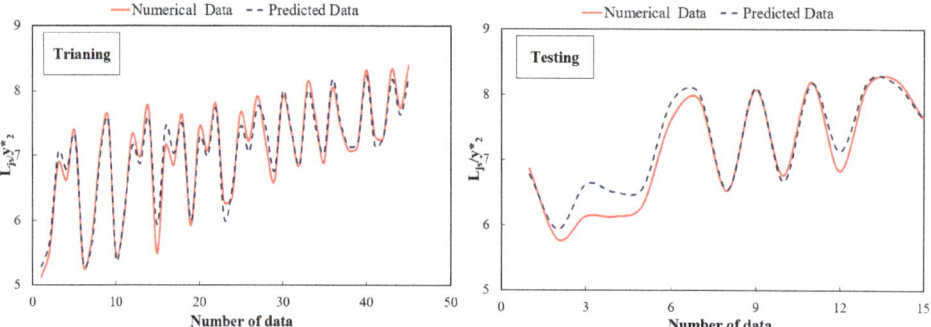

Figure 16. Comparison of FLOW-3D® and SVM model values to estimate the L_{js}/y_2^*.

3.5. The Energy Dissipation (ΔE/E₁)

The energy dissipation of hydraulic jumps based on free and submerged is calculated as follows by Pourabdollah et al. [17]:

$$\left(\frac{\Delta E}{E_1}\right)_f = \left(\frac{E_1 - E_2}{E_1}\right)_f = \left(\frac{(y_1 + V_1^2/2g) - (y_2 + V_2^2/2g)}{y_1 + V_1^2/2g}\right)_f \tag{24}$$

$$\left(\frac{\Delta E}{E_1}\right)_s = \left(\frac{E_3 - E_4}{E_3}\right)_s = \left(\frac{(y_3 + V_1^2/2g) - (y_4 + V_4^2/2g)}{y_3 + V_1^2/2g}\right)_s \tag{25}$$

E_1, E_2, E_3, and E_4 are specific energies upstream and downstream of the free and submerged jumps, respectively (see Figure 1). According to the results of the FLOW-3D®, the $\Delta E/E_1$ increases with increasing the Fr_1. The highest $\Delta E/E_1$ occurs with $T/I = 0.50$ in the free and submerged jumps compared to other distances between the roughnesses of the corresponding T/I ratios [26,27]. Determining the amount of $\Delta E/E_1$ that occurs due to hydraulic jumps will lead to the stilling basin's more efficient and economical design. The results of predicting energy dissipation due to free jump $(\Delta E/E_1)_f$ and submerged jump $(\Delta E/E_1)_S$ are presented in Table 8. The results showed that for energy dissipation for $(\Delta E/E_1)_f$, the SVM model with $R^2 = 0.9848$ and RMSE = 0.0313, and for the testing phase $(\Delta E/E_1)_S$, $R^2 = 0.9843$ and RMSE = 0.0238, these were recognized as the best models. Therefore, the best prediction with the least possible error among the three models is obtained by the SVM model.

Table 8. Prediction results for the energy dissipation ($\Delta E/E_1$).

		Training				Testing		
$(\Delta E/E_1)_f$	R^2	RMSE	NRMSE (%)	MAPE (%)	R^2	RMSE	NRMSE (%)	MAPE (%)
GEP	0.980	0.029	6.76	6.59	0.977	0.040	10.19	16.04
RF	0.855	0.069	16.5	25.61	0.801	0.072	16.32	21.22
SVM	0.985	0.027	6.37	6.58	0.984	0.031	7.80	9.04
$(\Delta E/E_1)_S$	R^2	RMSE	NRMSE (%)	MAPE (%)	R^2	RMSE	NRMSE (%)	MAPE (%)
GEP	0.980	0.025	7.22	8.85	0.969	0.033	10.07	11.63
RF	0.912	0.051	12.94	13.83	0.916	0.047	13.53	13.43
SVM	0.985	0.022	6.22	6.43	0.984	0.023	7.25	9.05

Two radar graphs of the R^2 and RMSE of energy dissipation due to free and submerged jumps are presented for the testing phase (Figure 17). Radar graphs can show the accuracy of predictions of different models compared to each other. It can be seen that the SVM model has provided acceptable performance and has a much better prediction than the GEP and RF models. Furthermore, because RMSE values are small and their changes are not visible in the graph, by multiplying the RMSE by 10, the range of changes became broader and more precise.

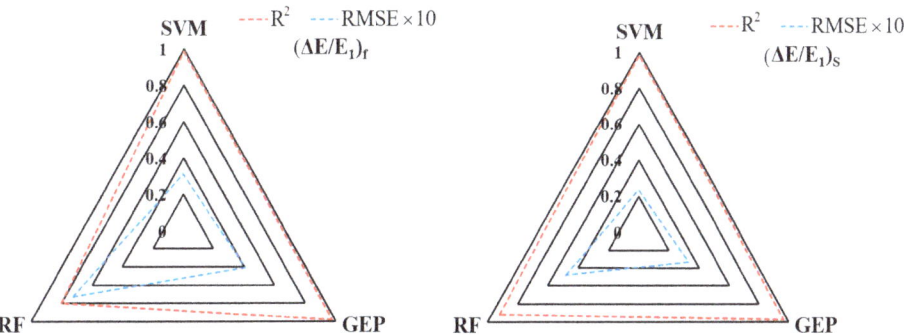

Figure 17. Radar graphs of R^2 and RMSE for energy dissipation due to free and submerged jumps in the testing phase.

The distribution graph between numerical and predicted values is plotted for the best energy dissipation model due to free and submerged jumps (Figures 18 and 19). Changes in energy dissipation for jumps during the testing and training phase indicate good agreement and overlap between the values of the numerical model and the predicted. According to the figure, it can be seen that the data of the numerical model had less dispersion with the predicted data. In other words, the output data are very well matched to each other. Additionally, during the model simulation process, the network training did not fail, and the training values were always higher than the testing.

The following equations show the relationship between the $\Delta E/E_1$ and Fr_1 with a correlation coefficient equal to 0.963 and 0.946, respectively, for the free and submerged jumps:

$$\left(\frac{\Delta E}{E_1}\right)_f = -0.009 Fr_1^2 + 0.184 Fr_1 - 0.177 \tag{26}$$

$$\left(\frac{\Delta E}{E_1}\right)_s = -0.007 Fr_1^2 + 0.146 Fr_1 - 0.143 \tag{27}$$

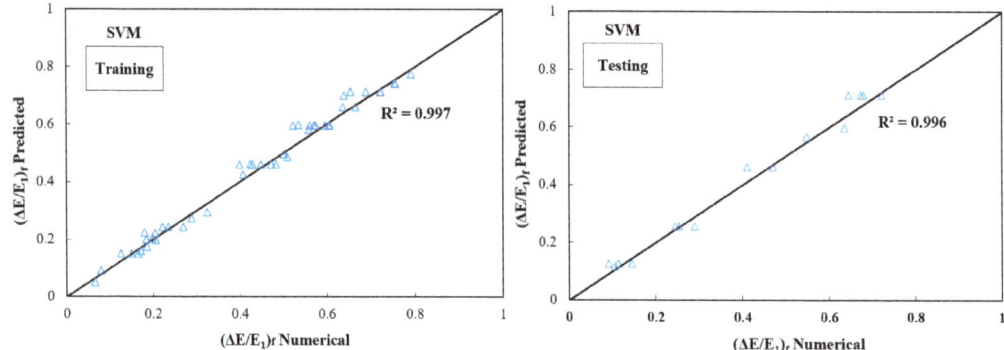

Figure 18. FLOW-3D® model versus SVM predicted for the free jump.

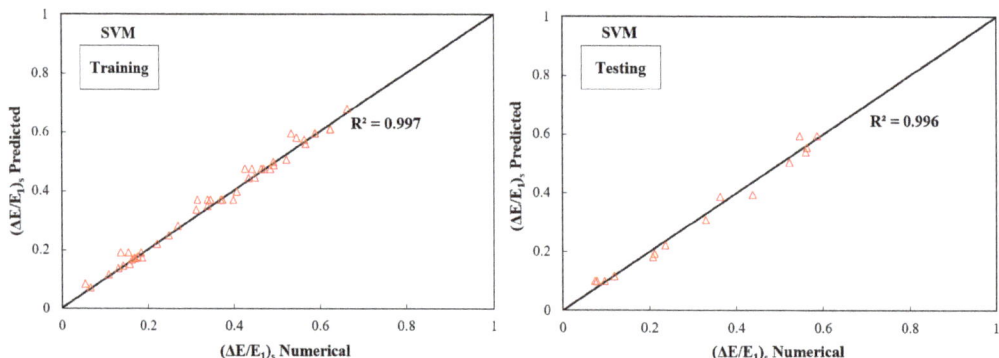

Figure 19. FLOW-3D® model versus SVM predicted for the submerged jump.

3.6. Sensitivity Analysis

Sensitivity analysis is the best solution to achieve the effectiveness of the input variables of a statistical model in a study. Sensitivity analysis is used when sufficient inputs are changed in an organized statistical model to observe the effects of the presence or absence of these variables in the predictive output model. The present study omitted the one-by-one parameters to predict the submerged depth ratio (y_3/y_1). The parameter that had the most impact was identified, and its results are presented in Table 9.

Table 9. Sensitivity analysis results for the submerged depth ratio (y_3/y_1).

		Training				Testing			
Input parameter	Omitted parameter	R^2	RMSE	NRME (%)	MAPE (%)	R^2	RMSE	NRMSE (%)	MAPE (%)
$Fr_1, SF, T/I$	-	0.999	0.163	1.99	1.63	0.998	0.280	3.25	5.08
$SF, T/I$	Fr_1	0.787	1.639	24.33	34.63	0.731	2.417	34.21	36.69
$Fr_1, T/I$	SF	0.986	0.479	7.12	10.83	0.984	0.664	9.41	15.9
Fr_1, SF	T/I	0.989	0.415	6.17	9.85	0.988	0.548	7.76	14.66

It can be seen that the best result for predicting the effective parameter of (y_3/y_1) is when all three parameters of Fr_1, T/I, and SF are involved in the prediction. The Fr_1 has the greatest effect on predicting the (y_3/y_1) based on sensitivity analysis. By omitting this parameter, the prediction accuracy is significantly reduced. The SF and T/I are also involved in the study of (y_3/y_1), but the impact of each is less than the Fr_1.

4. Conclusions

This paper presented and discussed the characteristics of free and submerged hydraulic jumps on the triangular bed roughness in various roughness arrangements of the corresponding T/I ratios with the CFD techniques and compared the prediction of this numerical model with the help of artificial intelligence methods (SVM, GEP, and RF). To simulate the free flow surface, the Volume of Fluid (VOF) method, and the turbulence, the RNG k-ε model in FLOW-3D® software was used. Key findings of the comparative analysis are given below:

1. By comparing the results of the two experiments (physical and numerical), the FLOW-3D® software can accurately predict the characteristics of free and submerged hydraulic jumps. The overall mean value of relative error between numerical results and experimental data is 4.1%, which confirms the numerical model's ability to predict the characteristics of the free and submerged jumps.
2. The SVM model with the RMSE = 0.2075 and R^2 = 0.9966 for the training phase and RMSE = 0.2990 and R^2 = 0.9960 for the testing phase in predicting the y_2/y_1 is the best model and close to the FLOW-3D® result.
3. For the y_3/y_1, the SVM model with values of RMSE = 0.3391 and R^2 = 0.9964 for the testing phase is close to the FLOW-3D® model. The SVM model also performed better in predicting y_4/y_1 and had very little error. After the SVM model, the GEP model also provided acceptable results in estimating (y_3/y_1) and (y_4/y_1).
4. The SVM model demonstrated better statistical criteria among other models (i.e., GEP and RF) and has high accuracy in predicting the relative length of free and submerged hydraulic jumps. Furthermore, the best result for predicting the L_{jf}/y_2^* in the optimal gamma is 10 (γ = 10) and the L_{js}/y_2^* in the optimal gamma is 0.60 (γ = 0.60).
5. For energy dissipation due to $(\Delta E/E_1)_f$ and $(\Delta E/E_1)_S$, for the testing phase, SVM model with R^2 = 0.9848 and RMSE = 0.031 as well as R^2 = 0.9843 and RMSE = 0.0238 were recognized as the best models, respectively.
6. The Fr_1 has the greatest effect on predicting the (y_3/y_1) based on sensitivity analysis. By omitting this parameter, the prediction accuracy is significantly reduced. The SF and T/I are also involved in the (y_3/y_1), but the impact of each is less than the Fr_1.
7. Relationships with good correlation coefficients for the mentioned parameters in free and submerged hydraulic jumps were presented based on numerical results.

Finally, the methodology presented in this study and the solution-oriented result contributes to helping hydraulic engineers to design and construct cost-effective spillways, stilling basins, and other hydraulic structures that experience hydraulic jumps. Indeed, the accurate estimation of the hydraulic jump length, especially in high head spillways, reduces construction operations' volume and ultimately reduces the project's overall cost of the stilling basin built to dissipate the hydraulic jumps.

Author Contributions: Conceptualization, M.D., A.G., M.B., M.A. and A.K.; methodology, M.D., A.G., M.B., M.A. and A.K.; software, M.D., A.G., M.B., M.A. and A.K.; validation, M.D., A.G., M.B., M.A. and A.K.; formal analysis, M.D., A.G., M.B., M.A. and A.K.; investigation, M.D., A.G., M.B., M.A. and A.K.; resources, M.D., A.G., M.B., M.A. and A.K.; data curation, M.D., A.G., M.B., M.A. and A.K.; writing—original draft preparation, M.D., A.G., M.B., M.A. and A.K.; writing—review and editing, M.D., A.G., M.B., M.A. and A.K.; visualization, M.D., A.G., M.B., M.A. and A.K.; supervision, M.D., A.G., M.B., M.A. and A.K.; project administration, M.D., A.G., M.B., M.A. and A.K. All authors have read and agreed to the published version of the manuscript.

Funding: This research received no external funding.

Institutional Review Board Statement: Not applicable.

Informed Consent Statement: Not applicable.

Data Availability Statement: Data are contained within the article.

Acknowledgments: Alban Kuriqi acknowledge the support of the Portuguese Foundation for Science and Technology (FCT) through the project PTDC/CTA-OHR/30561/2017 (WinTherface).

Conflicts of Interest: The authors declare no conflict of interest.

Notation

The following symbols are used in this paper:

Q	Discharge ($L^3 T^{-1}$)
d	Gate opening (L)
E_1, E_2	Specific energy at the beginning and after the free jump (L)
E_3, E_4	Specific energy at the beginning and after the submerged jump (L)
ΔE	Energy dissipation (L)
y_1	Inlet depth of the hydraulic jump (L)
y_2	Sequent depth of the free jump (L)
y_3	Submerged depth (L)
y_4	Tailwater depth (L)
L_{jf}	Length of the free jump (L)
L_{js}	Length of the submerged jump (L)
u_1	Inlet horizontal velocity (LT^{-1})
g	Gravitational acceleration (LT^{-2})
I	Distance of triangular roughness (L)
T	Roughness height (L)
Fr_1	Inlet Froude number (-)
Re_1	Inlet Reynolds number (-)
SF	Submergence factor (-)
t	Time (T)
p	Pressure ($ML^{-1}T^{-2}$)
F	Fraction function
ρ	Mass density of water (ML^{-3})
ν	Kinematic viscosity of water (LT^{-1})
μ	Dynamic viscosity of fluid ($ML^{-1}T^{-1}$)
k	Turbulence kinetic energy ($L^2 T^{-3}$)
ε	Turbulence dissipation rate ($L^2 T^{-3}$)
μ_{eff}	Effective viscosity ($ML^{-1}T^{-1}$)
G_k	The generation of turbulent kinetic energy caused by the average velocity gradient
G_b	The generation of turbulent kinetic energy caused by buoyancy
S_k, S_ε	Source terms
SVM	Support Vector Machine
GEP	Gene Expression Programming
RF	Random Forest
R^2	Correlation coefficient
RMSE	Root Mean Square Error
NRMSE	Normalized Root Mean Square of Error
MAPE	Mean Absolute Percentage Error

References

1. Ebrahimi, S.; Salmasi, F.; Abbaspour, A. Numerical study of hydraulic jump on rough beds stilling basins. *J. Civ. Eng. Urban.* **2013**, *3*, 19–24.
2. Chanson, H. *Hydraulics of Open Channel Flow*; Elsevier: Amsterdam, The Netherlands, 2004.
3. McCorquodale, J.A.; Khalifa, A.M. Submerged radial hydraulic jump. *J. Hydraul. Div.* **1980**, *106*, 355–367. [CrossRef]
4. Smith, C.D. The submerged hydraulic jump in an abrupt lateral expansion. *J. Hydraul. Res.* **1989**, *27*, 257–266. [CrossRef]
5. Graber, S.D.; Ohtsu, I.; Yasuda, Y.; Ishikawa, M. Submerged Hydraulic Jumps below Abrupt Expansions. *J. Hydraul. Eng.* **2001**, *127*, 84–85. [CrossRef]
6. Vallé, B.L.; Pasternack, G.B. Submerged and unsubmerged natural hydraulic jumps in a bedrock step-pool mountain channel. *Geomorphology* **2006**, *82*, 146–159. [CrossRef]
7. Dey, S.; Sarkar, A. Characteristics of turbulent flow in submerged jumps on rough beds. *J. Eng. Mech.* **2008**, *134*, 49–59. [CrossRef]
8. Tokyay, N.; Evcimen, T.; Şimşek, Ç. Forced hydraulic jump on non-protruding rough beds. *Can. J. Civ. Eng.* **2011**, *38*, 1136–1144. [CrossRef]

9. Samadi-Boroujeni, H.; Ghazali, M.; Gorbani, B.; Nafchi, R.F. Effect of triangular corrugated beds on the hydraulic jump characteristics. *Can. J. Civ. Eng.* **2013**, *40*, 841–847. [CrossRef]
10. Ead, S.; Rajaratnam, N. Hydraulic jumps on corrugated beds. *J. Hydraul. Eng.* **2002**, *128*, 656–663. [CrossRef]
11. Carollo, F.G.; Ferro, V.; Pampalone, V. Hydraulic jumps on rough beds. *J. Hydraul. Eng.* **2007**, *133*, 989–999. [CrossRef]
12. Pagliara, S.; Lotti, I.; Palermo, M. Hydraulic jump on rough bed of stream rehabilitation structures. *J. Hydro-Environ. Res.* **2008**, *2*, 29–38. [CrossRef]
13. Abbaspour, A.; Hosseinzadeh Dalir, A.; Farsadizadeh, D.; Sadraddini, A.A. Effect of sinusoidal corrugated bed on hydraulic jump characteristics. *J. Hydro-Environ. Res.* **2009**, *3*, 109–117. [CrossRef]
14. Chanson, H. Momentum considerations in hydraulic jumps and bores. *J. Irrig. Drain. Eng.* **2012**, *138*, 382–385. [CrossRef]
15. Ahmed, H.M.A.; El Gendy, M.; Mirdan, A.M.H.; Ali, A.A.M.; Haleem, F.S.F.A. Effect of corrugated beds on characteristics of submerged hydraulic jump. *Ain Shams Eng. J.* **2014**, *5*, 1033–1042. [CrossRef]
16. Palermo, M.; Pagliara, S. Semi-theoretical approach for energy dissipation estimation at hydraulic jumps in rough sloped channels. *J. Hydraul. Res.* **2018**, *56*, 786–795. [CrossRef]
17. Pourabdollah, N.; Heidarpour, M.; Koupai, J.A. Characteristics of free and submerged hydraulic jumps in different stilling basins. *Water Manag.* **2020**, *173*, 121–131. [CrossRef]
18. Habibzadeh, A.; Rajaratnam, N.; Loewen, M. Characteristics of the flow field downstream of free and submerged hydraulic jumps. *Water Manag.* **2019**, *172*, 180–194. [CrossRef]
19. Gharangik, A.M.; Chaudhry, M.H. Numerical simulation of hydraulic jump. *J. Hydraul. Eng.* **1991**, *117*, 1195–1211. [CrossRef]
20. Ma, F.; Hou, Y.; Prinos, P. Numerical calculation of submerged hydraulic jumps. *J. Hydraul. Res.* **2001**, *39*, 493–503. [CrossRef]
21. Mousavi, S.N.; Júnior, R.S.; Teixeira, E.D.; Bocchiola, D.; Nabipour, N.; Mosavi, A.; Shamshirband, S. Predictive Modeling the Free Hydraulic Jumps Pressure through Advanced Statistical Methods. *Mathematics* **2020**, *8*, 323. [CrossRef]
22. Abbaspour, A.; Farsadizadeh, D.; Dalir, A.H.; Sadraddini, A.A. Numerical study of hydraulic jumps on corrugated beds using turbulence models. *Turk. J. Eng. Environ. Sci.* **2009**, *33*, 61–72.
23. Chern, M.-J.; Syamsuri, S. Effect of corrugated bed on hydraulic jump characteristic using SPH method. *J. Hydraul. Eng.* **2013**, *139*, 221–232. [CrossRef]
24. Bayon, A.; Valero, D.; García-Bartual, R.; Vallés-Morán, F.J.; López-Jiménez, P.A. Performance assessment of OpenFOAM and FLOW-3D in the numerical modeling of a low Reynolds number hydraulic jump. *Environ. Model. Softw.* **2016**, *80*, 322–335. [CrossRef]
25. Nikmehr, S.; Aminpour, Y. Numerical Simulation of Hydraulic Jump over Rough Beds. *Period. Polytech. Civ. Eng.* **2020**, *64*, 396–407. [CrossRef]
26. Ghaderi, A.; Dasineh, M.; Aristodemo, F.; Ghahramanzadeh, A. Characteristics of free and submerged hydraulic jumps over different macroroughnesses. *J. Hydroinform.* **2020**, *22*, 1554–1572. [CrossRef]
27. Ghaderi, A.; Dasineh, M.; Aristodemo, F.; Aricò, C. Numerical Simulations of the Flow Field of a Submerged Hydraulic Jump over Triangular Macroroughnesses. *Water* **2021**, *13*, 674. [CrossRef]
28. Karbasi, M.; Azamathulla, H.M. GEP to predict characteristics of a hydraulic jump over a rough bed. *KSCE J. Civ. Eng.* **2016**, *20*, 3006–3011. [CrossRef]
29. Roushangar, K.; Ghasempour, R. Explicit prediction of expanding channels hydraulic jump characteristics using gene expression programming approach. *Hydrol. Res.* **2017**, *49*, 815–830. [CrossRef]
30. Roushangar, K.; Ghasempour, R. Evaluation of the impact of channel geometry and rough elements arrangement in hydraulic jump energy dissipation via SVM. *J. Hydroinform.* **2018**, *21*, 92–103. [CrossRef]
31. Roushangar, K.; Homayounfar, F. Prediction Characteristics of Free and Submerged Hydraulic Jumps on Horizontal and Sloping Beds using SVM Method. *KSCE J. Civ. Eng.* **2019**, *23*, 4696–4709. [CrossRef]
32. Naseri, M.; Othman, F. Determination of the length of hydraulic jumps using artificial neural networks. *Adv. Eng. Softw.* **2012**, *48*, 27–31. [CrossRef]
33. Nasrabadi, M.; Mehri, Y.; Ghassemi, A.; Omid, M.H. Predicting submerged hydraulic jump characteristics using machine learning methods. *Water Supply* **2021**. [CrossRef]
34. Huš, M.; Grilc, M.; Pavlišič, A.; Likozar, B.; Hellman, A. Multiscale modelling from quantum level to reactor scale: An example of ethylene epoxidation on silver catalysts. *Catal. Today* **2019**, *338*, 128–140. [CrossRef]
35. Hager, W.H.; Bremen, R. Classical hydraulic jump: Sequent depths. *J. Hydraul. Res.* **1989**, *27*, 565–585. [CrossRef]
36. Fürst, J.; Halada, T.; Sedlář, M.; Krátký, T.; Procházka, P.; Komárek, M. Numerical Analysis of Flow Phenomena in Discharge Object with Siphon Using Lattice Boltzmann Method and CFD. *Mathematics* **2021**, *9*, 1734. [CrossRef]
37. Hirt, C.W.; Nichols, B.D. Volume of fluid (VOF) method for the dynamics of free boundaries. *J. Comput. Phys.* **1981**, *39*, 201–225. [CrossRef]
38. Nazari-Sharabian, M.; Nazari-Sharabian, A.; Karakouzian, M.; Karami, M. Sacrificial piles as scour countermeasures in river bridges a numerical study using flow-3D. *Civ. Eng. J.* **2020**, *6*, 1091–1103. [CrossRef]
39. Abbasi, S.; Fatemi, S.; Ghaderi, A.; Di Francesco, S. The Effect of Geometric Parameters of the Antivortex on a Triangular Labyrinth Side Weir. *Water* **2021**, *13*, 14. [CrossRef]
40. Chiu, C.-L.; Fan, C.-M.; Tsung, S.-C. Numerical Modeling for Periodic Oscillation of Free Overfall in a Vertical Drop Pool. *J. Hydraul. Eng.* **2017**, *143*, 04016077. [CrossRef]

41. Wang, Y.; Wang, W.; Hu, X.; Liu, F. Experimental and numerical research on trapezoidal sharp-crested side weirs. *Flow Meas. Instrum.* **2018**, *64*, 83–89. [CrossRef]
42. Ghaderi, A.; Daneshfaraz, R.; Dasineh, M.; Di Francesco, S. Energy dissipation and hydraulics of flow over trapezoidal–triangular labyrinth weirs. *Water* **2020**, *12*, 1992. [CrossRef]
43. Ghaderi, A.; Abbasi, S. CFD simulation of local scouring around airfoil-shaped bridge piers with and without collar. *Sādhanā* **2019**, *44*, 216. [CrossRef]
44. Ghaderi, A.; Dasineh, M.; Abbasi, S.; Abraham, J. Investigation of trapezoidal sharp-crested side weir discharge coefficients under subcritical flow regimes using CFD. *Appl. Water Sci.* **2019**, *10*, 31. [CrossRef]
45. Ghaderi, A.; Abbasi, S.; Abraham, J.; Azamathulla, H.M. Efficiency of Trapezoidal Labyrinth Shaped stepped spillways. *Flow Meas. Instrum.* **2020**, *72*, 101711. [CrossRef]
46. Daneshfaraz, R.; Aminvash, E.; Ghaderi, A.; Abraham, J.; Bagherzadeh, M. SVM Performance for Predicting the Effect of Horizontal Screen Diameters on the Hydraulic Parameters of a Vertical Drop. *Appl. Sci.* **2021**, *11*, 4238. [CrossRef]
47. Daneshfaraz, R.; Bagherzadeh, M.; Esmaeeli, R.; Norouzi, R.; Abraham, J. Study of the performance of support vector machine for predicting vertical drop hydraulic parameters in the presence of dual horizontal screens. *Water Supply* **2021**, *21*, 217–231. [CrossRef]
48. Thakur, B.; Kalra, A.; Ahmad, S.; Lamb, K.W.; Lakshmi, V. Bringing statistical learning machines together for hydro-climatological predictions—Case study for Sacramento San joaquin River Basin, California. *J. Hydrol. Reg. Stud.* **2020**, *27*, 100651. [CrossRef]
49. Roushangar, K.; Koosheh, A. Evaluation of GA-SVR method for modeling bed load transport in gravel-bed rivers. *J. Hydrol.* **2015**, *527*, 1142–1152. [CrossRef]
50. Roushangar, K.; Alami, M.T.; Shiri, J.; Asl, M.M. Determining discharge coefficient of labyrinth and arced labyrinth weirs using support vector machine. *Hydrol. Res.* **2017**, *49*, 924–938. [CrossRef]
51. Ferreira, C. Gene Expression Programming in Problem Solving. In *Soft Computing and Industry: Recent Applications*; Roy, R., Köppen, M., Ovaska, S., Furuhashi, T., Hoffmann, F., Eds.; Springer: London, UK, 2002; pp. 635–653.
52. Borrelli, A.; De Falco, I.; Della Cioppa, A.; Nicodemi, M.; Trautteur, G. Performance of genetic programming to extract the trend in noisy data series. *Phys. A Stat. Mech. Its Appl.* **2006**, *370*, 104–108. [CrossRef]
53. Majedi-Asl, M.; Daneshfaraz, R.; Fuladipanah, M.; Abraham, J.; Bagherzadeh, M. Simulation of bridge pier scour depth base on geometric characteristics and field data using support vector machine algorithm. *J. Appl. Res. Water Wastewater* **2020**, *7*, 137–143.
54. Antoniadis, A.; Lambert-Lacroix, S.; Poggi, J.-M. Random forests for global sensitivity analysis: A selective review. *Reliab. Eng. Syst. Saf.* **2021**, *206*, 107312. [CrossRef]
55. Pavlišič, A.; Pohar, A.; Likozar, B. Comparison of computational fluid dynamics (CFD) and pressure drop correlations in laminar flow regime for packed bed reactors and columns. *Powder Technol.* **2018**, *328*, 130–139. [CrossRef]
56. Pavlišič, A.; Huš, M.; Prašnikar, A.; Likozar, B. Multiscale modelling of CO_2 reduction to methanol over industrial $Cu/ZnO/Al_2O_3$ heterogeneous catalyst: Linking ab initio surface reaction kinetics with reactor fluid dynamics. *J. Clean. Prod.* **2020**, *275*, 122958. [CrossRef]
57. French, R.H. *Open-Channel Hydraulics*; McGraw-Hill: New York, NY, USA, 1985.

Article

A Review of the Modification Strategies of the Nature Inspired Algorithms for Feature Selection Problem

Ruba Abu Khurma [1], Ibrahim Aljarah [1,*], Ahmad Sharieh [1], Mohamed Abd Elaziz [2,3,4], Robertas Damaševičius [5,*] and Tomas Krilavičius [5]

[1] King Abdullah II School for Information Technology, The University of Jordan, Amman 11942, Jordan; ruba_abukhurma@yahoo.jo (R.A.K.); sharieh@ju.edu.jo (A.S.)
[2] Faculty of Computer Science and Engineering, Galala University, Suez 435611, Egypt; abd_el_aziz_m@yahoo.com
[3] Artificial Intelligence Research Center (AIRC), Ajman University, Ajman P.O. Box 346, United Arab Emirates
[4] Department of Mathematics, Faculty of Science, Zagazig University, Zagazig 44519, Egypt
[5] Department of Applied Informatics, Vytautas Magnus University, 44404 Kaunas, Lithuania; tomas.krilavicius@vdu.lt
* Correspondence: i.aljarah@ju.edu.jo (I.A.); robertas.damasevicius@vdu.lt (R.D.)

Abstract: This survey is an effort to provide a research repository and a useful reference for researchers to guide them when planning to develop new Nature-inspired Algorithms tailored to solve Feature Selection problems (NIAs-FS). We identified and performed a thorough literature review in three main streams of research lines: Feature selection problem, optimization algorithms, particularly, meta-heuristic algorithms, and modifications applied to NIAs to tackle the FS problem. We provide a detailed overview of 156 different articles about NIAs modifications for tackling FS. We support our discussions by analytical views, visualized statistics, applied examples, open-source software systems, and discuss open issues related to FS and NIAs. Finally, the survey summarizes the main foundations of NIAs-FS with approximately 34 different operators investigated. The most popular operator is chaotic maps. Hybridization is the most widely used modification technique. There are three types of hybridization: Integrating NIA with another NIA, integrating NIA with a classifier, and integrating NIA with a classifier. The most widely used hybridization is the one that integrates a classifier with the NIA. Microarray and medical applications are the dominated applications where most of the NIA-FS are modified and used. Despite the popularity of the NIAs-FS, there are still many areas that need further investigation.

Keywords: feature selection; evolutionary algorithms; nature inspired algorithms; meta-heuristic optimization; computational intelligence; soft computing

Citation: Abu Khurma, R.; Aljarah, I.; Sharieh, A.; Abd Elaziz, M.; Damaševičius, R.; Krilavičius, T. A Review of the Modification Strategies of the Nature Inspired Algorithms for Feature Selection Problem. *Mathematics* **2022**, *10*, 464. https://doi.org/10.3390/math10030464

Academic Editors: Freddy Gabbay and Ripon Kumar Chakrabortty

Received: 2 December 2021
Accepted: 21 January 2022
Published: 31 January 2022

Publisher's Note: MDPI stays neutral with regard to jurisdictional claims in published maps and institutional affiliations.

Copyright: © 2022 by the authors. Licensee MDPI, Basel, Switzerland. This article is an open access article distributed under the terms and conditions of the Creative Commons Attribution (CC BY) license (https://creativecommons.org/licenses/by/4.0/).

1. Introduction

As data accumulate rapidly in databases and data warehouses, a dimensionality problem becomes the main challenge for machine learning tasks (e.g., classification or clustering) [1]. Many negative effects may result from scaling up the dimensionality of a data set. These include the existence of irrelevant and redundant features that may adversely affect the learning algorithm or cause data over-fit [2]. Thus, the development of effective data mining techniques becomes an urgent necessity in various fields such as medicine [3], bioinformatics [4], text mining [5], image processing [6], design of smart infrastructures and smart homes [7], financial estimation [8,9], coastal engineering [10], and sustainability [11]. Their significance depends on their ability to turn huge amounts of data into an acceptable form. This will simplify knowledge discovery and make huge data sets more understandable, analyzable, and predictable.

Feature Selection (FS) is a pre-processing data mining technique for dimensionality reduction [12]. In recent years, research in FS has been rapidly developed in line with the era of big data and huge data sets. This subject has attracted the attention of researchers who have become more interested in developing novel FS techniques and improving current technologies [13]. FS manages the dimensionality problem by finding the most representative feature subset. The essence of FS is to choose features that are highly correlated to the class concept (relevant features) and weakly correlated with each other (complementary features/not redundant) [14]. Removing irrelevant and redundant features from a data set will cause improvements in different directions. For the modeling process, it will promote the generalization process. This will improve the quality of the generated model, so it becomes less complicated and more understandable. As a result, the inductive learner will be more efficient. FS is categorized based on the evaluation strategy into filters and wrappers [15]. The main difference between them depends on integrating a learning algorithm in the evaluation stage. Wrappers use learning algorithms to evaluate the selected feature subset. Hence, wrappers are more accurate and more expensive. In contrast, filters do not rely on learning algorithms, but use some data proprieties for evaluation. Examples of filters include univariate and multivariate filters. The main difference between them is that univariate filters rank a single feature to evaluate its performance, while multivariate filters evaluate the entire feature subset, which includes a set of feature as a combination. The generation of a feature subset in multivariate filters depends on the search strategy and the staring point of generation such as: Forward selection, backward elimination, bidirectional selection, and heuristic feature subset selection. Forward selection starts with an empty feature subset and then adds features, backward selection starts with the whole feature subset and eliminates one or more features from the set, and bidirectional search starts from both sides from an empty feature subset and from the whole feature subset at the same time [16], F-statistic [17], and information gain [18].

FS is not only a variable shrinkage process, and the target is not just to perform arbitrary cardinality reduction for a data set. FS is a multi-objective optimization problem which searches for the (near) optimal subset of features in terms of certain evaluation criteria. The main target of the FS problem is to find trade-offs between various conflicting objectives [19]. FS tries to achieve the minimum number of selected features with maximum performance [20]. Relative to search space, FS is considered a combinatorial nondeterministic polynomial-time-hard (NP-hard) problem. The reason being that it has a large search space that needs exponential running time to traverse exhaustively all the generated subsets of features [12]. The 2^N run time complexity will grow exponentially with increasing the value of N which represents the number of dimensions (features/variables) in a data set. This means that the traditional brute force methods are too impractical to be applied and other advanced search methods should be used.

Meta-heuristic search techniques are promising alternative solutions. They observed superior performance in various optimization scenarios. Potentially, they have a great opportunity to be suitable solutions for the FS problem. Meta-heuristics includes Nature Inspired Algorithms (NIAs), which are further divided into two main subcategories, namely Swarm Intelligence (SI) and Evolutionary Algorithms (EA) [21]. Both categories simulate the public behavior and biological evolution of agents in nature, respectively. Examples on EAs are: Genetic Algorithms (GA) [22] and Differential Evolution (DE) [23]. The SI category includes other types of algorithms such as Particle Swarm Optimization (PSO) [24], Ant Colony Optimization (ACO) [25], the Artificial Bee Colony (ABC) algorithm [26], memetic algorithms [27], artificial ecosystem-based optimization [28], marine predators algorithm [29], polar bear optimization [30], and red fox optimization [31].

Despite the effectiveness of nature-inspired Algorithms (NIAs) in solving the FS problem, finding the optimal solution is still not guaranteed. The main challenges that affect meta-heuristic optimization are stagnation in local minima, premature convergence, parameter tuning, exploitation and exploration imbalance, the diversity problem, dynamicity, multi-objectivity, constraints, and uncertainty [32].

Several kinds of modifications were proposed in the literature to enhance the performance of NIAs in optimization. Examples of these modification techniques include a new operator, hybridization [33], updated mechanism, new initialization strategy, new fitness function, new encoding schemes, modified population structure, multi-objectives, state flipping [34,35], and parallelism [36]. Each modification addresses the weakness of the NIA algorithm in some issues without harming the essence of the algorithm and its logic. The research field of NIA-FS has witnessed considerable development. To show the expansion of the NIAs-FS models in the literature, Figure 1 illustrates the correspondence between the year and number of publications that combine modified NIAs with FS. In the first years, research was volatile, and there were also years of research disruptions. Since 2006, the number of publications has remarkably increased to reach its peak in 2018. Furthermore, the research in this area has become very effective in the last five years. An intensive search for surveys in this area found that there are very limited NIAs-FS surveys [20]. Some FS surveys did not refer to meta-heuristics at all, but focused on other issues such as data perspectives [19], supervised/unsupervised FS approaches [15], and other FS surveys were tailored to specific applications or limited to certain domains [37]. The analysis of FS surveys showed that either they briefly refer to the meta-heuristic FS or they do not refer to them at all. To our knowledge, there is no survey about a modified NIAs-FS. This finding was one of the main motivations for this work. Unlike the previous FS surveys, FS will not be discussed in isolation from other related issues. The main objective is to bridge the gap in FS surveys by providing a review of the important aspects and design issues of NIAs-based FS approaches. The main modification strategies that have been adopted to enhance NIA for solving FS problem are categorized and discussed.

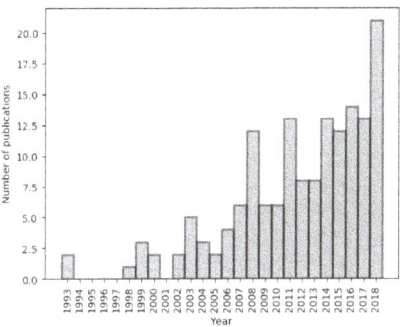

Figure 1. Development of research field regarding Nature Inspired Algorithms (NIA) modifications for tackling Feature Selection (FS).

In this review, a set of research questions will be asked and answered:
1. What is the current status of modified NIAs-FS research?
2. What are the important aspects and design issues regarding building NIA for tackling FS?
3. What are the modifications that were applied on NIA for tackling FS and in what domains were they applied?
4. Are there current open-source software systems that apply a modified NIA-FS?

Based on the aforementioned research questions, we have constructed this review based on three primary issues:
- Theoretical aspects of modified NIAs-FS provide detailed coverage for three main subjects: Meta-heuristic optimization, the FS problem, and modifications on meta-heuristic to enhance meta-heuristics for FS;
- Applied aspects of modified NIAs-FS presents different applications of modified NIAs-FS;
- Technical aspects of modified NIAs-FS presents a new developed FS tool, named Evolopy-FS.

The review will refer to various well-regarded publishers such as ACM, Elsevier, Springer, IEEE, World scientific, Hindawi, and others. Figure 2a shows the number of publications for each NIA in main publishers regarding modifications for tackling FS. Figure 2b shows the number of citations for popular NIAs articles in the main publishers regarding modifications for FS.

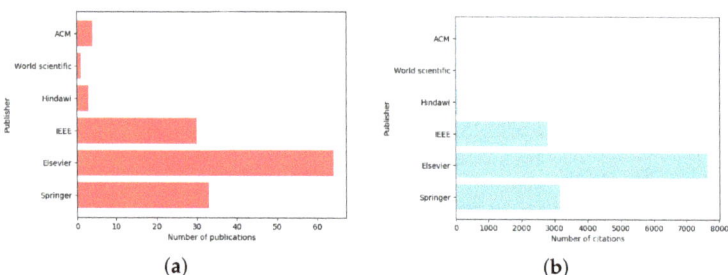

Figure 2. Statistics of the number of publications and citations for papers on NIAs modifications for FS. (**a**) Statistics of publications on modified NIAs-FS; (**b**) Statistics of citations for papers about modified NIAs-FS.

A description of meta-heuristic optimization is presented in Section 3. Section 4 discusses the problem of feature space symmetry in datasets and the need for feature selection as a disentanglement of symmetry. Section 4 discusses the feature selection problem and its related issues. A review of different NIAs-FS modification techniques is presented in Section 5. Section 6 highlights the main applications on modified NIAs-FS. An assessment of NIA-FS is provided in Section 7. Finally, in Section 8, the outlook for the NIA-FS research field and possible future directions are discussed.

2. Feature Selection as a Task of Disentangling the Symmetry of Feature Space

The aim of supervised machine learning is to estimate a function f that fits well with the features of training data and allows to predict the outputs on previously unseen inputs. The number of samples required for training grows exponentially with the dimension of a feature space, which is known as the "curse of dimensionality" [38]. To approximate a Lipschitz-continuous function composed of Gaussian kernels placed in the quadrants of a d-dimensional unit hypercube (blue) with error ϵ, one requires $\mathcal{O}(1/\epsilon^d)$ samples [39].

Intuitively, a symmetry of an object is a transformation that leaves certain properties of the object invariant. For example, translation and rotation are symmetries of objects, which do not change their representations [40]. The geometric structure of the feature space imposes the structure on the class of functions f that we are trying to learn. One can have invariant functions that are unaffected by the action of the group, i.e., $f(\rho(g)x) = f(x)$ for any $g \in \mathcal{G}$ and x, here $\mathcal{G}$ is the symmetry group, g is the symmetrical transformation in the feature space, $\rho(g)$ is the group representation, and x is an input in the space of input signals $\mathcal{G}(\Omega)$ that acts as a point in the feature space. Such symmetrical transformations (e.g., translation, rotation, shifting) are commonly used for data (image) augmentation to increase the number of data instances for effective training of machine learning models.

The goal of feature selection is to eliminate uninformative and/or redundant features from the feature space, leaving only relevant (i.e., predictive) features [41]. Feature selection seeks to decrease M to M' and $M' \ll M$ for a dataset with N samples and M dimensions (or features). In other words, feature selection produces a disentangled representation [40] with respect to a particular decomposition of a feature space with some symmetry group, which may be useful for subsequent tasks, such as reduced complexity of training a machine learning classifier. Such disentanglement, in fact, is performed by a neural network as a part of the classification process by learning the weights of a network nodes [42], which produce asymmetric activations for separation of classes.

The redundant features are characterized by a high level of inter-correlation. Such correlated features result in the symmetrical distribution of instances in feature space.

Feature selection aims to reduce feature dimensionality by reducing the symmetry in feature space. The resulting distribution of classes in the lower-dimensional feature space should be as asymmetrical as possible to allow for easy separability of classes [43]. Furthermore, a strong correlation in features might result in numerous near-optimal feature subsets, making traditional feature selection approaches unstable and lowering the trust in selected features [44]. As many different feature space decompositions are possible, the problem of finding an optimal feature subspace in a high-dimensional feature space is known to be NP-hard [45]. In this paper, the nature-inspired meta-heuristic optimization algorithms are studied for solving the feature selection problem.

3. Meta-Heuristic Optimization

Meta-heuristic algorithms are characterized by flexibility, simplicity, low cost in computations, and they are derivation-free methods. The principle of meta-heuristics is reasonability vs completeness. In other words, it gives up completeness for providing approximated solutions for complex unsolved problems. Meta-heuristics are further categorized based on the number of candidate solutions encountered during the optimization process into the trajectory and population.

3.1. Trajectory-Based Optimization

A trajectory algorithm begins with one random solution and it tries to optimize the solution until a stop condition is satisfied. The computation overhead is reduced significantly because only one solution is being improved and evaluated during the optimization process. Equation (1) expresses the number of function evaluations needed in trajectory algorithms where T is the number of iterations:

$$\#Evaluations(in\ trajectory\ based) = 1 \times T. \tag{1}$$

Trajectory algorithms are local search techniques. They depend on making a few changes in the components of the current solution to find a better one. A potential solution is picked, and its neighboring solutions are checked if they are better. Local search implies searching within a limited region (exploitation). This process suffers from a potential entrapment in local minima because of the diversity weakness and a lack of information exchange. Examples of trajectory algorithms are Simulated Annealing (SA) [46] and Tabu search (TS) [47].

3.2. Population-Based Optimization

A population algorithm begins with a set of randomly generated solutions and tries to enhance them during the optimization process. Each candidate solution fluctuates outward or converges toward the best solution following a certain mathematical framework. The predominance of these algorithms is because of their simplicity and flexibility. Simplicity means that they are built upon simple methodologies and are evolved from simple concepts. They can be adopted to deal with real-world problems without structural modifications. All that is required is an accurate representation of the problem and the structure of the optimizer is left untouched. Population algorithms are more efficient in mitigating local minima compared with trajectory algorithms because more individuals and more information are shared between them. However, multiplicity in solutions increases the computation burden because more evaluations are required. The number of calls for a fitness function is driven by the number of individuals and the number of iterations. Equation (2) identifies the number of function evaluations in population algorithms where N is the number of individuals and T is the number of iterations:

$$\#Evaluations(in\ population\ based) = N \times T. \tag{2}$$

A population algorithm begins with the initialization step where a set of candidate solutions are generated. The solution is a candidate or possible solution if it satisfies the constraints of the problem. The next step is the evaluation of individuals. The evaluation is carried out using a specified fitness function and in terms of predefined evaluation criteria. The fitness function is called for each individual so that each individual gets a fitness value. After evaluating the individuals, the update process refines and improves current solutions. This requires updating the positions of individuals in the search space. This iterative process of evaluating and updating individuals continues until a predefined criterion is satisfied and the global optimal solution is best approximated.

Population-based algorithms compromise of NIAs that are the result of the union of nature with different scientific fields including physics, biology, mathematics, and engineering. Computer science utilized these relations between science and nature and turned it into a well-defined discipline for optimizing different challenging problems. NIAs are categorized based on the source of inspiration into EA- and SI-based algorithms [21].

3.2.1. Evolution-Based Optimization (EA)

This category includes different computational systems that share in their emulation for the biological evolution. EAs model the natural cellular processes such as reproduction, mutation, recombination, and selection.

EAs typically designed by generating a population of possible solutions $\vec{I_1}, \vec{I_2}, \vec{I_3} \ldots \vec{I_{n-1}}, \vec{I_n}$ called chromosomes. Each chromosome is split into smaller units called genes. The length of the chromosome (#genes) determines the dimensionality of a problem. The relation between gene, chromosome, and population can be expressed as *gene* $\subset$ *chromosome* $\subset$ *population*. Most of the current evolutionary frameworks implement the chromosome and population as 1-d array (vector) and 2-d array, respectively. Equation (3) identifies the individual I_i with a length d and Equation (4) identifies a population P where each individual represents a row in a matrix. Each solution is evaluated by a certain object function $\vec{O_1}, \vec{O_2}, \vec{O_3} \ldots \vec{O_{n-1}}, \vec{O_n}$ to determine its quality and decide if it is fitted or unfitted. The highest evaluated solution (best individual) is preserved at each iteration. The unfitted solutions (worst individuals) are candidates to be replaced by newly-generated offsprings. This allows the average fitness value to increase dramatically throughout iterations. Common EA examples are GA [22] and DE [23]. GA undoubtedly is the most widespread and typical example of EAs:

$$I_i = \begin{bmatrix} x_i^1 & x_i^2 & x_i^3 & \cdots & x_i^{d-1} & x_i^d \end{bmatrix} \quad (3)$$

$$P = \begin{bmatrix} x_1^1 & x_1^2 & x_1^3 & \cdots & x_1^{d-1} & x_1^d \\ x_2^1 & x_2^2 & x_2^3 & \cdots & x_2^{d-1} & x_2^d \\ \vdots & \vdots & \vdots & \ddots & \vdots & \vdots \\ x_n^1 & x_n^2 & x_n^3 & \cdots & x_n^{d-1} & x_n^d \end{bmatrix}. \quad (4)$$

3.2.2. Swarm-Based Optimization (SI)

SI algorithms have a common behavior that is very similar to the social behavior of creators. The Swarm system comprises an abundant number of agents that are distributed in the environment to achieve a global target. Intelligence can be seen in the actions of agents to coexist. The main characteristics of swarm systems are adaptability, self-organization, distributed control, scalability, and flexibility [20]. The most common SI examples are Particle Swarm Optimization (PSO) [24] and Ant Colony Optimization (ACO) [25]. A PSO source of inspiration are flocks of birds that search for food. The search procedure is guided by two main factors: Pbest and gbest. Pbest represents the best experience that was gained by the previous particle itself. Gbest represents the best individual in the whole swarm. Particles also have a position and velocity that are both updated in each iteration.

3.3. Challenges of Meta-Heuristic Optimization

Despite the efficiency of meta-heuristics in tackling challenging optimization problems, some obstacles impact their performance. These include dynamicity, multi-objectivity, constraint, and uncertainty. For multi-objectivity, there are multiple conflicting objectives to be optimized until trade-offs (Pareto optimal set) are achieved. The search space is quite more complex. The optimization problem becomes highly challenging when the number of objectives becomes larger than four [48]. Many objective fields have emerged to deal with these cases. Constraints of real problems create gaps in the search space by dividing it into feasible and infeasible regions. Feasible regions satisfy the constraints while infeasible regions violate these constraints [49]. Accordingly, the optimization algorithm should follow certain mechanisms to become closer to the promising region and avoid the infeasible region until an optimal solution is found. The other main issue of meta-heuristic optimization is uncertainties. For example, the global solution frequently changes its position in the search space, which requires more attention from the optimization algorithm. Some operators are used for registering the history and memorizing the locations of the global optima all the time. Other severe challenges are related to the problem search space, such as the existence of many holes or valleys that lead to stagnation in local minima, discontinuities in a search space, the location of global optima that comes onto the boundary of a search space (the boundary of constraints), and the isolation of global optima [32].

Population algorithms are characterized by two conflicting milestones that are called exploration (diversification) and exploitation (intensification) [32]. In exploration, the candidate solutions churn and change violently, which leads one to examine more regions and to find diverse solutions. Exploitation changes gently and causes a less sudden stir for the candidate solutions. GA realizes these processes through crossover and mutation operators. Crossover intermixes a combination of solutions while mutation squeezes certain regions and searches locally. PSO configures the inertia weight operator by large values for more explorations and selects small values for more exploitation.

The main challenges of exploration and exploitation include: Firstly, since they have conflicting purposes, increasing any of the causes decreasing the other. Secondly, a transition between these two milestones is not defined because the search spaces of the optimization problems are usually unknown. Thirdly, performing pure exploration causes less accuracy in approximating an optimal solution because different regions are being explored without a focus on a certain promising region. Performing pure exploitation gives rise to entrap in local optima. Fulfilling a balance between exploration and exploitation may produce better results and increases the chance of being close to the optimal solution. Recently, this idea has become an active research problem. Several types of research have tried to attain balance by integrating several random and adaptive operators in the structure of the algorithms.

4. Feature Selection

This section introduces FS in two parts: The dimensionality problem and the FS system based on the NIA search strategies.

4.1. Dimensionality Problem

Due to the incremental growth of information and the abundance of data, data sets have increased in both data samples (number of instances) and dimensions (number of features). As a result of the increased dimensionality, different negative effects were embedded in data mining tasks. One of these problems is called the curse of dimensionality, which describes the status of data as it becomes sparser in large dimensionality space [12]. This raises the need for more instances for the training of the classifier, which increases the learning time. Learning algorithms were designed to build their models based on rules inferred from a small number of dimensions. Learning algorithms cannot generalize well in a large dimensionality space. High dimensionality implies the existence of noisy features, such as redundant and irrelevant features that mask the informative features and mislead the classifier and cause data to overfit. An overfitting [2] problem occurs when a classifier overtrained on the data and learned all examples, including outliers. Considering that noise and random fluctuations as related concepts will cause building complex models; logically, learning from relevant features allows the classifier to be more accurate. Another negative effect of increasing dimensions is the increased demand for specialized devices such as large memory storage and high-speed processors, which increases cost.

4.2. FS Preliminaries

Features are defined as measurable properties of the observation under study. The complexity of the problem is determined by its features. In real-world applications, the discovery of relevant features is a big challenge. In 1997, the first papers about relevance and feature selection were published [14]. Feature relevance can be formalized as follows. Let $1 \leq i \leq n$, E_i be the domain of feature x_i, $X = x_1, x_2, \cdots x_n$ be the set of all features. $E = E_1 \times E_2 \cdots \times E_n$ is the instance space from which instances derive their values. Each instance can be represented as a point in space and the distribution of these data points has a probability P. If we consider the class (label) space to be T, then we can define an objective function c as a relation that maps an instance S to a specified label/class in labels space T as: $c: E \to T$. Arguably, a data set with $|S|$ number of instances is the result of sampling $|S|$ times from E with a probability P and get label from T. An x_i in X is a relevant feature with respect to class concept if there exist two instances (A and B) in E, which only differ in their assignment to xi (all their feature values are the same except those for feature xi) and $c(A) \neq c(B)$. In contrast, a variable with no correlation or weak correlation with the target concept is called an irrelevant feature. Other types of noisy features are redundant features. These are features that are highly related and connected with other features and add nothing new regarding the classification decision.

In the literature, FS was defined in different ways, which are all close in meaning and intuition [15]. FS is a searching process that tries to find the subset of features which is the best one to describe the data. According to relevance discovery, FS aims to determine the most meaningful subset of features, which has the largest relevance and minimum redundancy. Even though those features are fewer than the original features, but they carry the maximum discriminate information. Classically, FS selects a subset of M features from a set of N features where $M < N$ and the value of an evaluation function is optimized over all subsets of size M. The essence of FS is to select or discard features intelligibly in such a way, the resulting class distribution is as close to the class distribution with the complete set of features. In another meaning, FS is not a technique for only reducing data set cardinality, but it should find a trade-off and a balance between different conflicting objectives. As a multi-objective optimization problem, there are two primary objectives to be optimized. These objectives are the performance and the number of selected features. These are conflicting objectives because the optimization algorithms require getting the maximum performance and the minimum number of selected features.

Typically, the standard process of FS consists of four primary stages of subset generation, subset evaluation, stopping criterion, and results validation [15].

Regarding the subset generation and search procedure, FS is considered an NP-hard problem. When the number of features equals n, the search space comprises 2^n subsets of features. Using brute search methods such as a huge search space needs an exponential running time to traverse all the candidate subsets of features.

Concerning subset evaluation, there are different methods to assess the goodness of a feature subset such as filters and wrappers. A stop criterion is a condition that halts the FS process and prevents the infinite loop. For example, the search completion (all feature subsets have been examined), the learning performance reached its highest limit, the subset of features with a specified size is obtained, the pre-defined number of iterations is reached, the occurrence of conversion situation in which results become stable, and no further enhancement is achieved. A direct way to validate the obtained results is based on prior knowledge from a domain. Unfortunately, this features knowledge is usually unavailable so other methods have to be used instead. FS could be validated by comparing the system performance using the whole subset of features with its performance using the selected features. FS has many advantages that positively affect the data mining task, including improving the quality of the generated model, speeding the learning time of the classifier, enhancing the ease of reading the data set, and reducing the need for more hardware resources.

4.3. NIAs for Feature Selection

Two important points should be focused on: The representation of a solution and the evaluation for it. Normally, a feature subset is represented by a binary vector. The dimensionality of the problem is equal to the number of features in the data set. If the gene value is set to 1, this indicates that the feature is selected, otherwise, it is not selected. The quality of a feature subset is evaluated based on two contradictory objectives: The classification accuracy (minimum error rate) and the minimal number of selected features simultaneously. These two criteria are represented in one fitness function that is shown in Equation (5), where $\alpha \gamma_R(D)$ is the error rate of the classification produced by a classifier, $|R|$ is the number of selected features in the reduced data set, and $|C|$ is the number of features in the original data set, and $\alpha \in [0,1]$, $\beta = (1 - \alpha)$ are two parameters for representing the significance of classification and length of feature subset according to recommendations:

$$Fitness = \alpha \gamma_R(D) + \beta \frac{|R|}{|C|}. \tag{5}$$

5. NIAs FS Modifications

This section highlights the main modification techniques applied in the literature to enhance the NIAs as wrappers FS. By referring to 156 articles in the domain of modified NIAs-FS, it can be noticed that the modification techniques can be classified into nine categories as depicted in Figure 3: New operators, hybridization, update mechanism, modified population structure, different encoding scheme, new initialization, new fitness function, multi-objective, and parallelism.

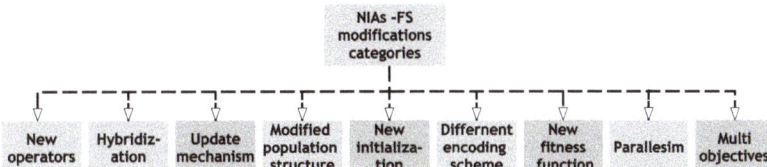

Figure 3. NIAs FS modifications categories

5.1. New Operators

This modification depends on integrating a new operator in the original NIA structure to achieve certain targets, such as improving the algorithm performance, increasing the diversity among the population, enhancing the exploitation and exploration processes, facilitating the sharing of information between population's individuals, repositioning of the worst individuals in the population, and performing a search along various vectors in search space [36]. In literature, several operators have been used to enhance NIAs wrappers. Some of these operators are discussed next.

5.1.1. Chaotic Maps

The denotation of chaos means a state of disorder. In mathematics, it is a formula that describes a dynamic system with time dependence. The chaotic system has a high level sensitivity to its initial conditions. This behavior implies that even a simple modification in the initial conditions will lead to big changes in the outcomes. Although the chaotic system is deterministic and does not incorporate any randomness but the results are not always predictable [50].

Chuang in [51] used two kinds of chaotic maps and integrated them with Binary Particle Optimization (BPSO), namely logistic maps and tent maps. Equation (6) describes how the logistic map is written in mathematics (general formula), where X_n is a number between 0 and 1 which represents the ratio of the current population size to the maximum population size and μ is a constant value between 0 and 4. Equation (7) describes how Chuang exploited Equation (6) to modify the inertia weight value where w is the inertia value between (0,1) and t is the number of iteration. The same thing was followed to apply the tent map chaotic map. Equation (8) is the general mathematical formula and Equation (9) is the modified version of inertia weight using a tent map. Using large values for inertia weight facilitates more exploration while selecting small values facilitates more exploitation. Hence, chaos theory could be used for balancing the two types of search in the search space. Besides, the study contributed that Chaos Binary Particle Swarm Optimization (CBPSO) with a tent map achieved a higher classification accuracy than CBPSO with a logistic map:

$$X_{n+1} = \mu X_n (1 - X_n) \quad (6)$$

$$w(t+1) = 4.0 * w(t)(1 - w(t)) \quad (7)$$

$$f(x) = X_{n+1} = \begin{cases} \mu X_n, & \text{if } X_n < 0.5 \\ 1 - \mu X_n, & \text{otherwise} \end{cases} \quad (8)$$

$$w(t+1) = \begin{cases} w(t)/0.7, & \text{if } w(t) < 0.7 \\ 10/3 w(t)(1 - w(t)), & \text{otherwise.} \end{cases} \quad (9)$$

In the same year, Chuang presented another model for FS [52]. The proposed model was a filter-wrapper approach based on using a correlation-based filter (CFS) and Taguchi chaotic BPSO (TCBPSO). In [53], chaotic was applied with BPSO for FS in text clustering. Ahmad in [54] used chaotic maps as modifications for the SSA algorithm. He replaced the C3 random parameter with chaotic sequences, namely logistic map, piecewise map, and tent map. It was clear the impact of chaotic maps in improving the SSA. In the same year, the influence of chaotic operators on SSA was investigated in [55]. The experiments proved that the logistic map achieved a better performance for the SSA algorithm over nine chaotic maps. The chaotic multiverse optimization (MVO) FS model was proposed in [56] to cope with some limitations of MVO. Tent, logistic, singer, sinusoidal, and piecewise chaotic maps were used. The results showed that the logistic chaotic maps were the best, which increased the MVO performance more than other maps. Sayed in [57] developed a new wrapper FS approach based on the Whale Optimization Algorithm (WOA) and chaotic theory named CWOA. He used 10 chaotic maps namely chebyshev, circle, guass/mouse, iterative, logistic, piecewise, sine, singer, sinusoidal, and tent. The results showed that a circle chaotic maps was the best among other chaotic. In [3], a model based on chaotic Moth Flame Optimization (CMFO) and Kernel Extreme Learning Machine (KELM) was proposed. In [58], Sayed developed a new FS system composed of the Crow Search Algorithm (CSA) algorithm and chaos theory to enhance the performance and convergence speed of CSA. Lately in [59], a Binary Black Hole optimization Algorithm (BBHA) has been modified by embedding new chaotic maps embedded with the movement of stars in the BBHA. This model was called CBBA and uses 10 chaotic maps. The results of three chemical data sets demonstrated that CBBA outperformed the BBHA in terms of the number of selected features, classification performance, and computational time.

5.1.2. Rough Set

Rough Set (RS) was first described by Zdzislaw Pawlak at the beginning of the 1980s [60]. This is a mathematical concept related to topological operations. In mathematics, RS is a theory that tries to find two approximate sets for the original conventional set (crisp set). The first RS gives the lower approximation for the crisp set which compromises the elements that surely belong to the target subset. The second RS gives the upper approximation of the crisp set which compromises the elements that possibly belong to the target subset. The pair of rough sets are themselves either crisp sets or fuzzy sets. Rather than belonging or not belonging in relation to the elements as in crisp sets, the fuzzy sets depend on the membership function for gradual assessments of the elements. Unlike the fuzzy sets, RSs depend on finding the positive region, not the membership function for dealing with uncertainties and vagueness. The RS has many advantages, including the approximation of concepts, reduction of spaces, discovering the equivalence relations, and finding the minimal sets of data in vague and uncertain domains. In FS, the RS tries to define the attribute dependency. Zainal in [61] proposed the RS-PSO model for a better representation of data. Another RS-PSO-FS model was proposed in [62] based on Relative Reduct (PSO-RR) and PSO-based Quick Reduct (PSO-QR). Both tools depend on the dependency measure for comparing sets of attributes. In [63], the authors proposed a model for FS in nominal data sets based on BCS and Rough Sets. Another CS model was introduced in [64] by incorporating the RS with different classifiers. In [65], a new model was developed based on two incremental techniques (QuickReduct and CEBARKCC). Quick reduct and CEBARKCC are two filtering methods where the former one is a rough set-based filter that simulates the forward generation method and the latter is a conditional entropy-based method. These two methods were integrated with the Ant Lion Optimization (ALO) algorithm to improve the initial population quality. The RS-FA model was developed in [66]. Hassanien in [67] developed a new system based on rough set and MFO. Lately, in [68], a hybrid model called BPSOFPA composed of Flower Pollination Algorithm (FPA) and PSO was also developed. BPSOFPA was integrated with the RS approach for the FS problem. Ropiak in [69] integrate RSs with deep learning as rough mereological granular computing.

5.1.3. Selection Operators

Inspired from Darwin's theory [70], which explained the evolution and changes in species through the natural selection mechanism, the genetic algorithm incorporated selection operators to select some individuals from the population for later breeding. A conventional strategy to implement the selection is using the fitness values of the solutions. In other methods, these fitness values are normalized by finding the summation of them then divide the fitness of each individual by this summation. Another method sorts all individuals in the population according to their fitness values in descending order. The selection mechanism was applied in other studies by finding the accumulated fitness for each individual so that the final individual fitness value is one [71]. All such methods become computationally expensive and may negatively impact the performance of GA when the population becomes larger. Other methods of selection which are widely implemented with GA are Tournament Selection (TS) and Roulette Wheel Selection (RWS). The stochastic nature of these methods makes them simpler in implementation and better in performance than the aforementioned methods. TS is the most applied selection operator with GA because of its simplicity. It selects randomly a set of solutions from the population then the best one is used for breeding the successive generation. In RWS, the mechanism differs in that no agent in the population is discarded. The RWS strategy depends on creating something like a roulette where all fitness scores of the individuals are represented as areas or sectors on this roulette. The individual with a large fitness value well reserve a large sector on the roulette, which shows a larger probability for selection. Individuals with small fitness scores will reserve small areas on the roulette. In RWS, the final selection for the agent is done by rotating the roulette and the selected individual is the one where the point stayed when the roulette had stopped. Mafarja in [46] developed a new model that combines TS with the WOA optimizer to enhance the exploration of the search. One year later, Mafarja presented in [72] an FS model based on the Grasshopper Optimisation Algorithm (GOA) algorithm with RWS and TS. Mafarja in the same year developed a new wrapper FS model based on WOA along with studying the effect of TS and RWS [73].

In [26], the selection operators were incorporated to improve the ABC optimizer. In [74], the method compromised of a DE optimizer and RWS structure for the selection of the Wavelet Packet Transform.

5.1.4. Sigmoidal Function

A sigmoid function is a mathematical function that falls under the S-shaped family and is considered a special case of a more general function called a logistic function, which has the mathematical formula defined by Equation (10), where e is the natural logarithm base (Euler's number), $x0$ is the sigmoid's midpoint, L is the sigmoid's maximum value, and k is the logistic growth rate of the curve [75]. The sigmoid function formula is defined by Equation (11):

$$f(x) = L/(1 + exp^{-k(x-x0)}) \tag{10}$$

$$S(x) = 1/(1 + \exp(-x)) = \exp(x)/(\exp(x) + 1). \tag{11}$$

The sigmoid function has some special characteristics including the monotonic behavior, which means that the function is defined on all real numbers but the output of the function is increasing either from 0 to 1 or from -1 to 1. Moreover, the sigmoid function is differentiable and has a bell-shaped first derivative where the derivative at each point is a non negative value. There are several variations of the sigmoid function such as hyperbolic tangent, arctangent function, and algebraic functions which are respectively defined by Equations (12)–(14). The sigmoid function is widely applied as the activation function of a Neural Network (NN). Other useful usage of the sigmoid function is that it is used as a discretization method to convert a continuous space into a binary one, such an application is a feature selection application:

$$f(x) = tanh(x) = (e^x - e^{-x})/(e^x + e^{-x}) \tag{12}$$

$$f(x) = arctan(x) \tag{13}$$

$$f(x) = x/\sqrt{(1 + x^2)}. \tag{14}$$

For solving the FS problem, Aneesh developed a modified BPSO called Accelerated BPSO (ABPSO). The strategy for accelerating the particles was using a new velocity update function based on a sigmoidal function [76]. In [6], the sigmoidal function was used with BGWO in solving FS. In [77], different transfer functions that map continuous solutions to binary ones were applied in combination with the CS algorithm. The CS-sigmoid and CS-hyperbolic tangent was performed on five data sets. In [78], the effect of different transfer functions on the Bat optimization (BA) algorithm was studied. Sigmoid and hyperbolic tangent functions were used to analyze their influence on FS. The results proved that the sigmoid function was better than the hyperbolic function in feature reduction for almost all data sets. Mafarja, in [79], presented new versions of the Grasshopper Optimization Algorithm (GOA) based on sigmoid and V-shaped TFs in the context of FS.

5.1.5. Transfer Functions

Transfer functions (TFs) are mathematical formulas that play a significant role in mapping a continuous search space to discrete search space. The discrete search space could be viewed as a hyper-cube in which solutions move in different directions within its boundaries by flipping their bit values. TFs are one of the most efficient ways that could be utilized to covert continuous meta-heuristic algorithms into their corresponding binary versions [80]. The mathematical formulations of these TFs can be found in [80]. The update procedure in a binary meta-heuristic algorithm is switching solutions elements between 0 and 1 based on certain mapping formula TFs that links the original continuous update procedure with a new binary update procedure. TFs in a close meaning define the probability of updating each element (gene/feature) in a solution to be either selected 1 or not selected 0.

Equations (15) and (16) define the general update formulas of a solution using S-TFs and V-TFs, respectively, where $X_i^d(t+1)$ represents the *ith* element (gene/feature value) in the X solution (feature subset) at dimension d (feature number/index) in iteration $t+1$, rand $\in [0, 1]$, which was generated using a random probability distribution:

$$X_i^d(t+1) = \begin{cases} 0, & \text{if } rand < S\_TF(X_i^d(t+1)) \\ 1, & \text{if } rand \geqslant S\_TF(X_i^d(t+1)) \end{cases} \quad (15)$$

$$X_{t+1} = \begin{cases} X_t, & \text{if } rand < V\_TF(X_{t+1}) \\ \neg X_t, & \text{if } rand \geqslant V\_TF(X_{t+1}). \end{cases} \quad (16)$$

These can be reformulated to preserve the concepts of searching using any specific meta-heuristic algorithm. As an example, PSO was converted by Kennedy and Eberhart [81] from a real algorithm to a binary algorithm. The PSO binary conversion started by employing a sigmoid function to convert the velocity values into probability values bounded in the interval [0,1] as in Equation (17), where $T(v_i^d(t))$ indicates the velocity of particle i at dimension d in iteration t. In the next step, the computed probabilities are used to update the position vector using Equation (18). To preserve the PSO continuous searching method and keep the concepts of pbest/gbest, the TF gives a high probability for switching gene values for those genes having high-velocity values since they are far away from the best solution. Small probability is given for genes having small velocity values since they are considered close to the best solution [80]:

$$T(v_i^d(t)) = 1/(1 + e^{-v_i^d(t)}) \quad (17)$$

$$X_i^d(t+1) = \begin{cases} 0, & \text{if } rand < TF(v_i^d(t+1)) \\ 1, & \text{if } rand \geqslant TF(v_i^d(t+1)). \end{cases} \quad (18)$$

In the literature, there were several studies that adopted TFs operators with FS problem. Mirjalili in [80] improved the performance of BPSO by using TFs, S-shaped, and V-shaped transfer. The results of V-TFs improved the performance of BPSO more than S-TFs. In [82], a new wrapper was developed by modifying the Salp Swarm Algorithm (SSA) using TFs. The proposed approach achieved significant superiority over other competitive approaches in 90% of the data sets. Mafarja in [83] presented a new wrapper FS method based on a modified Dragonfly Algorithm (DA) using time-varying S-shaped and V-shaped TFs. Recently, in the context of Internet of Things (IoT) attack, a new wrapper-based approach using the WOA was developed. The augmented WOA used both V-shaped and S-shaped transfer functions.

5.1.6. Crossover

In living things, the chromosomal crossover is a recombination process that occurs between non-sister chromatids to exchange the genetic material during recombination (sexual reproduction). This process ends in the production of new recombinant chromosomes. Faraway from the biological chromosomal crossover in the genetic algorithm and evolutionary computation, this process was inspired to exchange information between solutions in the population and generating new offsprings in the next generation. In the genetic algorithm, recombination (crossover) is defined as a stochastic operator that enforces the diversity in the population by exchanging (swapping) the bits after a random cutting point (crossover point) between the parents' vectors (selected individuals) to produce new children (offsprings). Equation (19) shows how a crossover operator is used to combine solutions where ⋈ is an operator that performs the crossover scheme on the two binary solutions X_i and X_{i-1}. In a binary space, the crossover can be realized by exchanging the binary bits of two solutions to obtain an intermediate solution. Equation (20) shows that the crossover mechanism switches between two input vector with the same probability, where Xd is the value of the dth dimension in the yielded vector after applying the crossover operator on X_i and X_{i-1}:

$$X_i^{t+1} = \bowtie(X_i, X_{i-1}) \tag{19}$$

$$X^d = \begin{cases} X_1^d, & \text{if } rand \geqslant .5 \\ X_2^d, & \text{otherwise.} \end{cases} \tag{20}$$

In [84], a crossover operator was applied in combination with the sigmoid function to modify a Binary Grey Wolf Optimizer (BGWO). The BGWO1 approach was used to convert the Continuous version of GWO (CGWO) into the binary version. The first steps toward the three best solutions are converted into binary, then a random crossover is applied among them to find the updated position. The results of the approach positively affected the performance of GWO. In [82], the crossover operator was applied to improve the Salp Swarm Algorithm (SSA) optimizer in solving the FS problem. The crossover job was to increase the diversity of the model and improve the exploration process of the search space. In [73], the study incorporated many modifications strategies with WOA. Solving the limitations of the WOA represented by local minima and slow convergence was the priority. The crossover was used for achieving this target. Mafarja in [79] applied multiple operators with GOA. The combination operator together with the mutation was applied in his approach to BGOA-M for achieving more exploration.

5.1.7. Mutation

In the organism, the mutation is an error that occurs during DNA replication (meiosis). The error specifically results from a permanent deletion, insertion, or alternation on the DNA segment (nucleotide sequence of the genome). Even though this is a small genome error, it causes abnormal changes in the characteristics of an organism. Evolutionary and genetic algorithms inspired the same idea to make changes and increase the diversity in the population. The advantages of mutation come from preventing solutions becoming similar and thus ensuring the evolution does not stop. Mutation operators alter one or more gene values (a bit in chromosome vector) which causes the solution to be changed from its previous state. Besides diversity, the mutation could contribute to mitigating the local minima problem. Equation (21) identifies the mutation process where $Xi(t+1)^d$ is the ith element at the dth dimension in the X_i solution,

$$X_i^d(t+1) = \begin{cases} 0, & \text{if } rand \geqslant .5 \\ 1, & \text{otherwise.} \end{cases} \tag{21}$$

In [85], a Particle Swarm Optimization (PSO) applied mutation to a solution was conducted after it was updated. A probability commonly $1/n$ indicates one bit of the solution will be muted (flipped). The model proved the effectiveness of the suggested modified PSO-FS model. In [53], the authors developed a hybrid intelligent algorithm that combined mutation with the BPSO and other operators to solve FS in the text clustering. The model attained a higher clustering accuracy and improved the convergence speed of BPSO. In [79], the mutation operator was applied with the GOA optimizer. The BGOA-M approach achieved superiority in comparison with other approaches compared. In [86], an Improved Harris Hawks Optimization (IHHO) was proposed based on elite opposite-based learning, mutation neighborhood search, and rollback strategies to increase the search performance.

5.1.8. Levy Flight

Levy flight has its source from chaos theory. It describes a random walk that follows a heavy-tailed probability distribution. This probability distribution represents the step-lengths that take place either on a discrete grid or continuous space. In mathematics, according to a central limit theorem, the steps from the original point of a random walk follow a stable distribution which could be modeled using equations of Levy flights. Investigators in nature found that Levy flights can describe the animals hunting patterns especially when the prey is sparsely distributed and not easily detected as opposed to Brownian motion, which can only approximate the prey place when the hunting is near an abundant and predictable prey [87]. In [64], a novel Cuckoo Search (CS) algorithm was developed using the Levy flight with the rough sets. He applied his idea by integrating the Levy flight random probability distribution in the equation that generates new solutions as shown in Equation (22) where $\oplus$ denotes the entry-wise multiplication, α is the step size, $\alpha > 0$, and $Levy(\lambda)$ is the Levy distribution which is described in Equation (23). In [88], Levy flight was used in combination with transfer functions to enhance the performance of the MFO algorithm and increase diversity:

$$X_i^{t+1} = X_i^t + \alpha \oplus Levy(\lambda) \tag{22}$$

$$Levy \sim u = t^{-\lambda} \qquad 0 < \lambda < 3. \tag{23}$$

5.1.9. Other Operators

A local search operator was incorporated with GA to mitigate the weakness of standard GA in fine-tuning near the local minima [89]. In [90], local search was used to improve the BPSO. A new local search and gbest resetting strategy called PSO-LSRG was proposed in [24] to facilitate the exploitation. A Uniform Combination (UC) operator was used in [80] to improve the performance of BPSO. Later, UC was adopted in [91] to balance the exploitation and exploration of bones PSO. The DE evolutionary operator was used in [5] to solve the local optima in standard WOA. The DE evolutionary includes mutation, crossover, and selection operators. Boolean algebra (and operator) was used in BPSO [92]. The bacterial evolutionary algorithm and PSO algorithm, both with a plain and a memetic variant complemented with gradient-based local search and fuzzy logic numbers were used in [93] for solving various resource allocation problems.

A catfish strategy was applied in [94] to improve the performance of BPSO based on introducing new particles into the search space when there is no improvement in the searching process. For example, when the gbest is unchanged over a consecutive number of iterations. The catfish particles replace the particles with the worst fitness and initialize a new search from the extreme positions of the search space. Feature subset ranking was introduced in [95]. The idea was to compute the significance of each feature according to its classification accuracy and compute the accuracy for some combinations of these ranks, then the BPSO wrapper approach was used to search on the top-ranked features subsets instead of the whole features.

A Gaussian operator was introduced in [96] and the idea was that FS is highly influenced by features interaction. The highly relevant features with a class label may have high interactions with other features which makes them redundant. On the other hand, irrelevant features concerning a class label may have small interactions with other features. As feature interaction is a challenge to classification and FS, a statistical clustering method based on Gaussian distribution was adopted. It groups homogeneous features based on the interactions between features then the PSO algorithm selects one feature from each cluster. Threshold was adopted in [97]. The idea was to set a nonzero value for a threshold based on the number of trails BPSO were run. The significance of a particular dimension is measured based on the frequency of appearance for that dimension in the gbest vector in all runs. The final gbest after thresholding will contain the most recurrent features.

Zhang in [91], used the Gaussian sampling to compute the positions of particles which is based on pbest and gbest instead of velocity. Another operator was incorporated, called reinforced memory. Reinforced memory is based on the idea of enhancing the probability of survival for outstanding genes. These are the important features with high fitness value in the current iteration. Consequently, the update of the local leaders (pbest) of each particle will avoid the gene degradation and preserve it in the next iteration. Hamming distance was used in [98] to replace the Euclidean distance in BPSO. Particularly, it was used to measure the distance between two binary vectors based on the Exclusively-OR (XOR) operator and count the number of ones in the resulting vector. In [99], a new model called Hybrid Particle Swarm Optimization Local Search (HPSO-LS) was proposed based on using local search with correlation information. The correlation information was used to guide the local search in PSO. This was carried out by including the most dissimilar features (low correlated) as a feature subset in the newly generated particles. Consequently, similar features (highly correlated) have less chance to be selected as a feature subset. Moreover, HPSO-LS used a specific subset size determination scheme to allow PSO to search within the abounded region and find a smaller number of features.

Binary quantum was used in [100] to modify and improve the PSO. The idea was to perform a sampling around the personal best and compute the mean best of the sampled points then introduce this value in the BQPSO. For any bit position of the mean best, it will be equal to 1 if 1 appears more often than 0 in all the corresponding bit positions of all pbests. On the other hand, if the 1 and 0 have the same frequencies, then each element of the mbest is set randomly either to 0 or 1. A re-initialization strategy was applied on PSO-mGA in [101]. The idea was to use a small population (3–6 chromosomes) with a reinitialization strategy to achieve convergence. A non replaceable memory operator was added to keep the original swarm and remains intact with it during the optimization process. This will help in increasing the diversity of a swarm. Moreover, the nonreplaceable memory was used for maintaining a secondary swarm with a leader and followers. Zhang in [102] developed a new wrapper-based approach by utilizing the Firefly Algorithm (FA), Return-cost, Pareto dominance-based, and adaptive movement operator. A return-cost indicator was used to compute attractiveness. The firefly is cloned based on the return cost instead of the distance so that the firefly with a big return and small cost has a great chance to be cloned. A pareto dominance-based operator was added. Pareto dominance is commonly used in multi-objective optimization. It is a selection strategy used to search for the attractive one of a firefly based on the cost and return. Adaptive jump was used in place of the fixed uniform jump. It requires a change in the jump probability based on a linear function concerning the number of iterations to allow for more exploration.

In [103], a greedy search was used to enhance the local search. Three modified versions of the Lion Algorithm (LA) (Lion M1, Lion M2, and Lion M1+M2) were proposed to improve the local search. Mafarja, in [72], applied a new methodology based on BGOA and Evolutionary Population Dynamics operator (EPD). EPD depends on making a local change in the population instead of external force. This idea comes from the theory of Self-organized Criticality (SOC). Hancer, in [26], developed a new version of the DisABC algorithm for FS by introducing a DE-based neighborhood mechanism into the similarity-based search of DisABC. DE evolutionary operators were also used in [5] for solving the problem of local optima in native WOA. These include mutation, crossover, and selection operators. Khushaba in [74] developed a new modified FS method called DEFS using a repair mechanism. The repair mechanism was based on feature distribution measures and the RWS structure. A new model was developed in [104] based on GA and m-features (OR operator). The OR operator performed a search space reduction and improved GA performance and convergence. Zeng in [105] developed a novel GA with a new population structure and a new operator called dynamic neighboring. Dynamic neighboring is a new selection strategy that was used to boost the capabilities of GA for the FS problem. In [106], Guo proposed a new repair operator that allowed GA to transform feature subsets from arbitrary combinations to valid combinations that conform to the feature model constraints and domain-specific objective function.

5.2. Hybridization

Hybridization means the integration of over one algorithm to build a powerful predictive framework that combines the power of the integrated algorithms. The expectation of combining the complementary features of different optimization strategies is to achieve a better performance compared with implementing them separately as pure paradigms.

There are several categories of NIAs hybridization techniques that were investigated in the literature such as combining NIA with other NIA or combining NIA with other algorithmic components from different areas of optimizations, such as with tree search, dynamic programming, and constraint programming [107].

5.2.1. NIA-NIA Hybridization

In mimetic models, a single solution algorithm is embedded in the population's structure algorithm to enhance the local search and exploitation of the search space. These algorithms are implemented in two search stages. In the first stage, the algorithm captures a global view of the search space. In the second stage, the algorithm focuses on the most promising area to perform a successive process of local search. As exploration/exploitation balance is guaranteed using these models and the premature conversion is avoided. In [4], Zawbaa developed a novel hybrid GWO-ALO system that exploits the GWO global search ability and Ant Lion Optimization algorithm (ALO) local search performance. In [65], Mafarja developed a hybrid model based on BALO and hill-climbing techniques called HBALO. A new hybrid algorithm was presented in [108] by combining the Clonal Selection Algorithm (CSA) with the Flower Pollination Algorithm (FPA). CSA was good in exploitation, while FPA was good in exploration via Levy flight. In [109], the Mine Blast Algorithm (MBA) was used to support the exploration phase. MBA was integrated with simulated annealing to optimize a local search in the exploitation phase to get closer to the optimal solutions. Ibrahim in [110] designed a hybrid SSA-PSO model. He integrated the update strategy of PSO into the structure of SSA so that the update for the current population was done by using either the SSA or PSO depending on the quality of the fitness function. PSO-mGA (micro Genetic Algorithm) model was presented in [101]. The ACO-DE model was developed in [23]. A novel SA-MFO model was presented by Sayed in [111]. The use of SA was to make the conversion rate slower, to reach to the global optima, and escape the local minima. A new MFO-based hybrid model was developed in [112] by combining MFO and Levy FA (LFA) algorithms. The other target of NIA-NIA hybridization is to refine the best solutions by implementing the NIAs sequentially as a pipeline where the operators of the first algorithm applied first then the operators of the other integrated algorithms are applied sequentially. These models often suffer from being slow in the search process. This hybridization strategy was applied in [113] to develop the PSO-GA model. In [46], the WOA-SA model was developed. In WOASA-1 (Low-Level Team-work Hybrid (LTH)) SA was used as an operator in WOA to enhance the exploitation. In WOASA-2 (High-Level Relay Hybrid (HRH)) SA was used after WOA to enhance the final solution. In 2020 [114], SA was hybridized with the HHO algorithm and AND and OR bitwise operations. SA was used to flee the HHO optimizer from local minima in the feature search space. A new hybrid binary version of the Bat Algorithm (BA) is suggested to solve feature selection problems. In [115], BA was hybridized with an enhanced version of the DE algorithm to reach the global solution. Hybridizing different NIAs to perform parallel exploration for the search space was also a primary target for other studies. Each algorithm generates its initial population and iteratively explores and evaluates the feature subsets. Using this strategy increases the speed of the search process. ACO-GA is an example of these hybrid models [116,117]. Recently, in [118], an enhanced hybrid approach using GWO and WOA was proposed to alleviate the drawbacks of both algorithms.

Another target for NIA-NIA hybridization is to enhance the initialization of the search using different NIAs. In these models, one algorithm is used to generate the initial solutions. Then the other combined algorithm is used to update these solutions. An example of these models is GA-IGWO presented in [119]. In [120], the hybridization of two Immune Firefly Algorithms (IFA1 and IFA2) was proposed. In IFA1, the FFA and Artificial Immune System (AIS) are used simultaneously to increase the global search of fireflies and select the best feature subset. IFA2 was used to study the influence the initial population on the searching progress of the AIS algorithm.

5.2.2. NIA-Classifier Hybridization

Hybridizing different classifiers such as SVM, Artificial Neural Network (ANN), aided Radial Basis Function (RBF), Optimum Path Forest (OPF), bagging, and Bayesian statistical with NIA for evaluating the solutions. Since classifiers have different capabilities regarding the training speed, computation complexity, and generalization capability; many studies investigated their influence when used in the wrappers framework. Other studies tried to make simultaneous FS and parameter optimization to enhance the performance of a classifier. NIA in these hybrid models works as a tuner to optimize the training parameters set up and select the optimal feature subset. In [121], a new wrapper approach was built to perform parallel FS and optimization for SVM parameters by exploiting the merits of MVO. Another hybrid model was presented in [122] for optimizing the SVM parameters simultaneously with selecting the best feature subsets using a GOA optimizer.

5.2.3. NIA-Filter (Wrapper-Filter) Hybridization

The filter-wrapper hybrid model is applied in two ways. First, a filter is applied to eliminate redundant and irrelevant features, minimize the dimensionality, and produce a reduced data set that is ready to be used by a wrapper. The second way to apply the filter-wrapper model is to use the filter in the structure of a wrapper to evaluate the generated features subsets. In [123], the Information gain and correlation-based were integrated with BPSO in models called IG-IBPSO and CB-IBPSO, respectively to solve FS. In [17], a MSPSO-F-score was developed. A mutual information filter was integrated with PSO and presented as a model called MI-PSO in [124]. PSO-MI and PSO-Intropy were developed in [125]. CS-MI was developed in [126]. BALO with QuickReduct and CEBARKCC filtering approaches were developed in [65]. In [5], IWOA-IG was developed. The ACO-MI model was presented in [127]. ACO with the multivariate filter was presented in [16]. The GA-MI model was presented in [128], GA-IG in [18,129], and GA-entropy in [130]. In [131], Relief-f was used with DE to rank the most significant features. Lately, in [132], an Embedded Chaotic Whale Survival Algorithm (ECWSA) has been proposed as a wrapper process and a filter method. In [133], an efficient hybrid model based on a combining filter and evolutionary wrapper approach was proposed for sentiment analysis of various topics on Twitter. The classification system was based on a SVM classifier and two FS methods using the ReliefF and MVO algorithms. Authors in [134] proposed a filter wrapper approach using a Sequential Floating Forward Search (SFFS) to acquire features for activity recognition. The model was validated using a benchmark dataset with a multiclass Support Vector Machine (SVM). The results show that the system is affected even with limited hardware resources.

5.3. Update Mechanism

The update modification aims to achieve a balance in exploration/exploitation processes. The update strategy is performed by either enhancing the update process of individuals or dynamically control the NIA parameters. A new variant of ACO was presented in [25,135]. The update strategy used performance and the number of selected features as heuristic information for ACO with no need for prior information about features. In [136], the gbest was updated based on some conditions. This strategy determines when to reset the gbest based on several epochs (iterations) in which the value of the gbest did not change. The same strategy was applied in [24,90]. Martinez, in [137], claimed that the initialization procedure and the update of all particles are not beneficial in high dimensional space. Hence, only a small subset of particles is randomly selected to be updated. The update for a particle is carried out by filling it with active features from the current particle, local best, and global best. This strategy was applied to the original PSO to get a new variant called CuPSO.

In [138], a new rule to update particle's positions was proposed. Instead of the original rule in BPSO that lies in giving equal probabilities to either selecting or not selecting a feature. $P(x_i^d(t) = 0) = P(x_i^d(t) = 1) = 0.5$ where $x_i^d(t)$ is the gene in the d dimension of the position vector at iteration t. The new rule was introduced to increase the probability of $x_i^d(t+1) = 0$ and reduce the probability of $x_i^d(t+1) = 1$. The idea in [139] is that pbest is usually updated based on the fitness value. However, if the new position has the same fitness value as the current pbest, then the pbest will not be updated even if the new solution corresponds to a smaller feature subset. This is a limitation of PSO. The proposed PSO was to update pbest and gbest into two stages where the priority is given first for the classification accuracy. Next, if the new particle position has the same performance as the current pbest but the number of features is smaller, then in this case, pbest will be updated and replaced by the new position.

In [96], the objective was to update PSO based on a clustering approach. The new GPSO uses Gaussian distribution. The idea was to group homogeneous features based on interactions between features, then PSO is used to select one representative feature from each cluster. Mafarja in [140] proposed five update strategies for the inertia weight (w) parameter. Linear, non-linear, coefficient, decreasing, oscillating, and logarithm were applied. His idea was based on applying an exploration operator more than exploitation at the beginning of the search then search those regions carefully to find the global optima. The conclusion was that the gradual decrease for the inertia weight (w) either linearly or non linearly improves BPSO. Mafarja in [141] studied the influence of the inertia weight (w) parameter on the performance of BPSO. He suggested the adaptive change for the exploration and exploitation by using a rank-based for updating the inertia weight (w) parameter. The same author presented in [83] the time-varying update strategy to improve the performance of the DA optimizer. In [142], Aljarah applied several asynchronous update strategies to solve the FS problem. An adaptive update strategy based on a descending linear function was used to update the SSA c1 parameter.

Recently in [143], a Binary DA (BDA) was proposed with new mechanisms to update its main coefficients. The main target is to apply the survival-of-the-fittest principle using different functions such as linear, quadratic, and sinusoidal. Three variants of BDA were introduced and compared with the standard DA. The new variants are linear-BDA, quadratic-BDA, and sinusoidal-BDA. Recently, in [144], a time-varying number of leaders and followers in a binary SSA (TVBSSA) with Random Weight Network (RWN) was proposed. In 2020, the CSA algorithm was enhanced in [145] using three enhancement strategies to solve the FS problem: Adaptive awareness probability to balance exploration and exploitation, dynamic local neighborhood to improve local search, and proposing a global search strategy to increase the global exploration of the crow.

In [146], an enhanced Binary Global Harmony Search algorithm, called IBGHS, was proposed to solve FS problems. An improved step is proposed to enhance the global search ability. In [147], a new update strategy based on ranking of the individuals was proposed. Each moth in the MFO algorithm is given a rank based on its fitness value. Therefore, a moth with a small fitness value will have a high rank so that there will be a great change in its position. On the other hand, a moth with a high fitness value will have a small rank so that there will be a small change in its position. This adaptive update strategy enhanced the performance of the optimizer. In [148], a time varying flame strategy was proposed to enhance the MFO algorithm. The number of flames represents the number of the best solution that decreases gradually across iterations. Different mathematical formulas were experimented with to decide the best formula that ensures exploitation around the best solution in the late stages.

5.4. Modified Population Structure

Zeng in [105] developed a novel GA with a dynamic chain-like agent population structure. CAGA aimed to enhance the population structure and diversity. This was better than the lattice-like agent population structure where agents do genetic operations just with neighboring agents. In [101], Mistry used a new population structure for PSO-mGA. He used a small-population secondary swarm strategy. A secondary swarm performs a collaborative role to avoid stagnation and overcome premature convergence.

5.5. Different Encoding Scheme

Galbally in [149], tried to minimize the verification error rate in the online signature system. Different encoding schemes were used, including binary and integer coding. GA with binary coding was used to search the complete search space. On the other hand, GA with integer coding was used for searching a subset of the search space. GA with an optimized descriptor weight or/and optimal descriptor subset was developed in [150] over MPEG-7. There were three different encoding schemes: A real-coded chromosome for weight optimization, binary-coded chromosome for the selection of optimal feature descriptor subset, and bi-coded chromosomes for simultaneous weight optimization and optimal feature descriptor selection. A new ensemble classifier was proposed in [151]. It was based on AdaBoost learning and parallel GA. A hybrid model parallel-GA-AdaBoost with different encoding schemes BGAFS and BCGAFS was proposed.

5.6. New Initialization

In [53], authors developed a hybrid model based on BPSO to solve the FS problem. A new initialization strategy called Opposition-based Learning (OBL) was proposed. The OBL strategy was used to enhance the initialization of particles and enforce diversity among solutions by considering the solution as well as its opposite solution simultaneously. OBL was used also to generate the opposite position of the gbest particle to get rid of the stagnation case. A novel framework based on IGWO and Kernel Extreme Learning Machine (KELM) was developed in [119]. In the GA-IGWO-KELM model, GA was applied first to generate high quality and diversified initial positions, then GWO was used to update the positions of the individuals in the discrete search space. Tubishat, in [5] developed a hybrid model called IWOA-SVM-IG. The OBL strategy was applied for increasing the level of diversity in the initial solutions generated by standard WOA. In [152], a quasi-oppositional learning-based Multi-Verse Optimization (MVO) algorithm was used to improve the initial setting up of solutions.

5.7. New Fitness Function

Chakraborty [153], proposed the PSO algorithm where the fitness evaluation of each particle is based on ambiguity. The new fuzzy evaluation function was used to measure the fuzziness of a fuzzy set. The best feature was represented with minimum intraclass ambiguity as well as maximum interclass ambiguity. In [154], GA was proposed with Fisher's Linear Discriminant function in a model called GA-FLD. The new evaluation function estimates the probability distribution of the class in the N-dimensional feature space. It uses also the cardinality of the feature subset using covariance matrices which is an extension of FLD. This method was used to measure the statistical proprieties of the feature subset. Authors in [53] developed BPSO with a new fitness function based on dynamic inertia weight. High inertia weights are assigned to particles with low fitness values to facilitate more exploration of the search space. Low inertia weights are assigned to particles with high fitness to facilitate more exploitation. In [6], GWO was modified using several fitness functions. The fitness functions were accuracy, Hausdorff distance, Jeffries–Matusita (JM) distance, the weighted sum of the accuracy and Hausdorff, and the weighted sum of the accuracy and JM. In [155], different fitness functions were used to enhance the performance of the MFO algorithm. The best fitness function was the one that was applied across two-stages. The first stage optimizes the classification performance only while the second stage takes into consideration the number of genes. The results show that the proposed fitness functions can achieve better classification results compared with the fitness function that takes into account only the classification performance.

5.8. Multi Objective

Zio [156], developed a system for nuclear plants based on GA to select among the several measured plant parameters. The first approach applied was single objective GA with fuzzy k-Nearest Neighbor classifier (KNN) then multi-objective approaches were applied. Mandal, in [157] developed a prediction system based on a multi-objective PSO that satisfies the Pareto front and makes a trade-off between the non-dominated solutions based on different objectives. The proposed multi-objective PSO FS algorithm performed a dual-task where the first objective was maximizing the mutual information between a feature and class label (relevance) and the second objective was minimizing the mutual information among the features (redundancy). A Dynamic Locality Multi-Objective SSA for FS was proposed in [158]. In [159], a multi-objective FS method was proposed based on bacterial foraging optimization. In [160], a multi-objective PSO modified by Levy Flight was proposed for intrusion detection in Internet of Things (IoT). RWS mechanism was used to remove redundant features and information exchange mechanisms to avoid local minima. A systematic review of the multi-objective FS problem that covered the related studies in the period (2012, 2019) was introduced in [161].

5.9. Parallelism

In [162], Punch applied a wrapper FS based on GA to biological datasets. 5KNN was modified to work on weighted features (multiplied by weights according to their importance). The new approach was applied to a parallel distributed machine (Sparc and HP). A new ensemble classifier was proposed in [151]. It was based on AdaBoost learning and parallel GA. A parallel version of GA was applied on 16 processors with a master-slave paradigm and KNN was used as a base classifier. Ghamisi in [163] applied the parallelism strategy on PSO. Darwinian PSO (DPSO) was based on running many PSO algorithms simultaneously. Each algorithm runs as a different swarm on the same problem. A natural selection process was applied by rewarding the swarm that got better results and extending its particles' life so that new descendants were spawned. On the other hand, the swarm with suboptimal results (stagnate) was punched so its search area was discarded and its life was reduced by deleting its particles.

6. NIAs FS Applications

This section provides an extensive discussion on the use of modified NIA algorithms in different applications.

6.1. Microarray Gene Expression Classification

In [164], a hybrid model of GA and SVM was developed to perform FS and kernel parameter optimization. GA-SVM is a recommended approach for FS especially when the kernel parameters are optimized and the number of selected features is not known beforehand. Huang, in [128], developed a new GA-based wrapper approach. He adopted two stages of optimizations. The outer optimization stage (global search) applied a fitness function based on mutual information between actual classes and predicted classes. The second stage (the inner optimization) implements a local search (filter manner) based on feature ranking. A gene selection approach based on ACO was developed in [165]. A high-dimensional multi-class cancer gene expression (GCM) and colon cancer data sets were used. The comparisons were conducted with several rank-based models. The simulated results proved the validity of the proposed ACO approach for FS in high dimensional data sets.

A reliable FS technique was developed in [136] for selecting relevant features in the gene expression data set. The proposed methodology was IBPSO-KNN. The results of the accuracy increased by 2.85% compared with other methods in the literature. Yang [92], presented a new modified model for BPSO and applied it over six multi-category cancer-related human gene expression data set. Yang [18] developed a hybrid filter wrapper method for FS in microarray data sets using GA and IG. The ranking of features was performed using a decision tree. Experiments showed that the IG-GA algorithm simplified the number of gene expression levels and either achieved higher accuracy or used fewer features compared to other methods. A hybrid filter-wrapper model based on Information Gain (IG), Correlation-based (CFS), and IBPSO was proposed in [123]. Kabir [166] developed a new hybrid model based on GA, NN, MI, and local search operators. A new PSO model that has the capability of discovering biomarkers from microarray data was designed in [137].

Chuang [52] developed a hybrid model for FS and classification of large-dimensional microarray data sets. Mohammad [138] developed a diagnostic medical model based on IBPSO to find the least possible number of discriminative genes. One year later, Kabir developed an ACO-based FS model in [167]. The ACOFS target was to select the salient features with the smallest size. The model combined the ACO, neural network, filter, and included an update for the rules-based on subset size determination scheme. In [24], the PSO variant was superior to other methods in terms of performance, a number of features, and cost.

A new filter-based approach based on the CS optimizer, mutual Information filter (MI), entropy, and Artificial Neural Network (ANN) classifier was proposed in [126]. The entropy and mutual information were applied in the fitness function to calculate the relevance and redundancy for the feature subsets. Banka developed a new modified version of the PSO algorithm in [98]. Three benchmark data sets were used for colon cancer, defused B-cell lymphoma, and leukemia. The model achieved a minimal number of features and a higher classification accuracy. In [100], a model for cancer gene selection and cancer classification was developed based on BQPSO and SVM with LOOCV. Five DNA microarray data sets were used. Experiments showed better results for BQPSO/SVM compared with BPSO/SVM and GA/SVM in terms of accuracy, robustness, and the number of genes selected. Zawbaa [4], handled the complexity of the FS problem in data sets with large dimensionality and few numbers of instances by developing a novel hybrid system called GWO-ALO. A total of 27 different microarray and image processing data sets were used. Some of the data sets were very complex with 50,000 features and less than 200 instances. The experiments showed promising results when compared with GA and PSO. Ibrahim, in [168], developed a novel wrapper approach based on combining SVM with the GOA optimizer, and then he applied the hybrid model on three biomedical data sets from Iraqi cancer patients and UCI.

6.2. Facial Expression Recognition

A new modified ACO-based FS approach without a need for prior knowledge about features was presented in [25]. The experiments were applied to an ORL gray-scale face image database. The same author proposed after one year another ACO-based FS approach [135], which showed superior performance compared with GA-based and other ACO FS approaches. Aneesh, in [76], proposed a new face recognition technology using a modified version of BPSO, called Accelerated BPSO (ABPSO). ORL database images taken at the AT&T Laboratories and Cropped Yale B database-4 were used in the experiments. A biometric technique for Face Recognition (FR) based on BPSO was developed in [97]. Seven benchmark databases, namely, Cambridge ORL, UMIST, Extended YaleB, CMUPIE, Color FERET, FEI, and HP were used in the experiments.

Zhang [112] developed a facial recognition system based on the MFO-LFA-SA hybrid model to avoid premature stagnation and to guide the search procedure towards global optima. MFO logarithmic spiral search behavior increased the exploitation power meanwhile the LFA used the attractiveness function for more exploration in the search space. The SA empowered the exploitation around the most promising solution. Experiments used frontal-view images extracted from CK+JAFFE, MMI, and BU-3DFE. MFO-LFA FS outperformed other facial expression recognition models. Mistry [101] incorporated several update mechanisms in one model including the hybridization of a PSO- and mGA- (micro Genetic Algorithm), modified population structure, new velocity update strategy, diversity maintenance strategy, and a sub-dimension-based regional facial feature search strategy. Cross-domain images from the extended Cohn Kanade and MMI benchmark databases were used in the experiments besides multiple classifiers including NN with back-propagation, a multi-class SVM, and ensemble classifiers.

In [169], a system for Facial Emotion Recognition (FER) was developed based on GWO-NN. The hybridization was used to tune the weights with less training error, then it classified the emotions from the selected features. The proposed FER system was evaluated using the JAFFE and Cohn–Kanade database and the results showed higher accuracy compared with conventional methods.

6.3. Medical Applications

A new recognition system for skin tumor diagnosis was developed by handels in [170]. A GA algorithm was used to extract the most suitable features from 2D images that characterize the structure of the skin surface. NN with back-propagation was used as a learning paradigm that was trained using the selected feature sets. Different network topologies and parameter settings were investigated for optimization purposes and GA was compared with heuristic greedy algorithms. The GA skin tumor achieved the highest classification performance of 97.7%. An optimized mass detection system for digitized mammograms was developed by Zheng [171]. A GA-BBN hybrid model was used to classify positive and negative regions for masses depicted in digitized mammograms. The results showed that GA achieved the same ratio of feature reduction in comparison with the exhaustive search but reduced the total computation time by a factor of 65. In [113], a hybrid PSO-GA FS system was developed to improve the cancer classification performance and reduce the cost of medical diagnoses. Chakraborty [153] proposed a modified version of PSO using a new fuzzy evaluation.

In [172], different hybridization models were developed using the GA algorithm with different neural classifiers to get the best feature subset while preserving accuracy. A comparison was conducted between GA-KNN, GA-BP-NN, GA-RBF-NN, and GA-LQV-NN. The results showed that GA with neural classifiers were more robust and effective. In [173], Babaoglu investigated the effectiveness of both BPSO and GA as FS models for determining the existence of Coronary Artery Disease (CAD). BPSO-SVM and GA-SVM were applied on a data set obtained from patients who had performed Exercise Stress Testing (EST) and coronary angiography. The results showed that the BPSO-FS method was more successful than GA-FS and SVM on determining CAD. An automatic breast cancer diagnosis framework was designed by Ahmad [174]. The developed hybrid Genetic Algorithm Multilayer Perceptron (GA-MLP) model performed simultaneous FS and parameter optimization of ANN.

Three different variations of the backpropagation training algorithm, namely the resilient backpropagation (GAANN-RP), Levenberg Marquardt (GAANN-LM), and Gradient Descent with momentum (GAANN-GD) were investigated. The Wisconsin Breast Cancer Database (WBCD) was used. The experiments showed that the best accuracy was achieved by the RP. Sheikhpour developed a hybrid model to distinguish between benign and malignant breast tumors [175].

PSO-KDE was used to minimize the kernel density estimation error and avoid the time needed by the surgical biopsy. The Wisconsin Breast Cancer Data set (WBCD) and Wisconsin Diagnosis Breast Cancer Database (WDBC) were used. Sayed [176] developed an automatic system based on MFO for Alzheimer's Disease (AD) diagnosis. It was able to distinguish three kinds of classes including Normal, AD, and Cognitive Impairment. A benchmark data set consisted of 20 patients from the National Alzheimer's Coordinating Center (NACC). Experiments showed that the SVM-polynomial kernel function was the best one in terms of accuracy precision, recall, and f-score. A novel medical diagnosis framework based on IGWO and KELM was developed in [119].

The model was investigated on Parkinson's and breast cancer disease data sets. The comparison was performed between IGWO-KELM, GWO-KELM, and GA-KELM. The experimental results proved that the proposed method was better than the other two competitive counterparts. One year later, Sayed developed a new approach for mitosis detection in breast cancer histopathology slide images based on the MFO FS algorithm [177]. MFO was used to extract the best discriminating features of mitosis cells such as statistical, shape, texture, and energy then the selected features were used to feed the Classification and Regression Tree (CART) to make classification into either mitosis and non-mitosis.

Wang [3] developed an efficient medical diagnosis tool based on CMFO and KELM to minimize the number of features and to perform parameters optimization for KELM.

6.4. Handwritten Letter Recognition

The target in [178] was to study which one of the machine learning algorithms had the right bias to solve specific natural language processing tasks. GA achieved the best results on a language processing WSD data set. In [89], authors developed a hybrid GA to mitigate the weakness of standard GA in fine-tuning near the local minima. The proposed approach was validated using a data set gained by extracting the gray-mesh features from the CENPARMI handwritten numeral samples. Galbally [149] tried to find a way to minimize the verification error rate in the online signature verification system. A GA-based approach with new modification was proposed. Experiments were conducted on the MCYT signature database with 330 users and 16,500 signatures. The new approach showed remarkable performance in all the carried out experiments. Zeng [105] developed a novel GA with a dynamic chain-like agent population structure and dynamic neighboring competitive selection strategy. He used a letter-recognition database from UC Irvine (UCI). The experimental results showed that the feature subset generated from CAGA achieved a higher classification rate, more stability, and lower classification complexity in comparison with the other four GAs. A novel FS algorithm based on ACO was presented in [179] to improve the performance of the algorithm in text categorization. Comparisons were conducted with GA, information gain, and Chi Square test (CHI) on the Reuters-21578 data set. The proposed approach proved its superiority concerning the Reuters-21578 data set. In [129], Principal Component Analysis (PCA) was used with the IG filter method and GA optimizer in a model called IG-GA-PCA. In the first stage, the IG method was applied to rank the terms of the document according to their importance. In the second stage, GA and PCA FS and feature extraction methods were applied separately to the ranked terms. Experiments used both Reuters-21578 and Classic3 data sets. The experiments showed that the IG-GA-PCA model could achieve high categorization results as measured by precision, recall, and F-measure. In [154], a GA-FLD-based FS approach was used in order to find features subsets that could optimally discriminate samples from different classes without prior knowledge about features dimensionality. Another modification based on fitness function were also proposed. Three standard databases of handwritten digits and one of handwritten letters were used in the experiments. In [53], authors developed a hybrid intelligent algorithm using BPSO and other operators to solve the FS problem in the text clustering. A new initialization strategy, new fitness function, and new operator were proposed. The Reuters-21578, Classic4, and WebKB benchmark text data sets were used. The results showed higher clustering accuracy and improved the convergence speed of BPSO. Ewees, in [180] introduced a new approach for Arabic handwritten letter recognition (AHLR) called MFO-AHLR. A data set for Arabic handwritten letter images (CENPARMI) was used. Results showed that MFO-AHLR achieved a 99.25% accuracy, which was the highest ratio achieved among all AHLR approaches. Tubishat, in [5], developed a novel hybrid model for Arabic SA. The targets of the study were to mitigate the limitations of the WOA such as local minima, slow convergence diversity, and over-fitting problems. A hybrid model IWOA-SVM-IG was applied over four Arabic benchmark data sets for sentiment analysis. IWOA was compared with six well-known optimization algorithms and two deep learning algorithms, namely Convolution NN (CNN) and Long Short-term Memory (LSTM). The results showed that the IWOA algorithm outperformed all other algorithms.

6.5. Hyper Spectral Images Processing

Tackett in [181] worked on extracting the statistical features from a large noisy US Army NVEOD Terrain Board imagery database using GP. In [182], a new model was proposed based on GA, Bayesian classification, and a new proposed fitness function to discriminate the targets from clutters in SAR images. Jarvis [183] developed a novel approach based on GA and DFA for the selection of important discriminatory variables from Fourier Transform Infrared (FT-IR) spectroscopic data. The GA achieved 16% reduction in the model error. The GA-SVM model for hyper-spectral data classification was proposed in [184]. The proposed GA-SVM was tested on an HYPERION hyper-spectral image. Experiments demonstrated that the number of bands was reduced from 198 to 13, while accuracy increased from 88.81% to 92.51%. A GA-based image annotation system with optimized descriptor weights or/and optimal descriptor subset over MPEG-7 was developed in [150]. The Corel image database consisted of 2000 images with 20 categories used. Experiments showed that the binary-coded GA and the bi-coded GA improved the accuracy of the image annotation system by 7%, 9%, and 13.6%, respectively compared to the commonly used methods.

A new ensemble classifier was proposed in [151]. It was based on AdaBoost and parallel GA in the context of the FS problem for image annotation in MPEG-7 standard. The experiments were performed over 2000 classified Corel images. In [185], a new approach based on GA, SVM, MI, and BB was developed to search for the best combination of bands in the hyper spectra images. MI was used as a pre-processing step for band grouping based on the correlation between bands and classes. GA-SVM was used to search for the optimal combinations of bands that increase accuracy. A post-processing step based on BB was used to filter out those irrelevant band groups. Ghamisi [163] applied the FODPSO SVM approach to determine the most informative bands in the Hekla and Indian Pines hyper-spectral data set using the parallelism modification technique. In the same year, Ghamisi [186] presented a new hybrid approach based on GA, PSO, and SVM. His target was to detect roads from a background in complex urban images. He integrated the standard velocity and update rules of PSO with selection, crossover, and mutation from GA. In [6], Medjahed developed a novel GWO framework for Pavia and AVIRIS hyper-spectral images data sets.

6.6. Protein and Related Genome Annotation

In [116], a new FS model based on ACO-GA was proposed. Both ACO and GA generated the feature subsets in parallel then the generated subsets were evaluated by a certain fitness function. ACO used GA operators to update the solutions. The GPCR-PROSITE dataset and ENZYME-PROSITE challenging protein sequences data sets were used. Mandal [157], developed a prediction system to identify the possible subcellular location of a protein-based on a multi-objective PSO.

6.7. Biochemistry and Drug Design

Raymer [187] developed a system that integrates FE, FS, and classifier training using GA and KNN. This approach was applied in biochemistry and drug design for the identification of favorable water-binding sites on protein surfaces. The approach was validated using protein water interactions from a biochemistry field. Another model was developed by Salcedo [104]. The proposed FS model was based on GA and m-features operator (OR operator). The new approach was evaluated using two machine learning classification problems; the first one used two artificial data sets and the second one was a real application in molecular bioactivity for a drug design taken from the ones used in the KDD Cup. THe m-features operator improved the GA performance over the other existing approaches.

6.8. Electroencephalogram (EEG) Application

Palani [188] used GA and Fuzzy ARTMAP (FA) NN for FS. GA-FA-NN was used with the VEP data which was recorded from 10 alcoholics and 10 controls. The target was to classify alcoholics and controls, using multi-channel EEG signals. The discriminatory spectral bands reduced from 7 to 2. The identification of useful spectral power ratios produced better performance. In [189], a hybrid GA-SVM model was used to extract the favorable patterns from noisy multidimensional time series obtained from EEG which are a base for Brain-computer Interfaces (BCIs). The data set was collected by a procedure in which subjects were placed in a dim, sound controlled room. The proposed nonlinear system was better than other linear approaches with a slight difference. A novel ACO-DE FS system called ANTDE was presented in [23]. It could cope with the limitations of ACO regarding the sequential generation for solutions. ANTDE was used in EEG and Myoelectric Control (MEC) biosignal applications. Wang [130] developed a BCI system using a hybrid model GA-SVM-entropy and 28 EEG channels. Noori designed an effective BCI in [190]. He used a new version of GA based on SVM to get smaller optimal features from functional Near-infrared Spectroscopy signals (fNIRS). The experiments were established by recruiting seven subjects who do not have any psychological disorder. Subjects were seated in a quiet room and asked to relax to settle down their responses before beginning to perform mental arithmetic tasks for a certain period.

6.9. Financial Prediction

In [191], a new financial prediction model was proposed. A hybrid model SVM-GA was evaluated using 15 business data sets. Each data set consisted of 186 sampled firms. GA-SVM achieved a prediction accuracy of up to 95.56% for all the tested business data. In [192], the authors developed a hybrid fuzzy-GA approach for stock selection. The fuzzy-based scoring mechanism was applied for scoring a set of stocks then the topmost stocks were selected. GA applied for performing a dual job of FS and parameter optimization. The constituent stocks of the 200 largest market capitalization listed in the Taiwan Stock Exchange were used in the experiments.

6.10. Software Product Line Estimation

Oliveira [22] investigated the use of the GA method for simultaneous FS and parameters optimization of Support Vector Regression (SVR) when applied for software effort estimates. GA, SVR, MLP, and model trees were used. Six benchmark data sets of software projects, namely, Desharnais, NASA, COCOMO, Albrecht, Kemerer and Koten, and Gray were used in the experiments. In [106], Guo presented a new methodology for FS in the software application. The target of the new modified GA was optimizing FS in a Software Product Line (SPLs) to find a feature subset with an optimal product capability subject to feature model constraints and resource constraints. The results showed that GA FS algorithms produced a system with high performance and in 45–99% less time than existing heuristic FS techniques.

6.11. Spam Detection in Emails

Temitayo [193] developed a new approach for the classification of emails, either spam or legitimate. GA was used to perform simultaneous FS and parameter optimization. The hybrid GA-SVM spam detection model was evaluated using a Spam Assassin (6000 emails) data set. Experiments showed that GA-SVM improved the results compared with SVM by achieving a higher recognition rate with only a few feature subsets. In [85], a mutation-based BPSO FS model was developed in an email application. A data set of 6000 emails manually collected during the year 2012 was used. The proposed was able to effectively reduce the false-positive error. In [194], a hybrid GA-RWN was used for identifying the most relevant features in spam emails and automatic tuning for the hidden neurons. The GA-RWN achieved promising results according to the spam detection rate and optimization for the configuration of its core classifier. Lately, in [195], a novel Northern Bald Ibis Algorithm (NOA) was used with a SVM classifier to get an optimal feature subset of the Enron-spam dataset.

6.12. Other Various Applications

Zio [156] developed an efficient transient diagnosis system for nuclear power plants based on GA to select among the several measured plant parameters. In [61], the target was addressing the problem of a lengthy Intrusion Detection (ID) process based on attributes of network packets. Rough-PSO was used and evaluated using the KDDCup 1999 data set. An automatic FS model that can choose the most relevant features from password typing patterns was designed in [196]. The data sets were captured on a Sun Sparc-Station by a program in an X window environment in which the keystroke duration times were measured. Rodrigues in [197] proposed a CS-OPF model for theft detection in power distribution systems. The proposed model was evaluated using two data sets from a Brazilian electrical power company. Experiments proved the robustness of the CS-OPF model by increasing the theft recognition up to 40%. Zhang [127] developed a new forecaster FS model based on combining MI and ACO. The ACO-MI model was applied on forecasters data sets at the Australian Bureau of Meteorology. A system for diagnosing different types of fault in a gearbox was designed in [198]. Hassanien [67] developed an automatic tomato disease detection system based on integrating rough set with MFO.

6.13. An Open Source Evolopy-FS Framework

EvoloPy-FS [199] is an open-source FS software tool developed by our team and it is publicly available on (www.evo-ml.com). It serves as an explicit white-box NIAs-FS optimization framework. The main objective was to support researchers from different disciplines with an easy-to-use, transparent, and automated NIAs-FS optimization tool. The framework contains severe recent NIAs algorithms written in Python and a set of different operators such as transfer functions (S-TFs and V-TFs). Moreover, the framework applies wrappers, filters and a hybrid filter-wrapper, different evaluation metrics, and allows for loading data from different resources. Evolopy-FS is a continuation of our path, which is building an integrated optimization environment. The work was started by EvoloPy [200] for global optimization problems then EvoloPy-NN for optimizing MLP and recently Evolopy-FS for optimizing the feature selection process. In [199], authors constructed the experiments based on 30 different well-regarded data sets from common repositories such as UCI and Kaggle. The comparisons were conducted between wrapper FS, filter FS, and hybrid filter-wrapper approaches. It was shown that wrapper and hybrid filter-wrapper were superior and more trustable in dealing with large dimensionality data sets. However, the filter approach was faster and generated results in a shorter time and fewer computational efforts.

7. Assessment and Evaluation of NIAs FS Modification Techniques

As discussed, NIAs-FS approaches achieved big contributions and clear success in solving the FS problem in different domains. This section presents the results from the analysis of modified NIAs-FS studies. Table 1 shows a summary of the main studies in the literature that adopted new operators as modification techniques for NIAs-FS, Table 2 shows a summary of the main studies in the literature that adopted hybridization modification technique for NIAs-FS, Table 3 shows a summary of main studies in the literature that adopted the remaining modifications techniques for NIAs-FS, Table 4 shows a summary of main modifications applied in the literature on main NIAs (applied/not applied), Table 5 shows a summary of main modifications applied in the literature on main NIAs (by numbers), Table 6 shows a summary of the main studies in the literature that applied modified NIAs FS in applications, and Table 7 shows a summary of modifications applied on NIAa-FS in the main applications. It was observed that 34 different operators were applied on NIAs wrappers in 48 different papers. Some references adopted over one operator in their work. As it is clear also from the list, the most applied operator is the chaotic map, which was applied in 10 references, then rough set in 6 references, then selection operators (RWS, TS) in 5, then S-shaped and V-shaped transfer functions and crossover in 4 references. The mutation was applied in 3 and UC, DE, and local search operators each were been adopted by 2 references. A single reference adopted the remaining operators. It was found that the PSO wrapper was the most modified optimizer using newly adopted operators for tackling the FS problem. It was modified using a new operator in 21 references. In addition, GA was modified in 6 references, WOA in 4, CS in 3, SSA in 3, GWO, GOA, and MFO each one was modified by a new operator in 2 references. For DA,

FFA, LA, BA, MVO, ABC, CSO, DE, and CSA, the number of references was 1. For FPA and ACO, no work applies new operators to their algorithms for solving the FS problem.

It was clear from the analysis that the hybridization modification technique was applied in 75 references to solve FS. This counting result shows that hybridization is the most widely applied modification technique to enhance NIAs wrappers in the FS domain. This high number of work comes from GA, which is the NIA that had the most number of works regarding wrapper hybridization. GA wasapplied hybridization in 38 different works, which is much higher than 6, the number of works that adopted new operators to GA. We can infer from this works count and from the contribution of studies that hybridization is the best suitable modification technique to be applied with GA. ACO also were hybridized in 7 references, while no work adopted a new operator to modify ACO. Conversely, PSO hybridization works were 11, which is less than 21, the number of works with new operators, thus we can again infer from these counts and the contribution of studies that adopting a new operator to PSO is more suitable than hybridization. It was also noticed that hybridization using different kinds of classifiers was the most prominent hybridization technique.

There are 49 studies that tried to investigate the influence of the classification technique on the performance of wrappers for optimizing FS, some of these studies applied simultaneous optimization for FS and a classification/prediction task by tuning the parameters of the classifier with applying FS. The next widespread hybridization technique is a filter-wrapper, which was applied in 14 studies and was very effective in dealing with large dimensionality feature space. Hybridization techniques that tried to balance the exploration/exploitation of the search space also were adopted by a considerable number of works. In summary, PSO and GA are the most widely modified NIAs-FS approaches. They were equally modified and used. Each one of them was adopted and modified for FS in 56 references of the gathered studies.

On the other side, regarding applications of NIAs-FS, it was evident that microarray gene expression classification is the most dominant application where NIAs-FS approaches were applied in 18 studies with a ratio 24% concerning other applications. The medical application was the second prominent application for applying NIAs-FS approaches with a ratio of 21%. The medical application includes different medical branches SONAR, tumor, mass, and various disease detection, medical diagnosis, medical data, and bio-signal analysis. Then, follows hyper-spectral image with a ratio of 17%, Arabic handwritten recognition with a ratio of 13%, facial expression recognition with a ratio of 9%, EEG application with a ratio of 7%, financial diagnosis with a ratio of 5%, and spam detection with a ratio of 4%. Furthermore, it is noticeable that GA is the most dominant NIA optimizer for optimizing FS in applications with a ratio of 45%. PSO is the second most widespread optimizer with a ratio of 26%, then ACO with a ratio of 11%, MFO with a ratio of 7%, GWO with a ratio of 6%, CS with a ratio of 2%, WOA with a ratio of 2%, and GOA with a ratio of 1%.

No Free Lunch (NFL) theorem [201] states no optimization algorithm can solve all the optimization problems equally. The success in solving a specific problem does not guarantee that the algorithm will perform similarly for other problems. On average, all the optimization algorithms perform equally. This theorem has motivated researchers to develop new algorithms or improve the existing ones to solve another wide area of optimization problems, such as feature selection. Researchers are advised to read the following references as they are the most cited papers after 2010 in the field of NIAs-FS: [17,51,80,90,94,124,202].

Table 1. Summary of main ro pub the tables, and please check if it is the background color can be deleted, same as follows. studies in the literature that adopted new operators as modification techniques for Nature Inspired Algorithms Feature Selection NIAs-FS.

New Operator	NIA Wrapper and No. of Publications	References	Total NIAs
Chaotic maps	PSO(4), SSA(2), MVO(1), CSA(1) WOA(1), MFO(1)	[3,51–58]	10
Rough set	PSO(2), CS(2), FA(1), MFO(1)	[61–64,66,67]	6
Selection operators (RWS,TS)	GOA(1), WOA(2), ABC(1), DE(1)	[26,46,72–74]	5
Sigmoidal function	PSO(1), GWO(1), CS(1), BA(1), GOA(1)	[6,76–79]	5
S-shaped and V-shaped TFs	PSO(1), SSA(1), DA(1), GOA(1)	[79,80,82,83]	4
Crossover	GWO(1), SSA(1), WOA(1), GOA(1)	[73,79,82,84]	4
Mutation	PSO(2), GOA(1)	[53,79,85]	3
Uniform Combination (UC)	PSO(2)	[80,91]	2
Local search	PSO(2)	[24,90]	2
DE evolutionary operators	WOA(1), ABC(1)	[5,26]	2
Boolean algebra operation	PSO(1)	[92]	1
Logistic regression	PSO(1)	[202]	1
Catfish strategy	PSO(1)	[94]	1
Feature subset ranking	PSO(1)	[95]	1
Statistical clustering	PSO(1)	[96]	1
Threshold	PSO(1)	[97]	1
Gaussian sampling	PSO(1)	[91]	1
Reinforced memory strategy	PSO(1)	[91]	1
XOR operator	PSO(1)	[98]	1
Correlation information	PSO(1)	[99]	1
Binary quantum	PSO(1)	[100]	1
Reinitialization strategy	PSO(1)	[101]	1
Non replaceable memory	PSO(1)	[101]	1
Levy flight	CS(1)	[64]	1
Return cost indicator	FFA(1)	[102]	1
Pareto dominance based	FFA(1)	[102]	1
Movement operator and adaptive jump	FFA(1)	[102]	1
Greedy search	ALO(1)	[103]	1
Evolutionary Population Dynamics (EPD)	GOA(1)	[72]	1
DE-based neighborhood mechanism	ABC(1)	[26]	1
Repair mechanism	DE(1)	[74]	1
m-features operator (OR operator)	GA(1)	[104]	1
Dynamic neighboring genetic	GA(1)	[105]	1
Repair operator	GA(1)	[106]	1

Table 2. Summary of main studies in the literature that adopted hybridization modification technique for NIAs-FS.

Target	Type	Strategy	Model	References	Total
Enhance the exploitation	NIA-NIA (Population-trajectory mimetic) SI-EA, SI-SI)	Global search followed by local search	GWO-ALO, WOA-SA, CSA-FPA, SSA-PSO, PSO-mGA, ACO-DE, MFO-SA, MFO-LFA	[4,23,46,101,108,110–112]	8
Refine the best solutions	NIA-NIA	Implementing NIAs sequentially as a pipeline. Apply operators of 1st algorithm then apply operators of the 2nd algorithm sequentially	WOA-SA, PSO-GA	[46,113]	2
Speed up the search process	NIA-NIA	Perform parallel exploration	ACO-GA	[116,117]	2
Enhance the initialization process	NIA-NIA	Generate initial solutions by one algorithm then update them using the other algorithm	GA-IGWO	[119]	1
Improve the training process. Improve the evaluation process. Reduce the computation complexity. Investigate the capability of different classifiers. Simultaneous parameters optimization and feature reduction. Study the influence of different evaluation strategies on wrappers performance.	NIA-Classifier	For the simultaneous parameter and FS optimization, NIA works as a tuner to optimize the training parameters set up, selecting the optimal feature subset by making new representations of an individual, in such a way the length of the individual equals the number of parameters and number of features and adjusts the values of genes by either doing real or binary conversion.	PSO-KDE, GA-IGWO-KELM, GWO-NN, CS-OPF, CSA-FPA-OPF, BBA-OPF, BBA-NaiveBayes, [MVO-RF, MVO-J48, MVO-Kstar, MVO-LMT], MVO-SVM, GOA-SVM, ACO-SVM, ACO-NN, CMFO-KELM, CSO-SVM, GA-KNN, [GA-NN, GA-BNN, GA-MLP, GA-BP-NN, GA-RBF-NN, GA-LQV-NN, GAANN-RP, GAANN-LM, GAANN-GD, GA-RWN], GA-C4.5, GA-SVM, GA-Bayesian, GA-FKNN, GA-adaboost, GA-SVR,	[3,18,22,22,56,77,78,108,117, 119,121,122,128,130,151, 156,164,166,166–169,171– 175,182,185,188–191,194,196– 198,203–208]	49

Table 2. *Cont.*

Target	Type	Strategy	Model	References	Total
Minimize the dimensionality of the large datasets, eliminate the redundant/irrelevant features, evaluate the generated features subsets	NIA-Filter (Wrapper-Filter)	Classically applied in two steps: Filtering of the features applied first then a wrapper is applied on the reduced dataset. Other studies tried to embed the filter in the structure of a wrapper in order to evaluate the generated features subsets	[IG-IBPSO, CFS-IBPSO], MSPSO-F-Score, MI-PSO, [BPSO-G (Intropy-PSO), BPSO-P (MI-PSO)], CS-MI-Entropy, LA-QuickReduct-CEBARKCC, IWOA-IG, ACO-MI, ACO-Multivariate filter, GA-MI, IG-GA, GA-Entropy	[5],[6–8],[4,6.5],[2]–[30]	14

Table 3. Summary of main studies in the literature that adopted the remaining modifications techniques for NIAs-FS.

Modification	NIA Wrapper and No. Publications	References	Total Works
Update mechanism	PSO(10), SSA(1), DA(1), ACO(5)	[24,25,53,83,90,96,135–139,141,142,179, 209–211]	17
Modified population structure	PSO(1), SSA(1), GA(1)	[101,105,142]	3
Different encoding scheme	PSO(3), GA(4)	[17,90,91,105,149,150,212]	7
New initialization	PSO(2), GWO(1)	[53,84,139]	3
New fitness function	PSO(2), GWO(1), FFA(1), WOA(1), GA(4)	[5,6,106,153,154,182,183,213,214]	9
Multi objective	PSO(4), ACO(1), GA(1)	[156,157,209,215–217]	6
Parallelism	PSO(2), GA(2)	[17,151,162,163]	4

Table 4. Summary of main modifications applied in the literature on main NIAs-FS (applied/not applied).

Modification	PSO	GWO	SSA	CS	DA	FFA	LA	BA	MVO	GOA	WOA	ACO	MFO	GA
New operator	✓	✓	✓	✓	✓	✓	✓	✓	✓	✓	✓	✗	✓	✓
Hybridization	✓	✓	✓	✓	✗	✗	✓	✓	✓	✓	✓	✓	✓	✓
Update mechanism	✓	✗	✓	✗	✓	✗	✗	✗	✗	✗	✗	✓	✗	✗
Modified population structure	✓	✗	✓	✗	✗	✗	✗	✗	✗	✗	✗	✗	✗	✓
Different encoding scheme	✓	✗	✗	✗	✗	✗	✗	✗	✗	✗	✗	✗	✗	✓
New initialization	✓	✓	✗	✗	✗	✗	✗	✗	✗	✗	✗	✗	✗	✗
New fitness function	✓	✓	✗	✗	✗	✓	✗	✗	✗	✗	✓	✗	✗	✓
Multi objective	✓	✗	✗	✗	✗	✗	✗	✗	✗	✗	✗	✓	✗	✓
Parallelism	✓	✗	✗	✗	✗	✗	✗	✗	✗	✗	✗	✗	✗	✓

Table 5. Summary of main modifications applied in the literature on main NIAs-FS (by number of studies).

Modification	PSO	GWO	SSA	CS	DA	FFA	LA	BA	MVO	GOA	WOA	ACO	MFO	GA	Total Studies/Modification
New operator	21	2	3	3	1	1	1	1	1	2	4	0	2	6	48
Hybridization	11	2	1	3	0	1	1	3	2	2	2	7	3	38	75
Update mechanism	10	0	1	0	1	0	0	0	0	0	0	5	0	0	17
Modified population structure	1	0	0	0	0	0	0	0	0	0	0	0	0	1	3
Different encoding scheme	3	0	0	0	0	0	0	0	0	0	0	0	0	4	7
New initialization	2	1	0	0	0	1	0	0	0	0	1	0	0	0	3
New fitness function	2	0	0	0	0	0	0	0	0	0	0	1	0	4	9
Multi objective	4	0	0	0	0	0	0	0	0	0	0	0	0	1	6
Parallelism	2	0	0	0	0	0	0	0	0	0	0	0	0	2	4
Total studies/FS-NIA	56	6	6	6	2	2	2	4	3	4	7	13	5	56	172

Modification	PSO	GWO	SSA	CS	DA	FFA	LA	BA	MVO	GOA	WOA	ACO	MFO	GA	Total Studies/Modification
New operator	21	2	3	3	1	1	1	1	1	2	4	0	2	6	48
Hybridization	11	2	1	3	0	1	1	3	2	2	2	7	3	38	75
Update mechanism	10	0	1	0	1	0	0	0	0	0	0	5	0	0	17
Modified population structure	1	0	0	0	0	0	0	0	0	0	0	0	0	1	3
Different encoding scheme	3	0	0	0	0	0	0	0	0	0	0	0	0	4	7
New initialization	2	1	0	0	0	1	0	0	0	0	1	0	0	0	3
New fitness function	2	0	0	0	0	0	0	0	0	0	0	1	0	4	9
Multi objective	4	0	0	0	0	0	0	0	0	0	0	0	0	1	6
Parallelism	2	0	0	0	0	0	0	0	0	0	0	0	0	2	4
Total studies/FS-NIA	56	6	6	6	2	2	2	4	3	4	7	13	5	56	172

Table 6. Summary of main studies in the literature that applied modified NIAs-FS in applications.

FS Application	NIA Wrapper and No. of Publications	References	Total NIAs
Microarray gene expression classification. (DNA micro array classification)	PSO(9), GWO(1), CS(1), GOA(1), ACO(2), GA(4)	[4,18,24,52,98,100,123, 126,128,136–138,164–168]	18
Facial expression recognition	PSO(3), GWO(1), ACO(2), MFO(1)	[25,76,97,101,112,135, 169]	7
Medical applications (SONAR, tumor, mass and various disease detection, medical diagnosis, medical data, and bio signal analysis)	PSO(4), GWO(1), WOA(1), ACO(1), MFO(3), GA(6)	[3,57,62,113,119,153, 172–177,211]	16
Handwritten letter recognition, sentiment analysis, language processing, signature verification system, and text categorization	PSO(1), WOA(1), ACO(1), MFO(1), GA(6)	[5,53,89,105,129,149, 154,178–180]	10
Hyper spectral images processing and classification.	PSO(3), GWO(2), GA(9)	[4,6,150,151,163,181–186,212]	13
Intrusion detection	PSO(1)	[61]	1
Protein and related genome annotation .	PSO(1), ACO(1)	[116,157]	2
Meteorology weather forecasting	ACO(1)	[127]	1
Biological application	GA(1)	[162]	1
Biochemistry and drug design	GA(2)	[104]	2
Electroencephalogram (EEG) signals application/Brain Computer Interface (BCI) system	ACO(1), GA(2)	[23,130,188–190]	5
Design an automatic FS model that can choose the most relevant features from password typing	GA(1)	[196]	1

Table 6. Cont.

FS Application	NIA Wrapper and No. of Publications	References	Total NIAs
Transient diagnosis system for nuclear power plants	GA(1)	[156]	1
Financial diagnosis /business crisis detection/stock price prediction	PSO(1), GA(3)	[191,192,202]	4
Diagnose different types of fault in a gearbox	GA(1)	[198]	1
Software Product Line (SPLs) and sw effort estimation	GA(2)	[22,106]	2
Theft detection in power distribution systems	CS(1)	[197]	1
Spam detection in emails	PSO(1), GA(2)	[85,193,194]	3
Transient diagnosis system for nuclear power plants	GA(1)	[156]	1
Automatic tomato disease detection system based	MFO(1)	[67]	1

Table 7. Summary of modifications applied on NIA-FS in the main applications.

App	Modification	Model	Classifier	Datasets	Dimension	Year	Ref
Microarray gene expression	Hybridization	GA-SVM	SVM	2	2000	2003	[164]
	Hybridization	GA-MI	SVM	13	2000	2007	[128]
	Update mechanism	IPSO	1-KNN	11	15,009	2008	[136]
	Update mechanism	BPSO	1-KNN	6	10,509	2008	[92]
	Hybridization	IBPSO-IG, IBPSO-CFS	1-KNN	6	11,225	2008	[123]
	Update mechanism	cuPSO	1-KNN	11	15,009	2010	[137]
	Hybridization	GA-IG	1-KNN	11	15,009	2010	[18]
	Hybridization New operator/chaotic	TCBPSO-CFC	1-KNN	10	9868	2011	[52]
	Update mechanism	IBPSO	SVM	10	12,600	2011	[138]
	Hybridization	GA-NN-MI	NN-BP	11	7129	2011	[166]
	Hybridization	ACOFS	NN	9	2000	2012	[167]
	Hybridization	CS-MI-Entropy	ANN	6	15,009	2014	[126]
	Update mechanism	PSO-LSRG	1-KNN	5	12,600	2014	[24]
	New operator/XOR	HDBPSO	1-KNN, 3-KNN, 5-KNN	3	7129	2015	[98]
	New operator/binary quantum	BQPSO-SVM	SVM	5	12,600	2016	[100]
	Hybridization	GWO-ALO	5-KNN	7	49151	2018	[4]
	Hybridization	GOA-SVM	SVM	3	17,678	2018	[168]
Medical application	Hybridization	GABPNN, GARBFNN, GALQVNN	BPNN, RBFNN, LQVNN	1	30	2006	[172]
	Hybridization	PSO-GA	SVM	3	7129	2008	[113]
	New fitness function	PSO-MLP	MLP	3	16	2008	[153]
	Hybridization	BPSO-FST, GA-FST	SVM	1	23	2010	[173]
	New operator/rough set	PSO-RR, PSO-QR	Naive Bayes, BayesNet, KStar	4	45	2014	[62]
	Hybridization	GAANNRP, GAANNLM, GAANNGD,	NNRP, NNLM, NNGD	1	10	2015	[174]
	Hybridization	PSO-KDE	KDE	2	32	2016	[175]
	Hybridization	IGWO-GA-KELM	KELM	2	32	2017	[119]
	New operator/chaotic maps	CMFOFS-KELM	KELM	1	22	2017	[3]

Table 7. Cont.

App	Modification	Model	Classifier	Datasets	Dimension	Year	Ref
Hyper spectral images processing	New fitness function hybridization	GA-MDLP-Bayesian	Bayesian	3	20	2003	[82]
	New fitness function	GA-DFA	Validation by projection	1	882	2004	[83]
	Hybridization	GA-SVM	SVM	1	198	2008	[84]
	Different encoding schemes	GA-KNN	KNN	1	25	2008	[156]
	Different encoding schemes	GA-KNN	KNN	1	25	2008	[212]
	Hybridization parallelism	Parallel-GA-Adaboost	Adaboost ensemble, KNN	1	25	2010	[15]
	Different encoding schemes	BGAFS, BCGAFS	Adaboost ensemble, KNN	1	25	2010	[15]
	Hybridization	MI-GA-SVM-BB	SVM,BB	2	202	2011	[85]
	Parallelism	FODPSO-SVM	SVM	2	220	2015	[163]
	Hybridization	HGAPSO-SVM	SVM	1	220	2015	[86]
	New fitness function	GWO-KNN	7-KNN	3	224	2016	[6]
	Hybridization	GWO-ALO	5-KNN	5	10,304	2018	[4]
Arabic HR	New operator/local search	HGA	1-KNN	1	16	2004	[89]
	Different encoding schemes	GA	No classifier	1	100	2007	[149]
	Update mechanism	ACO	KNN	1	7542	2009	[179]
	New population structure	CAGA	NN-BP	1	16	2009	[105]
	Hybridization	IG-GA-PCA	KNN, C4.5	4	7542	2011	[29]
	New fitness function	GA-FLD	KNN, MLP, SVM (RBF, Poly, Sigm)	4	780	2014	[154]
	New initialization+New operator/mutation	BPSO	No classifier/clustering problem	3	8830	2016	[53]
	Hybridization new operator/DE evolutionary	IWOA-IG-SVM	SVM	4	8057	2018	[5]
Face recognition	Update mechanism	ACO	KNN	1	400	2007	[25]
	Update mechanism	ACO	KNN	1	400	2008	[135]
	New operator/intelligent acceleration	ABPSO	Euclidean classifier	2	2204	2012	[76]
	New operator/threshold	BPSO	Euclidean classifier	7	16,380	2014	[97]
	Hybridization new operator update	PSO-mGA	NN-BP,SVM-RBF,ensembles	2	1280	2017	[10]
	Hybridization	GWO-NN	NN	2	486	2018	[69]
	New operators/return-cost	Rc-BBFA	1-KNN	10	1280	2016	[112]

255

8. Conclusions and Future Research Directions

In this study, a survey about modifications of NIAs for tackling the FS optimization problem is presented. The review is based on a solid theoretical, applied, and technical foundation. Three main research streams are identified in this review: Meta-heuristic optimization, feature selection, and modification on NIAs for tackling FS. This review aims to draw the map for researchers and guide them when creating new research in this area. This survey is based on 156 articles collected and studied on modifications of NIAs for solving the FS problem. The sources of the information search came mainly from six well-regarded scientific databases: Elsevier, Springer, Hindawi, ACM, World scientific, and IEEE. From the review, it can be seen that the NIAs algorithms have been extensively investigated over the past years to improve the FS problem. About 34 different operators were investigated. The most popular operator is chaotic maps. Hybridization is the most widely used modification technique. There are three types of hybridization: Integrating NIA with another NIA, integrating NIA with a classifier, and integrating NIA with a classifier. The most widely used hybridization is the one that integrates a classifier with the NIA. Microarray and medical applications are the dominated applications where most of the NIA-FS are modified and used. Despite the popularity of the NIAs-FS, there are still many areas that need further investigation:

- Until now, there are few works in the binary optimization field. Many new operators can be proposed to enhance the performance of binary optimizers in a binary space. This is an interesting research direction;
- The proposed enhanced binary versions of optimizers can be used as a data mining tool in various applications. There are some applications where the usage of modified NIAs-FS in them is still limited;
- It would also be interesting to look at the dimensionality and number of instances in data sets. Nowadays, the majority of FS works to address problems with dimensionality up to several thousand but the question that may arise is what will happen if the data sets scaled up to millions of features? There is a scalability gap that should be addressed in the future;
- There is still room for improvement through parallel NIAs-FS. This might be a fruitful direction for research;
- Hyper volume Pareto optimal dominance and many-objective optimization need further crucial investigation.

Based on the above trends, the size of the NIAs-FS research area can be recognized. Besides, it can be imagined that a thorough investigation and improvement of NIAs will improve the FS process in various high-dimensional areas. This review paper will be used to help researchers take an excellent view of the modification strategies in nature-inspired algorithms for tackling the feature selection problem.

Author Contributions: Conceptualization, R.A.K., I.A. and A.S.; methodology, R.A.K., I.A. and A.S.; formal analysis, R.A.K., I.A. and A.S.; resources, R.A.K.; validation, M.A.E., R.D. and T.K.; writing—original draft preparation, R.A.K., I.A. and A.S.; writing—review and editing, M.A.E., R.D. and T.K.; supervision, M.A.E.; funding acquisition, T.K. All authors read and agreed to the published version of the manuscript.

Funding: No funding received for this research.

Institutional Review Board Statement: Not applicable.

Informed Consent Statement: Not applicable.

Data Availability Statement: Not applicable.

Acknowledgments: Not applicable.

Conflicts of Interest: All authors declare that they have no conflict of interest.

References

1. Liu, H.; Yu, L. Toward integrating feature selection algorithms for classification and clustering. *IEEE Trans. Knowl. Data Eng.* **2005**, *17*, 491–502.
2. Hawkins, D.M. The problem of overfitting. *J. Chem. Inf. Comput. Sci.* **2004**, *44*, 1–12.
3. Wang, M.; Chen, H.; Yang, B.; Zhao, X.; Hu, L.; Cai, Z.; Huang, H.; Tong, C. Toward an optimal kernel extreme learning machine using a chaotic moth-flame optimization strategy with applications in medical diagnoses. *Neurocomputing* **2017**, *267*, 69–84.
4. Zawbaa, H.M.; Emary, E.; Grosan, C.; Snasel, V. Large-dimensionality small-instance set feature selection: A hybrid bio-inspired heuristic approach. *Swarm Evol. Comput.* **2018**, *42*, 29–42.
5. Tubishat, M.; Abushariah, M.A.; Idris, N.; Aljarah, I. Improved whale optimization algorithm for feature selection in Arabic sentiment analysis. *Appl. Intell.* **2018**, *49*, 1–20.
6. Medjahed, S.A.; Saadi, T.A.; Benyettou, A.; Ouali, M. Gray wolf optimizer for hyperspectral band selection. *Appl. Soft Comput.* **2016**, *40*, 178–186.
7. Makhadmeh, S.N.; Al-Betar, M.A.; Alyasseri, Z.A.A.; Abasi, A.K.; Khader, A.T.; Damaševičius, R.; Mohammed, M.A.; Abdulkareem, K.H. Smart home battery for the multi-objective power scheduling problem in a smart home using grey wolf optimizer. *Electronics* **2021**, *10*, 1–35.
8. Faris, H.; Abukhurma, R.; Almanaseer, W.; Saadeh, M.; Mora, A.M.; Castillo, P.A.; Aljarah, I. Improving financial bankruptcy prediction in a highly imbalanced class distribution using oversampling and ensemble learning: A case from the Spanish market. *Prog. Artif. Intell.* **2019**, *9*, 1–23.
9. Al-Madi, N.; Faris, H.; Abukhurma, R. Cost-Sensitive Genetic Programming for Churn Prediction and Identification of the Influencing Factors in Telecommunication Market. *Int. J. Adv. Sci. Technol.* **2018**, 13–28.
10. Pourzangbar, A.; Losada, M.A.; Saber, A.; Ahari, L.R.; Larroudé, P.; Vaezi, M.; Brocchini, M. Prediction of non-breaking wave induced scour depth at the trunk section of breakwaters using Genetic Programming and Artificial Neural Networks. *Coast. Eng.* **2017**, *121*, 107–118.
11. Okewu, E.; Misra, S.; Maskeliunas, R.; Damasevicius, R.; Fernandez-Sanz, L. Optimizing green computing awareness for environmental sustainability and economic security as a stochastic optimization problem. *Sustainability* **2017**, *9*, 18–57.
12. Xue, B.; Zhang, M.; Browne, W.N.; Yao, X. A survey on evolutionary computation approaches to feature selection. *IEEE Trans. Evol. Comput.* **2016**, *20*, 606–626.
13. Alweshah, M.; Al Khalaileh, S.; Gupta, B.B.; Almomani, A.; Hammouri, A.I.; Al-Betar, M.A. The monarch butterfly optimization algorithm for solving feature selection problems. *Neural Comput. Appl.* **2020**, *2020*, 1–15.
14. Kohavi, R.; John, G.H. Wrappers for feature subset selection. *Artif. Intell.* **1997**, *97*, 273–324.
15. Dash, M.; Liu, H. Feature selection for classification. *Intell. Data Anal.* **1997**, *1*, 131–156.
16. Tabakhi, S.; Moradi, P.; Akhlaghian, F. An unsupervised feature selection algorithm based on ant colony optimization. *Eng. Appl. Artif. Intell.* **2014**, *32*, 112–123.
17. Liu, Y.; Wang, G.; Chen, H.; Dong, H.; Zhu, X.; Wang, S. An improved particle swarm optimization for feature selection. *J. Bionic Eng.* **2011**, *8*, 191–200.
18. Yang, C.H.; Chuang, L.Y.; Yang, C.H. IG-GA: A hybrid filter/wrapper method for feature selection of microarray data. *J. Med Biol. Eng.* **2010**, *30*, 23–28.
19. Li, J.; Cheng, K.; Wang, S.; Morstatter, F.; Trevino, R.P.; Tang, J.; Liu, H. Feature selection: A data perspective. *ACM Comput. Surv. (CSUR)* **2018**, *50*, 94.
20. Brezočnik, L.; Fister, I.; Podgorelec, V. Swarm Intelligence Algorithms for Feature Selection: A Review. *Appl. Sci.* **2018**, *8*, 1521.
21. Diao, R.; Shen, Q. Nature inspired feature selection meta-heuristics. *Artif. Intell. Rev.* **2015**, *44*, 311–340.
22. Oliveira, A.L.; Braga, P.L.; Lima, R.M.; Cornélio, M.L. GA-based method for feature selection and parameters optimization for machine learning regression applied to software effort estimation. *Inf. Softw. Technol.* **2010**, *52*, 1155–1166.
23. Khushaba, R.N.; Al-Ani, A.; AlSukker, A.; Al-Jumaily, A. A combined ant colony and differential evolution feature selection algorithm. In *Proceedings of the International Conference on Ant Colony Optimization and Swarm Intelligence*; Springer: Berlin/Heidelberg, Germany, 2008; pp. 1–12.
24. Tran, B.; Xue, B.; Zhang, M. Improved PSO for feature selection on high-dimensional datasets. In *Proceedings of the Asia-Pacific Conference on Simulated Evolution and Learning*; Springer: Berlin/Heidelberg, Germany, 2014; pp. 503–515.
25. Kanan, H.R.; Faez, K.; Taheri, S.M. Feature selection using ant colony optimization (ACO): A new method and comparative study in the application of face recognition system. In *Proceedings of the Industrial Conference on Data Mining*; Springer: Berlin/Heidelberg, Germany, 2007; pp. 63–76.
26. Hancer, E.; Xue, B.; Karaboga, D.; Zhang, M. A binary ABC algorithm based on advanced similarity scheme for feature selection. *Appl. Soft Comput.* **2015**, *36*, 334–348.
27. Balázs, K.; Botzheim, J.; Kóczy, L.T. Comparative Investigation of Various Evolutionary and Memetic Algorithms. In *Computational Intelligence in Engineering*; Springer: Berlin/Heidelberg, Germany, 2010; pp. 129–140. doi:10.1007/978-3-642-15220-7_11.
28. Sahlol, A.T.; Elaziz, M.A.; Jamal, A.T.; Damaševičius, R.; Hassan, O.F. A novel method for detection of tuberculosis in chest radiographs using artificial ecosystem-based optimisation of deep neural network features. *Symmetry* **2020**, *12*, 11–46 .
29. Sahlol, A.T.; Yousri, D.; Ewees, A.A.; Al-qaness, M.A.A.; Damasevicius, R.; Elaziz, M.A. COVID-19 image classification using deep features and fractional-order marine predators algorithm. *Sci. Rep.* **2020**, *10*, 15–36.

30. Polap, D.; Woźniak, M. Polar bear optimization algorithm: Meta-heuristic with fast population movement and dynamic birth and death mechanism. *Symmetry* **2017**, *9*, 203.
31. Połap, D.; Woźniak, M. Red fox optimization algorithm. *Expert Syst. Appl.* **2021**, *166*, 114107.
32. Mirjalili, S. SCA: A sine cosine algorithm for solving optimization problems. *Knowl.-Based Syst.* **2016**, *96*, 120–133.
33. Jouhari, H.; Lei, D.; Al-qaness, M.A.A.; Abd Elaziz, M.; Damaševičius, R.; Korytkowski, M.; Ewees, A.A. Modified Harris Hawks optimizer for solving machine scheduling problems. *Symmetry* **2020**, *12*, 1460.
34. Ksiazek, K.; Połap, D.; Woźniak, M.; Damaševičius, R. Radiation heat transfer optimization by the use of modified ant lion optimizer. In Proceedings of the 2017 IEEE Symposium Series on Computational Intelligence, Honolulu, HI, USA, 27 November–1 December 2017; Volume 2018, pp. 1–7.
35. Damaševičius, R.; Maskeliūnas, R. Agent state flipping based hybridization of heuristic optimization algorithms: A case of bat algorithm and krill herd hybrid algorithm. *Algorithms* **2021**, *14*, 358.
36. Faris, H.; Aljarah, I.; Al-Betar, M.A.; Mirjalili, S. Grey wolf optimizer: A review of recent variants and applications. *Neural Comput. Appl.* **2018**, *30*, 413–435.
37. Hira, Z.; Gillies, D. A Review of Feature Selection and Feature Extraction Methods Applied on Microarray Data. *Adv. Bioinform.* **2015**, *2015*, 198363–198363.
38. Somol, P.; Grim, J.; Novovičová, J.; Pudil, P. Improving feature selection process resistance to failures caused by curse-of-dimensionality effects. *Kybernetika* **2011**, *47*, 401–425.
39. von Luxburg, U.; Bousquet, O. Distance-Based Classification with Lipschitz Functions. In *Learning Theory and Kernel Machines*; Springer Berlin/Heidelberg, Germany, 2003; pp. 314–328. doi:10.1007/978-3-540-45167-9_24.
40. Higgins, I.; Amos, D.; Pfau, D.; Racaniere, S.; Matthey, L.; Rezende, D.; Lerchner, A. Towards a Definition of Disentangled Representations. *arXiv* **2018**. arXiv:cs.LG/1812.02230.
41. Brank, J.; Mladenić, D.; Grobelnik, M.; Liu, H.; Mladenić, D.; Flach, P.A.; Garriga, G.C.; Toivonen, H.; Toivonen, H. Feature Selection. In *Encyclopedia of Machine Learning*; Springer: Boston, MA, USA, 2011; pp. 402–406. doi:10.1007/978-0-387-30164-8_306.
42. Achille, A.; Soatto, S. Emergence of Invariance and Disentanglement in Deep Representations. In Proceedings of the 2018 Information Theory and Applications Workshop (ITA), San Diego, CA, USA, 11–16 February 2018, doi:10.1109/ita.2018.8503149.
43. Ratner, A.J.; Ehrenberg, H.R.; Hussain, Z.; Dunnmon, J.; Ré, C. Learning to Compose Domain-Specific Transformations for Data Augmentation. *arXiv* **2017**,arXiv:stat.ML/1709.01643.
44. Khaire, U.M.; Dhanalakshmi, R. Stability of feature selection algorithm: A review. *J. King Saud Univ. Comput. Inf. Sci.* **2019**, *2019*, doi:10.1016/j.jksuci.2019.06.012.
45. Gheyas, I.A.; Smith, L.S. Feature subset selection in large dimensionality domains. *Pattern Recognit.* **2010**, *43*, 5–13, doi:10.1016/j.patcog.2009.06.009.
46. Mafarja, M.M.; Mirjalili, S. Hybrid Whale Optimization Algorithm with simulated annealing for feature selection. *Neurocomputing* **2017**, *260*, 302–312.
47. Glover, F. Tabu search-part I. *ORSA J. Comput.* **1989**, *1*, 190–206.
48. Mirjalili, S.Z.; Mirjalili, S.; Saremi, S.; Faris, H.; Aljarah, I. Grasshopper optimization algorithm for multi-objective optimization problems. *Appl. Intell.* **2018**, *48*, 805–820.
49. Mirjalili, S.; Lewis, A.; Mostaghim, S. Confidence measure: A novel metric for robust meta-heuristic optimisation algorithms. *Inf. Sci.* **2015**, *317*, 114–142.
50. Lu, H.; Wang, X.; Fei, Z.; Qiu, M. The effects of using Chaotic map on improving the performance of multiobjective evolutionary algorithms. *Math. Probl. Eng.* **2014**, *2014*, 924652 .
51. Chuang, L.Y.; Yang, C.H.; Li, J.C. Chaotic maps based on binary particle swarm optimization for feature selection. *Appl. Soft Comput.* **2011**, *11*, 239–248.
52. Chuang, L.Y.; Yang, C.S.; Wu, K.C.; Yang, C.H. Gene selection and classification using Taguchi chaotic binary particle swarm optimization. *Expert Syst. Appl.* **2011**, *38*, 13367–13377.
53. Bharti, K.K.; Singh, P.K. Opposition chaotic fitness mutation based adaptive inertia weight BPSO for feature selection in text clustering. *Appl. Soft Comput.* **2016**, *43*, 20–34.
54. Ahmed, S.; Mafarja, M.; Faris, H.; Aljarah, I. Feature selection using salp swarm algorithm with chaos. In *Proceedings of the 2nd International Conference on Intelligent Systems, Metaheuristics & Swarm Intelligence*; ACM: New York, NY, USA, 2018; pp. 65–69.
55. Sayed, G.I.; Khoriba, G.; Haggag, M.H. A novel chaotic salp swarm algorithm for global optimization and feature selection. *Appl. Intell.* **2018**, *48*, 1–20.
56. Ewees, A.A.; El Aziz, M.A.; Hassanien, A.E. Chaotic multi-verse optimizer-based feature selection. *Neural Comput. Appl.* **2017**, *31*, 991–1006.
57. Sayed, G.I.; Darwish, A.; Hassanien, A.E. A New Chaotic Whale Optimization Algorithm for Features Selection. *J. Classif.* **2018**, *35*, 300–344.
58. Sayed, G.I.; Hassanien, A.E.; Azar, A.T. Feature selection via a novel chaotic crow search algorithm. *Neural Comput. Appl.* **2017**, *31*, 171–188.
59. Qasim, O.S.; Al-Thanoon, N.A.; Algamal, Z.Y. Feature selection based on chaotic binary black hole algorithm for data classification. *Chemom. Intell. Lab. Syst.* **2020**, *15*, 104104.
60. Pawlak, Z. Rough sets. *Int. J. Comput. Inf. Sci.* **1982**, *11*, 341–356.

61. Zainal, A.; Maarof, M.A.; Shamsuddin, S.M. Feature selection using Rough-DPSO in anomaly intrusion detection. In *Proceedings of the International Conference on Computational Science and Its Applications*; Springer: Berlin/Heidelberg, Germany, 2007; pp. 512–524.
62. Inbarani, H.H.; Azar, A.T.; Jothi, G. Supervised hybrid feature selection based on PSO and rough sets for medical diagnosis. *Comput. Methods Programs Biomed.* **2014**, *113*, 175–185.
63. Alia, A.F.; Taweel, A. Feature selection based on hybrid Binary Cuckoo Search and rough set theory in classification for nominal datasets. *Algorithms* **2017**, *14*, 65.
64. El Aziz, M.A.; Hassanien, A.E. Modified cuckoo search algorithm with rough sets for feature selection. *Neural Comput. Appl.* **2018**, *29*, 925–934.
65. Mafarja, M.M.; Mirjalili, S. Hybrid binary ant lion optimizer with rough set and approximate entropy reducts for feature selection. *Soft Comput.* **2018**, *23*, 6249–6265.
66. Chen, Y.; Zhu, Q.; Xu, H. Finding rough set reducts with fish swarm algorithm. *Knowl.-Based Syst.* **2015**, *81*, 22–29.
67. Hassanien, A.E.; Gaber, T.; Mokhtar, U.; Hefny, H. An improved moth flame optimization algorithm based on rough sets for tomato diseases detection. *Comput. Electron. Agric.* **2017**, *136*, 86–96.
68. Tawhid, M.A.; Ibrahim, A.M. Hybrid binary particle swarm optimization and flower pollination algorithm based on rough set approach for feature selection problem. In *Nature-Inspired Computation in Data Mining and Machine Learning*; Springer: Berlin/Heidelberg, Germany, 2020; pp. 249–273.
69. Ropiak, K.; Artiemjew, P. On a Hybridization of Deep Learning and Rough Set Based Granular Computing. *Algorithms* **2020**, *13*, 63, doi:10.3390/a13030063.
70. Dennett, D.C. Darwin's dangerous idea. *Science* **1995**, *35*, 34–40.
71. Tanaka, M.; Watanabe, H.; Furukawa, Y.; Tanino, T. GA-based decision support system for multicriteria optimization. Systems, Man and Cybernetics. In *Proceedings of the Intelligent Systems for the 21st Century*; Springer: Berlin/Heidelberg, Germany, 1995; Volume 2, pp. 1556–1561.
72. Mafarja, M.; Aljarah, I.; Heidari, A.A.; Hammouri, A.I.; Faris, H.; AlaḾ, A.Z.; Mirjalili, S. Evolutionary population dynamics and grasshopper optimization approaches for feature selection problems. *Knowl.-Based Syst.* **2018**, *145*, 25–45.
73. Mafarja, M.; Mirjalili, S. Whale optimization approaches for wrapper feature selection. *Appl. Soft Comput.* **2018**, *62*, 441–453.
74. Khushaba, R.N.; Al-Ani, A.; Al-Jumaily, A. Feature subset selection using differential evolution and a statistical repair mechanism. *Expert Syst. Appl.* **2011**, *38*, 11515–11526.
75. Leibowitz, N.; Baum, B.; Enden, G.; Karniel, A. The exponential learning equation as a function of successful trials results in sigmoid performance. *J. Math. Psychol.* **2010**, *54*, 338–340.
76. Aneesh, M.; Masand, A.A.; Manikantan, K. Optimal feature selection based on image pre-processing using accelerated binary particle swarm optimization for enhanced face recognition. *Procedia Eng.* **2012**, *30*, 750–758.
77. Pereira, L.; Rodrigues, D.; Almeida, T.; Ramos, C.; Souza, A.; Yang, X.S.; Papa, J. A binary cuckoo search and its application for feature selection. In *Cuckoo Search and Firefly Algorithm*; Springer: Berlin/Heidelberg, Germany, 2014; pp. 141–154.
78. Rodrigues, D.; Pereira, L.A.; Nakamura, R.Y.; Costa, K.A.; Yang, X.S.; Souza, A.N.; Papa, J.P. A wrapper approach for feature selection based on bat algorithm and optimum-path forest. *Expert Syst. Appl.* **2014**, *41*, 2250–2258.
79. Mafarja, M.; Aljarah, I.; Faris, H.; Hammouri, A.I.; AlaḾ, A.Z.; Mirjalili, S. Binary grasshopper optimisation algorithm approaches for feature selection problems. *Expert Syst. Appl.* **2019**, *117*, 267–286.
80. Mirjalili, S.; Lewis, A. S-shaped versus V-shaped transfer functions for binary particle swarm optimization. *Swarm Evol. Comput.* **2013**, *9*, 1–14.
81. Kennedy, J.; Eberhart, R.C. A discrete binary version of the particle swarm algorithm. Systems, Man, and Cybernetics. In Proceedings of the Computational Cybernetics and Simulation, Orlando, FL, USA, 12–15 October 1997; Volume 5, pp. 4104–4108.
82. Faris, H.; Mafarja, M.M.; Heidari, A.A.; Aljarah, I.; AlaḾ, A.Z.; Mirjalili, S.; Fujita, H. An efficient binary Salp Swarm Algorithm with crossover scheme for feature selection problems. *Knowl.-Based Syst.* **2018**, *154*, 43–67.
83. Mafarja, M.; Aljarah, I.; Heidari, A.A.; Faris, H.; Fournier-Viger, P.; Li, X.; Mirjalili, S. Binary dragonfly optimization for feature selection using time-varying transfer functions. *Knowl.-Based Syst.* **2018**, *161*, 185–204.
84. Emary, E.; Zawbaa, H.M.; Hassanien, A.E. Binary grey wolf optimization approaches for feature selection. *Neurocomputing* **2016**, *172*, 371–381.
85. Zhang, Y.; Wang, S.; Phillips, P.; Ji, G. Binary PSO with mutation operator for feature selection using decision tree applied to spam detection. *Knowl.-Based Syst.* **2014**, *64*, 22–31.
86. Sihwail, R.; Omar, K.; Ariffin, K.A.Z.; Tubishat, M. Improved Harris Hawks Optimization Using Elite Opposition-Based Learning and Novel Search Mechanism for Feature Selection. *IEEE Access* **2020**, *8*, 121127–121145.
87. Sims, D.W.; Humphries, N.E.; Bradford, R.W.; Bruce, B.D. Lévy flight and Brownian search patterns of a free-ranging predator reflect different prey field characteristics. *J. Anim. Ecol.* **2012**, *81*, 432–442.
88. Khurmaa, R.A.; Aljarah, I.; Sharieh, A. An intelligent feature selection approach based on moth flame optimization for medical diagnosis. *Neural Comput. Appl.* **2020**, *33*, 7165–7204.
89. Oh, I.S.; Lee, J.S.; Moon, B.R. Hybrid genetic algorithms for feature selection. *IEEE Trans. Pattern Anal. Mach. Intell.* **2004**, *26*, 1424–1437.
90. Vieira, S.M.; Mendonça, L.F.; Farinha, G.J.; Sousa, J.M. Modified binary PSO for feature selection using SVM applied to mortality prediction of septic patients. *Appl. Soft Comput.* **2013**, *13*, 3494–3504.

91. Zhang, Y.; Gong, D.; Hu, Y.; Zhang, W. Feature selection algorithm based on bare bones particle swarm optimization. *Neurocomputing* **2015**, *148*, 150–157.
92. Yang, C.S.; Chuang, L.Y.; Ke, C.H.; Yang, C.H. Boolean binary particle swarm optimization for feature selection. In Proceedings of the Evolutionary Computation 2008, Hong Kong, China, 1–6 June 2008; pp. 2093–2098.
93. Danyadi, Z.; Foldesi, P.; Koczy, L.T. Solution of a fuzzy resource allocation problem by various evolutionary approaches. In Proceedings of the 2013 Joint IFSA World Congress and NAFIPS Annual Meeting (IFSA/NAFIPS), Edmonton, AB, Canada, 24–28 June 2013; doi:10.1109/ifsa-nafips.2013.6608504.
94. Chuang, L.Y.; Tsai, S.W.; Yang, C.H. Improved binary particle swarm optimization using catfish effect for feature selection. *Expert Syst. Appl.* **2011**, *38*, 12699–12707.
95. Xue, B.; Zhang, M.; Browne, W.N. Single feature ranking and binary particle swarm optimisation based feature subset ranking for feature selection. In *Proceedings of the Thirty-fifth Australasian Computer Science Conference*; Australian Computer Society, Inc.: Darlinghurst, Australia, 2012; Volume 122, pp. 27–36.
96. Lane, M.C.; Xue, B.; Liu, I.; Zhang, M. Gaussian based particle swarm optimisation and statistical clustering for feature selection. In *Proceedings of the European Conference on Evolutionary Computation in Combinatorial Optimization*; Springer: Berlin/Heidelberg, Germany, 2014; pp. 133–144.
97. Krisshna, N.A.; Deepak, V.K.; Manikantan, K.; Ramachandran, S. Face recognition using transform domain feature extraction and PSO-based feature selection. *Appl. Soft Comput.* **2014**, *22*, 141–161.
98. Banka, H.; Dara, S. A Hamming distance based binary particle swarm optimization (HDBPSO) algorithm for high dimensional feature selection, classification and validation. *Pattern Recognit. Lett.* **2015**, *52*, 94–100.
99. Moradi, P.; Gholampour, M. A hybrid particle swarm optimization for feature subset selection by integrating a novel local search strategy. *Appl. Soft Comput.* **2016**, *43*, 117–130.
100. Xi, M.; Sun, J.; Liu, L.; Fan, F.; Wu, X. Cancer Feature Selection and Classification Using a Binary Quantum-Behaved Particle Swarm Optimization and Support Vector Machine. *Comput. Math. Methods Med.* **2016**, *2016*, 3572705–3572705.
101. Mistry, K.; Zhang, L.; Neoh, S.C.; Lim, C.P.; Fielding, B. A micro-GA embedded PSO feature selection approach to intelligent facial emotion recognition. *IEEE Trans. Cybern.* **2017**, *47*, 1496–1509.
102. Zhang, Y.; Song, X.F.; Gong, D.W. A return-cost-based binary firefly algorithm for feature selection. *Inf. Sci.* **2017**, *418*, 561–574.
103. Lin, K.C.; Hung, J.C.; Wei, J.T. Feature selection with modified lions algorithms and support vector machine for high-dimensional data. *Appl. Soft Comput.* **2018**, *68*, 669–676.
104. Salcedo-Sanz, S.; Prado-Cumplido, M.; Pérez-Cruz, F.; Bousoño-Calzón, C. Feature selection via genetic optimization. In *Proceedings of the International Conference on Artificial Neural Networks*; Springer: Berlin/Heidelberg, Germany, 2002; pp. 547–552.
105. Zeng, X.P.; Li, Y.M.; Qin, J. A dynamic chain-like agent genetic algorithm for global numerical optimization and feature selection. *Neurocomputing* **2009**, *72*, 1214–1228.
106. Guo, J.; White, J.; Wang, G.; Li, J.; Wang, Y. A genetic algorithm for optimized feature selection with resource constraints in software product lines. *J. Syst. Softw.* **2011**, *84*, 2208–2221.
107. Blum, C.; Puchinger, J.; Raidl, G.R.; Roli, A. Hybrid metaheuristics in combinatorial optimization: A survey. *Appl. Soft Comput.* **2011**, *11*, 4135–4151.
108. Sayed, S.A.F.; Nabil, E.; Badr, A. A binary clonal flower pollination algorithm for feature selection. *Pattern Recognit. Lett.* **2016**, *77*, 21–27.
109. Alweshah, M.; Alkhalaileh, S.; Albashish, D.; Mafarja, M.; Bsoul, Q.; Dorgham, O. A hybrid mine blast algorithm for feature selection problems. *Soft Comput.* **2020**, *25*, 17–534.
110. Ibrahim, R.A.; Ewees, A.A.; Oliva, D.; Elaziz, M.A.; Lu, S. Improved salp swarm algorithm based on particle swarm optimization for feature selection. *J. Ambient. Intell. Humaniz. Comput.* **2019**, *10*, 3155–3169.
111. Sayed, G.I.; Hassanien, A.E. A hybrid SA-MFO algorithm for function optimization and engineering design problems. *Complex Intell. Syst.* **2018**, *4*, 195–212.
112. Zhang, L.; Mistry, K.; Neoh, S.C.; Lim, C.P. Intelligent facial emotion recognition using moth-firefly optimization. *Knowl.-Based Syst.* **2016**, *111*, 248–267.
113. Li, S.; Wu, X.; Tan, M. Gene selection using hybrid particle swarm optimization and genetic algorithm. *Soft Comput.* **2008**, *12*, 1039–1048.
114. Abdel-Basset, M.; Ding, W.; El-Shahat, D. A hybrid Harris Hawks optimization algorithm with simulated annealing for feature selection. *Artif. Intell. Rev.* **2020**, *54*, 593–637.
115. Ibrahim, A.M.; Tawhid, M.A. A New Hybrid Binary Algorithm of Bat Algorithm and Differential Evolution for Feature Selection and Classification. In *Applications of Bat Algorithm and Its Variants*; Springer: Berlin/Heidelberg, Germany, 2020; pp. 1–18.
116. Nemati, S.; Basiri, M.E.; Ghasem-Aghaee, N.; Aghdam, M.H. A novel ACO–GA hybrid algorithm for feature selection in protein function prediction. *Expert Syst. Appl.* **2009**, *36*, 12086–12094.
117. Wan, Y.; Wang, M.; Ye, Z.; Lai, X. A feature selection method based on modified binary coded ant colony optimization algorithm. *Appl. Soft Comput.* **2016**, *49*, 248–258.
118. Mafarja, M.; Qasem, A.; Heidari, A.A.; Aljarah, I.; Faris, H.; Mirjalili, S. Efficient hybrid nature-inspired binary optimizers for feature selection. *Cogn. Comput.* **2020**, *12*, 150–175.

119. Li, Q.; Chen, H.; Huang, H.; Zhao, X.; Cai, Z.; Tong, C.; Liu, W.; Tian, X. An Enhanced Grey Wolf Optimization Based Feature Selection Wrapped Kernel Extreme Learning Machine for Medical Diagnosis. *Comput. Math. Methods Med.* **2017**, *2017*, 9512741–9512741.
120. Kihel, B.K.; Chouraqui, S. Firefly optimization using artificial immune system for feature subset selection. *Int. J. Intell. Eng. Syst* **2019**, *12*, 337–347.
121. Faris, H.; Hassonah, M.A.; AlaM, A.Z.; Mirjalili, S.; Aljarah, I. A multi-verse optimizer approach for feature selection and optimizing SVM parameters based on a robust system architecture. *Neural Comput. Appl.* **2018**, *30*, 2355–2369.
122. Aljarah, I.; AlaM, A.Z.; Faris, H.; Hassonah, M.A.; Mirjalili, S.; Saadeh, H. Simultaneous feature selection and support vector machine optimization using the grasshopper optimization algorithm. *Cogn. Comput.* **2018**, *10*, 1478–495.
123. Yang, C.S.; Chuang, L.Y.; Ke, C.H.; Yang, C.H. A Hybrid Feature Selection Method for Microarray Classification. *IAENG Int. J. Comput. Sci.* **2008**, *35*, 285-290.
124. Unler, A.; Murat, A.; Chinnam, R.B. mr2PSO: A maximum relevance minimum redundancy feature selection method based on swarm intelligence for support vector machine classification. *Inf. Sci.* **2011**, *181*, 4625–4641.
125. Cervante, L.; Xue, B.; Zhang, M.; Shang, L. Binary particle swarm optimisation for feature selection: A filter based approach. In Proceedings of the Evolutionary Computation (CEC), Brisbane, Australia, 10–15 June 2012; pp. 1–8.
126. Moghadasian, M.; Hosseini, S.P. Binary cuckoo optimization algorithm for feature selection in high-dimensional datasets. In Proceedings of the International Conference on Innovative Engineering Technologies (ICIET2014), Bangkok, Thailand, 28–29 December 2014; pp. 18–21.
127. Zhang, C.K.; Hu, H. Feature selection using the hybrid of ant colony optimization and mutual information for the forecaster. Machine Learning and Cybernetics. In Proceedings of the 2005 International Conference on Machine Learning and Cybernetics, Guangzhou, China, 18–21 August 2005; Volume 3, pp. 1728–1732.
128. Huang, J.; Cai, Y.; Xu, X. A hybrid genetic algorithm for feature selection wrapper based on mutual information. *Pattern Recognit. Lett.* **2007**, *28*, 1825–1844.
129. Uğuz, H. A two-stage feature selection method for text categorization by using information gain, principal component analysis and genetic algorithm. *Knowl.-Based Syst.* **2011**, *24*, 1024–1032.
130. Wang, L.; Xu, G.; Wang, J.; Yang, S.; Guo, L.; Yan, W. GA-SVM based feature selection and parameters optimization for BCI research. Natural Computation (ICNC). In Proceedings of the 2011 Seventh International Conference on Natural Computation, Shanghai, China, 26–28 July 2011; Volume 1, pp. 580–583.
131. Zainudin, M.; Sulaiman, M.; Mustapha, N.; Perumal, T.; Nazri, A.; Mohamed, R.; Manaf, S. Feature selection optimization using hybrid relief-f with self-adaptive differential evolution. *Int. J. Intell. Eng. Syst* **2017**, *10*, 21–29.
132. Guha, R.; Ghosh, M.; Mutsuddi, S.; Sarkar, R.; Mirjalili, S. Embedded chaotic whale survival algorithm for filter-wrapper feature selection. *arXiv preprint* **2020**, arXiv:2005.04593.
133. Hassonah, M.A.; Al-Sayyed, R.; Rodan, A.; AlaM, A.Z.; Aljarah, I.; Faris, H. An efficient hybrid filter and evolutionary wrapper approach for sentiment analysis of various topics on Twitter. *Knowl.-Based Syst.* **2020**, *192*, 105353.
134. Ahmed, N.; Rafiq, J.I.; Islam, M.R. Enhanced human activity recognition based on smartphone sensor data using hybrid feature selection model. *Sensors* **2020**, *20*, 317.
135. Kanan, H.R.; Faez, K. An improved feature selection method based on ant colony optimization (ACO) evaluated on face recognition system. *Appl. Math. Comput.* **2008**, *205*, 716–725.
136. Chuang, L.Y.; Chang, H.W.; Tu, C.J.; Yang, C.H. Improved binary PSO for feature selection using gene expression data. *Comput. Biol. Chem.* **2008**, *32*, 29–38.
137. Martinez, E.; Alvarez, M.M.; Trevino, V. Compact cancer biomarkers discovery using a swarm intelligence feature selection algorithm. *Comput. Biol. Chem.* **2010**, *34*, 244–250.
138. Mohamad, M.S.; Omatu, S.; Deris, S.; Yoshioka, M. A modified binary particle swarm optimization for selecting the small subset of informative genes from gene expression data. *IEEE Trans. Inf. Technol. Biomed.* **2011**, *15*, 813–822.
139. Xue, B.; Zhang, M.; Browne, W.N. Novel initialisation and updating mechanisms in PSO for feature selection in classification. In *Proceedings of the European Conference on the Applications of Evolutionary Computation*; Springer: Berlin/Heidelberg, Germany, 2013; pp. 428–438.
140. Mafarja, M.; Sabar, N.R. Rank based binary particle swarm optimisation for feature selection in classification. In Proceedings of the 2nd International Conference on Future Networks and Distributed Systems, Amman, Jordan, 26–27 June 2018; pp. 1–6.
141. Mafarja, M.; Jarrar, R.; Ahmad, S.; Abusnaina, A.A. Feature selection using binary particle swarm optimization with time varying inertia weight strategies. In *Proceedings of the 2nd International Conference on Future Networks and Distributed Systems*; ACM: New York, NY, USA, 2018; p. 18.
142. Aljarah, I.; Mafarja, M.; Heidari, A.A.; Faris, H.; Zhang, Y.; Mirjalili, S. Asynchronous accelerating multi-leader salp chains for feature selection. *Appl. Soft Comput.* **2018**, *71*, 964–979.
143. Hammouri, A.I.; Mafarja, M.; Al-Betar, M.A.; Awadallah, M.A.; Abu-Doush, I. An improved Dragonfly Algorithm for feature selection. *Knowl.-Based Syst.* **2020**, *203*, 106131.
144. Faris, H.; Heidari, A.A.; AlaM, A.Z.; Mafarja, M.; Aljarah, I.; Eshtay, M.; Mirjalili, S. Time-varying hierarchical chains of salps with random weight networks for feature selection. *Expert Syst. Appl.* **2020**, *140*, 112898.
145. Ouadfel, S.; Abd Elaziz, M. Enhanced Crow Search Algorithm for Feature Selection. *Expert Syst. Appl.* **2020**, *159*, 113572.

146. Gholami, J.; Pourpanah, F.; Wang, X. Feature selection based on improved binary global harmony search for data classification. *Appl. Soft Comput.* **2020**, *93*, 106402.
147. Khurma, R.A.; Aljarah, I.; Sharieh, A. Rank based moth flame optimisation for feature selection in the medical application. In Proceedings of the 2020 IEEE Congress on Evolutionary Computation (CEC), Glasgow, UK, 19–24 July 2020; pp. 1–8.
148. Khurma, R.A.; Castillo, P.A.; Sharieh, A.; Aljarah, I. Feature Selection using Binary Moth Flame Optimization with Time Varying Flames Strategies. In *Proceedings of the 12th International Joint Conference on Computational Intelligence*; SciTePress: Setubal, Portugal, 2020; pp. 17–27, doi:10.5220/0010021700170027.
149. Galbally, J.; Fierrez, J.; Freire, M.R.; Ortega-Garcia, J. Feature selection based on genetic algorithms for on-line signature verification. In Proceedings of the IEEE Workshop on Automatic Identification Advanced Technologies, Alghero, Italy, 7–8 June 2007; pp. 198–203.
150. Lu, J.; Zhao, T.; Zhang, Y. Feature selection based-on genetic algorithm for image annotation. *Knowl.-Based Syst.* **2008**, *21*, 887–891.
151. Li, R.; Lu, J.; Zhang, Y.; Zhao, T. Dynamic Adaboost learning with feature selection based on parallel genetic algorithm for image annotation. *Knowl.-Based Syst.* **2010**, *23*, 195–201.
152. Hans, R.; Kaur, H. Quasi-opposition-Based Multi-verse Optimization Algorithm for Feature Selection. In *Proceedings of the First International Conference on Computing, Communications, and Cyber-Security (IC4S 2019)*; Springer: Berlin/Heidelberg, Germany, 2020; pp. 345–359.
153. Chakraborty, B. Feature subset selection by particle swarm optimization with fuzzy fitness function. In Proceedings of the 2008 International Conference on Intelligent System and Knowledge Engineering, ISKE, Xiamen, China, 17–19 November 2008; Volume 1, pp. 1038–1042.
154. De Stefano, C.; Fontanella, F.; Marrocco, C.; Di Freca, A.S. A GA-based feature selection approach with an application to handwritten character recognition. *Pattern Recognit. Lett.* **2014**, *35*, 130–141.
155. Khurma., R.A.; Castillo., P.A.; Sharieh., A.; Aljarah., I. New Fitness Functions in Binary Harris Hawks Optimization for Gene Selection in Microarray Datasets. In *Proceedings of the 12th International Joint Conference on Computational Intelligence*; SciTePress: Setubal, Portugal, 2020; Volume 1, pp. 139–146, doi:10.5220/0010021601390146.
156. Zio, E.; Baraldi, P.; Pedroni, N. Selecting features for nuclear transients classification by means of genetic algorithms. *IEEE Trans. Nucl. Sci.* **2006**, *53*, 1479–1493.
157. Mandal, M.; Mukhopadhyay, A.; Maulik, U. Prediction of protein subcellular localization by incorporating multiobjective PSO-based feature subset selection into the general form of Chous PseAAC. *Med Biol. Eng. Comput.* **2015**, *53*, 331–344.
158. Aljarah, I.; Habib, M.; Faris, H.; Al-Madi, N.; Heidari, A.A.; Mafarja, M.; Abd Elaziz, M.; Mirjalili, S. A Dynamic Locality Multi-Objective Salp Swarm Algorithm for Feature Selection. *Comput. Ind. Eng.* **2020**, *147*, 106628.
159. Niu, B.; Yi, W.; Tan, L.; Geng, S.; Wang, H. A multi-objective feature selection method based on bacterial foraging optimization. *Nat. Comput.* **2019**, *120*, 63–76.
160. Habib, M.; Aljarah, I.; Faris, H. A Modified Multi-objective Particle Swarm Optimizer-Based Lévy Flight: An Approach Toward Intrusion Detection in Internet of Things. *Arab. J. Sci. Eng.* **2020**, *45*, 6081–6108.
161. Al-Tashi, Q.; Abdulkadir, S.J.; Rais, H.M.; Mirjalili, S.; Alhussian, H. Approaches to Multi-Objective Feature Selection: A Systematic Literature Review. *IEEE Access* **2020**, *8*, 125076–125096.
162. Punch III, W.F.; Goodman, E.D.; Pei, M.; Chia-Shun, L.; Hovland, P.D.; Enbody, R.J. Further Research on Feature Selection and Classification Using Genetic Algorithms. *ICGA J.* **1993**, *93*, 557–564.
163. Ghamisi, P.; Couceiro, M.S.; Benediktsson, J.A. A novel feature selection approach based on FODPSO and SVM. *IEEE Trans. Geosci. Remote Sens.* **2015**, *53*, 2935–2947.
164. Frohlich, H.; Chapelle, O.; Scholkopf, B. Feature selection for support vector machines by means of genetic algorithm. In Proceedings of the 2003 15th IEEE International Conference, Sacramento, CA, USA, 3–5 November 2003; pp. 142–148.
165. Robbins, K.; Zhang, W.; Bertrand, J.; Rekaya, R. The ant colony algorithm for feature selection in high-dimension gene expression data for disease classification. *Math. Med. Biol. A J. IMA* **2007**, *24*, 413–426.
166. Kabir, M.M.; Shahjahan, M.; Murase, K. A new local search based hybrid genetic algorithm for feature selection. *Neurocomputing* **2011**, *74*, 2914–2928.
167. Kabir, M.M.; Shahjahan, M.; Murase, K. A new hybrid ant colony optimization algorithm for feature selection. *Expert Syst. Appl.* **2012**, *39*, 3747–3763.
168. Ibrahim, H.T.; Mazher, W.J.; Ucan, O.N.; Bayat, O. A grasshopper optimizer approach for feature selection and optimizing SVM parameters utilizing real biomedical data sets. *Neural Comput. Appl.* **2018**, *31*, 5965–5974.
169. Sreedharan, N.P.N.; Ganesan, B.; Raveendran, R.; Sarala, P.; Dennis, B. Grey Wolf optimisation-based feature selection and classification for facial emotion recognition. *IET Biom.* **2018**, *7*, 490–499.
170. Handels, H.; Roß, T.; Kreusch, J.; Wolff, H.H.; Poeppl, S.J. Feature selection for optimized skin tumor recognition using genetic algorithms. *Artif. Intell. Med.* **1999**, *16*, 283–297.
171. Zheng, B.; Chang, Y.H.; Wang, X.H.; Good, W.F.; Gur, D. Feature selection for computerized mass detection in digitized mammograms by using a genetic algorithm. *Acad. Radiol.* **1999**, *6*, 327–332.
172. Li, T.S. Feature selection for classification by using a GA-based neural network approach. *J. Chin. Inst. Ind. Eng.* **2006**, *23*, 55–64.
173. Babaoglu, İ.; Findik, O.; Ülker, E. A comparison of feature selection models utilizing binary particle swarm optimization and genetic algorithm in determining coronary artery disease using support vector machine. *Expert Syst. Appl.* **2010**, *37*, 3177–3183.

174. Ahmad, F.; Isa, N.A.M.; Hussain, Z.; Osman, M.K.; Sulaiman, S.N. A GA-based feature selection and parameter optimization of an ANN in diagnosing breast cancer. *Pattern Anal. Appl.* **2015**, *18*, 861–870.
175. Sheikhpour, R.; Sarram, M.A.; Sheikhpour, R. Particle swarm optimization for bandwidth determination and feature selection of kernel density estimation based classifiers in diagnosis of breast cancer. *Appl. Soft Comput.* **2016**, *40*, 113–131.
176. Sayed, G.I.; Hassanien, A.E.; Nassef, T.M.; Pan, J.S. Alzheimerś Disease Diagnosis Based on Moth Flame Optimization. In *Proceedings of the International Conference on Genetic and Evolutionary Computing*; Springer: Berlin/Heidelberg, Germany, 2016; pp. 298–305.
177. Sayed, G.I.; Hassanien, A.E. Moth-flame swarm optimization with neutrosophic sets for automatic mitosis detection in breast cancer histology images. *Appl. Intell.* **2017**, *47*, 397–408.
178. Daelemans, W.; Hoste, V.; De Meulder, F.; Naudts, B. Combined optimization of feature selection and algorithm parameters in machine learning of language. In *European Conference on Machine Learning*; Springer: Berlin/Heidelberg, Germany, 2003; pp. 84–95.
179. Aghdam, M.H.; Ghasem-Aghaee, N.; Basiri, M.E. Text feature selection using ant colony optimization. *Expert Syst. Appl.* **2009**, *36*, 6843–6853.
180. Ewees, A.A.; Sahlol, A.T.; Amasha, M.A. A Bio-inspired moth-flame optimization algorithm for Arabic handwritten letter recognition. Control, Artificial Intelligence, Robotics & Optimization (ICCAIRO). In Proceedings of the 2017 International Conference, Brussels, Belgium, 13–17 July 2017; pp. 154–159.
181. Tackett, W.A. Genetic Programming for Feature Discovery and Image Discrimination. *ICGA J.* **1993**, *1993*, 303–311.
182. Bhanu, B.; Lin, Y. Genetic algorithm based feature selection for target detection in SAR images. *Image Vis. Comput.* **2003**, *21*, 591–608.
183. Jarvis, R.M.; Goodacre, R. Genetic algorithm optimization for pre-processing and variable selection of spectroscopic data. *Bioinformatics* **2004**, *21*, 860–868.
184. Zhuo, L.; Zheng, J.; Li, X.; Wang, F.; Ai, B.; Qian, J. A genetic algorithm based wrapper feature selection method for classification of hyperspectral images using support vector machine. In *Proceedings of the Geoinformatics 2008 and Joint Conference on GIS and Built Environment: Classification of Remote Sensing Images*; International Society for Optics and Photonics: Bellingham, WA, USA, 2008; Volume 7147, p. 71471J.
185. Li, S.; Wu, H.; Wan, D.; Zhu, J. An effective feature selection method for hyperspectral image classification based on genetic algorithm and support vector machine. *Knowl.-Based Syst.* **2011**, *24*, 40–48.
186. Ghamisi, P.; Benediktsson, J.A. Feature selection based on hybridization of genetic algorithm and particle swarm optimization. *IEEE Geosci. Remote Sens. Lett.* **2015**, *12*, 309–313.
187. Raymer, M.L.; Punch, W.F.; Goodman, E.D.; Kuhn, L.A.; Jain, A.K. Dimensionality reduction using genetic algorithms. *IEEE Trans. Evol. Comput.* **2000**, *4*, 164–171.
188. Palaniappan, R.; Raveendran, P. Genetic Algorithm to select features for Fuzzy ARTMAP classification of evoked EEG. In Proceedings of the 2002 Asia-Pacific Conference, Taipei, Taiwan, China, 8 August 2002; Volume 2, pp. 53–56.
189. Garrett, D.; Peterson, D.A.; Anderson, C.W.; Thaut, M.H. Comparison of linear, nonlinear, and feature selection methods for EEG signal classification. *IEEE Trans. Neural Syst. Rehabil. Eng.* **2003**, *11*, 141–144.
190. Noori, F.M.; Qureshi, N.K.; Khan, R.A.; Naseer, N. Feature selection based on modified genetic algorithm for optimization of functional near-infrared spectroscopy (fNIRS) signals for BCI. In Proceedings of the 2016 2nd International Conference on IEEE, Chengdu, China, 26–27 October 2016; pp. 50–53.
191. Chen, L.H.; Hsiao, H.D. Feature selection to diagnose a business crisis by using a real GA-based support vector machine: An empirical study. *Expert Syst. Appl.* **2008**, *35*, 1145–1155.
192. Huang, C.F.; Chang, B.R.; Cheng, D.W.; Chang, C.H. Feature Selection and Parameter Optimization of a Fuzzy-based Stock Selection Model Using Genetic Algorithms. *Int. J. Fuzzy Syst.* **2012**, *14*, 1.
193. Temitayo, F.; Stephen, O.; Abimbola, A. Hybrid GA-SVM for efficient feature selection in e-mail classification. *Comput. Eng. Intell. Syst.* **2012**, *3*, 17–28.
194. Faris, H.; AlaM, A.Z.; Heidari, A.A.; Aljarah, I.; Mafarja, M.; Hassonah, M.A.; Fujita, H. An intelligent system for spam detection and identification of the most relevant features based on evolutionary random weight networks. *Inf. Fusion* **2019**, *48*, 67–83.
195. Saidala, R.K. Variant of Northern Bald Ibis Algorithm for Unmasking Outliers. *Int. J. Softw. Sci. Comput. Intell. (IJSSCI)* **2020**, *12*, 15–29.
196. Yu, E.; Cho, S. GA-SVM wrapper approach for feature subset selection in keystroke dynamics identity verification. In Proceedings of the International Joint Conference on IEEE, Portland, OR, USA, 20–24 July 2003; Volume 3, pp. 2253–2257.
197. Rodrigues, D.; Pereira, L.A.; Almeida, T.; Papa, J.P.; Souza, A.; Ramos, C.C.; Yang, X.S. BCS: A binary cuckoo search algorithm for feature selection. In Proceedings of the 2013 IEEE International Symposium on IEEE, Beijing, China, 19–23 May 2013; pp. 465–468.
198. Hajnayeb, A.; Ghasemloonia, A.; Khadem, S.; Moradi, M. Application and comparison of an ANN-based feature selection method and the genetic algorithm in gearbox fault diagnosis. *Expert Syst. Appl.* **2011**, *38*, 10205–10209.
199. Khurma, R.A.; Aljarah, I.; Sharieh, A.; Mirjalili, S. EvoloPy-FS: An Open-Source Nature-Inspired Optimization Framework in Python for Feature Selection. In *Evolutionary Machine Learning Techniques*; Springer: Berlin/Heidelberg, Germany, 2020; pp. 131–173.

200. Faris, H.; Aljarah, I.; Mirjalili, S.; Castillo, P.A.; Guervós, J.J.M. EvoloPy: An Open-source Nature-inspired Optimization Framework in Python. *IJCCI (ECTA)* **2016**, *1*, 171–177.
201. Wolpert, D.H.; Macready, W.G. No free lunch theorems for optimization. *IEEE Trans. Evol. Comput.* **1997**, *1*, 67–82.
202. Unler, A.; Murat, A. A discrete particle swarm optimization method for feature selection in binary classification problems. *Eur. J. Oper. Res.* **2010**, *206*, 528–539.
203. Nakamura, R.Y.; Pereira, L.A.; Costa, K.; Rodrigues, D.; Papa, J.P.; Yang, X.S. BBA: A binary bat algorithm for feature selection. In Proceedings of the 2012 25th SIBGRAPI Conference on Graphics, Patterns and Images, Ouro Preto, Brazil, 22–25 August 2012; pp. 291–297.
204. Taha, A.M.; Mustapha, A.; Chen, S.D. Naive bayes-guided bat algorithm for feature selection. *Sci. World J.* **2013**, *2013*, 325973.
205. Huang, C.L. ACO-based hybrid classification system with feature subset selection and model parameters optimization. *Neurocomputing* **2009**, *73*, 438–448.
206. Lin, K.C.; Chien, H.Y. CSO-based feature selection and parameter optimization for support vector machine. In Proceedings of the 2009 Joint Conferences on IEEE, Tamsui, Taiwan, China, 3–5 December 2009; pp. 783–788.
207. ElAlami, M.E. A filter model for feature subset selection based on genetic algorithm. *Knowl.-Based Syst.* **2009**, *22*, 356–362.
208. Huang, C.L.; Wang, C.J. A GA-based feature selection and parameters optimizationfor support vector machines. *Expert Syst. Appl.* **2006**, *31*, 231–240.
209. Zhang, Y.; Gong, D.W.; Cheng, J. Multi-objective particle swarm optimization approach for cost-based feature selection in classification. *IEEE/ACM Trans. Comput. Biol. Bioinform. (TCBB)* **2017**, *14*, 64–75.
210. Kashef, S.; Nezamabadi-pour, H. An advanced ACO algorithm for feature subset selection. *Neurocomputing* **2015**, *147*, 271–279.
211. Kashef, S.; Nezamabadi-pour, H. A new feature selection algorithm based on binary ant colony optimization. In Proceedings of the 2013 5th Conference on IEEE, Kuala Lumpur, Malaysia, 4–5 December 2013; pp. 50–54.
212. Zhao, T.; Lu, J.; Zhang, Y.; Xiao, Q. Feature selection based on genetic algorithm for cbir. In Proceedings of the Image and Signal Processing, 2008. CISP'08, Congress on IEEE, Sanya, China, 27–30 May 2008; Volume 2, pp. 495–499.
213. Xue, B.; Zhang, M.; Browne, W.N. New fitness functions in binary particle swarm optimisation for feature selection. In Proceedings of the 2012 IEEE Congress on IEEE, Brisbane, QLD, Australia, 10–15 June 2012; pp. 1–8.
214. Emary, E.; Zawbaa, H.M.; Ghany, K.K.A.; Hassanien, A.E.; Parv, B. Firefly optimization algorithm for feature selection. In Proceedings of the 7th Balkan Conference on Informatics Conference, ACM, Craiova, Romania, 2–4 September 2015; p. 26.
215. Xue, B.; Zhang, M.; Browne, W.N. Particle swarm optimization for feature selection in classification: A multi-objective approach. *IEEE Trans. Cybern.* **2013**, *43*, 1656–1671.
216. Xue, B.; Cervante, L.; Shang, L.; Browne, W.N.; Zhang, M. Binary PSO and rough set theory for feature selection: A multi-objective filter based approach. *Int. J. Comput. Intell. Appl.* **2014**, *13*, 1450009.
217. Doerner, K.; Gutjahr, W.J.; Hartl, R.F.; Strauss, C.; Stummer, C. Pareto ant colony optimization: A metaheuristic approach to multiobjective portfolio selection. *Ann. Oper. Res.* **2004**, *131*, 79–99.

MDPI
St. Alban-Anlage 66
4052 Basel
Switzerland
Tel. +41 61 683 77 34
Fax +41 61 302 89 18
www.mdpi.com

Mathematics Editorial Office
E-mail: mathematics@mdpi.com
www.mdpi.com/journal/mathematics

www.ingramcontent.com/pod-product-compliance
Lightning Source LLC
LaVergne TN
LVHW070509100526
838202LV00014B/1818